Deconstructing Olduvai

Vertebrate Paleobiology and Paleoanthropology Series

Edited by

Eric Delson
Vertebrate Paleontology, American Museum of Natural History,
New York, NY 10024, USA
delson@amnh.org

Ross D. E. MacPhee
Vertebrate Zoology, American Museum of Natural History,
New York, NY 10024, USA
macphee@amnh.org

Focal topics for volumes in the series will include systematic paleontology of all vertebrates (from agnathans to humans), phylogeny reconstruction, functional morphology, Paleolithic archaeology, taphonomy, geochronology, historical biogeography, and biostratigraphy. Other fields (e.g., paleoclimatology, paleoecology, ancient DNA, total organismal community structure) may be considered if the volume theme emphasizes paleobiology (or archaeology). Fields such as modeling of physical processes, genetic methodology, nonvertebrates or neontology are out of our scope.

Volumes in the series may either be monographic treatments (including unpublished but fully revised dissertations) or edited collections, especially those focusing on problem- oriented issues, with multidisciplinary coverage where possible.

Editorial Advisory Board

Published and forthcoming titles in this series are listed at the end of this volume.

Deconstructing Olduvai: A Taphonomic Study of the Bed I Sites

By

Manuel Domínguez-Rodrigo
Department of Prehistory
Complutense University, Madrid, Spain

Rebeca Barba Egido
Department of Prehistory
Complutense University, Madrid, Spain

and

Charles P. Egeland
Department of Anthropology
Indiana University, Bloomington, USA

Springer

A C.I.P. catalogue record for this book is available from the Library of Congress

ISBN 978-1-4020-6150-9 (HB)
ISBN 978-1-4020-6152-3 (e-book)

Published by Springer,
P.O. Box 17, 3300 AA Dordrecht, The Netherlands.

www.springer.com

Printed on acid-free paper

Cover illustration: Olduvai Gorge, photo by Manuel Domínguez-Rodrigo

This book is dedicated to the following people:

To Mary Leakey, whose dedication to the field and attention to detail at Olduvai Gorge created a truly impressive wealth of archaeological data, without which the present work would never be possible.
MDR dedicates this book to Mary, for her unlimited patience, support, and help. With love.
CPE dedicates this book to his family and to Amy, the love of his life.
RB dedicates this book to her father, Juan Carlos, who was not able to witness the final product.

Contents

Contributors

Manuel Domínguez-Rodrigo
Department of Prehistory
Complutense University
c/Prof. Aranguren s/n
28040 Madrid, Spain
Email: manueldr@ghis.ucm.es

Rebeca Barba Egido
Department of Prehistory
Complutense University
c/Prof. Aranguren s/n
28040 Madrid, Spain
Email: rbarbaegido@yahoo.es

Charles P. Egeland
Department of Anthropology
Indiana University
Bloomington, IN 47405, USA
Email: cegeland@indiana.edu

Elia Organista
Department of Prehistory
Complutense University
c/Prof. Aranguren s/n
28040 Madrid, Spain

Rafael Mora
División de Prehistoria
Facultad de Letras, Universidad Autónoma de Barcelona,
08193 Bellaterra, Barcelona, Spain
Email: rafael.mora@uab.es

Ignacio de la Torre
Institute of Archaeology
University College London
31–34 Gordon Square
WC1H 0PY, London, UK
Email: itorre@ucl.ac.uk

Acknowledgments

We are very thankful to Eric Delson and Ross MacPhee (editors of the Vertebrate Paleobiology and Paleoanthropology book series at Springer) for their help and advice during the completion of this book. We also thank the following for permission to reproduce sections of previous publications: Journal of World Prehistory (Springer), Journal of Human Evolution (Elsevier), and Palaeogeography, Palaeoclimatology, Palaeoecology (Elsevier). We thank Mary Prendergast for her support and help in editing various chapters of this book. We also thank the Graduate and Professional Student Organization and The College of Arts and Sciences, Indiana University, and the National Science Foundation (Grant no. 0603746).

Preface

Plio-Pleistocene sites are a rare occurrence in the archaeological record. When they are uncovered, the faunal materials so crucial to unlocking their behavioral meaning are often poorly preserved. For example, at Koobi Fora, Kenya, a prolific region that preserves several classic Plio-Pleistocene sites, many bones are affected by poor cortical surface preservation (Isaac, 1997). Such taphonomic vagaries limit the range of questions that can be addressed with these assemblages. In other instances, access to materials can be limited due to local politics or rivalries between individual research teams. As a result, many important assemblages either remain unstudied or have been interpreted without the advantage of a fully developed taphonomic framework, a situation that all but guarantees stagnant interpretations.

After more than half a century of systematic archaeological work in East Africa, most of what is known about hominid behavior during the Plio-Pleistocene still comes from the archaeological sites from Olduvai Gorge in Tanzania. The preeminent role of Olduvai Gorge in discussions of early hominid behavior stems from three major factors: (i) a long tradition of archaeological research spanning nearly five decades; (ii) the low-energy, largely undisturbed contexts in which the faunal remains are preserved, often with pristine cortical surface preservation; and (iii) the accessibility of the materials, which has allowed several researchers to examine the same sites. This combination of factors is unique in East African Plio-Pleistocene archaeology and has stimulated much debate over the socioeconomic function of early sites. Influential models of early hominid behavior in the late 1960s and early 1970s were based exclusively on information from Olduvai Gorge (Leakey, 1971). Although Isaac's (e.g., 1978) work at Koobi Fora expanded their application, the subsequent critiques, modifications, and reformulations of these models were based almost solely on studies from Olduvai (Binford, 1981, 1984; Bunn, 1981; Potts, 1982, 1988; Bunn and Kroll, 1986; Blumenschine, 1995; Rose and Marshall, 1996; Domínguez-Rodrigo, 1997a, 2002; Plummer, 2004). Having said that, it is also clear that continued work at Koobi Fora and recent research at sites like Kanjera, also in Kenya, promise to expand our knowledge of early hominid behavior well beyond Olduvai (Plummer, 2004).

Some of these studies have made a major academic impact and, given the importance of these issues for understanding human evolution, have been further popularized through books and TV documentaries. However, it is important to realize that the most debated issues, including the existence of a passive scavenging stage in the Plio-Pleistocene (as opposed to hunting or confrontational scavenging), the socioeconomic characterization of early sites (e.g., "Central Places" or "Stone Caches") and the behavioral complexity of early hominids relative to extant apes, rest on the assumption that the Olduvai sites, especially those from Bed I, are largely the result of hominid activity. Leakey's (1971) well-known interpretation of the Bed I assemblages as hominid campsites is a clear example. These interpretations were made when taphonomic applications in archaeology were still in their infancy, and researchers like Bunn (1982), Shipman (1986), and especially Potts (1982, 1988) undertook hands-on investigations of several Bed I sites to test the models of Leakey (1971) and Isaac (1978). Rose and Marshall's (1996) paper, which drew upon many of these seminal taphonomic studies, summarized the consensus view that hominids were the primary agents in the formation of these sites. (Binford's [1981] contention – based on outdated secondary data and comparisons with human and nonhuman bone accumulations – that hominids were marginal scavengers and contributed little if at all to site formation, appeared to have been safely refuted by the early 1990s.)

This volume provides a fresh look at this issue, and our analyses present a more complicated picture of site formation during Bed I times at Olduvai Gorge. Our major aim, therefore, is to demonstrate the importance of rigorous taphonomic analyses to the interpretation of archaeological sites. These investigations demonstrate that only the assemblage from Level 22 at the FLK locality (the Zinjanthropus Floor) is almost solely anthropogenic. The other Bed I sites from Olduvai represent exclusively carnivore-collected bone accumulations (FLK North North Levels 1–3), background scatters created in death arenas that favored bone accumulation (FLK North Level 6), or palimpsests in which carnivores and, to a much lesser extent, hominids, contributed to the accumulations (FLK North Levels 1–5 and DK Levels 1–3).

Given that FLK 22 (alternately termed FLK Zinj) is the only Bed I site that is largely the result of hominid activity, its role in understanding Plio-Pleistocene subsistence behavior is more important than ever. Although our study of this assemblage agrees with other analyses that argue for a significant hominid signal, it is also evident that previous estimates of carnivore involvement have been greatly overestimated because natural biochemical marks prevalent on many FLK 22 specimens have been misinterpreted as carnivore tooth marks. Our reexamination also shows that the frequencies and distribution of cut marks are consistent with the processing of completely fleshed carcasses. The systematic nature of these data also suggest that hominids were skilled butchers, indicating that these activities (disarticulating, defleshing, and demarrowing) were an integral part of hominid subsistence behavior during Bed I times. Although the data are not fully conclusive, we feel the possibility that hominids were actively hunting and accumulating carcasses at FLK 22 should again be seriously considered.

Many of these ideas will contradict the current paradigm, which assumes that many Bed I sites are hominid accumulations, that the central tenets of the home base hypothesis have been refuted and that passive scavenging was the major carcass acquisition strategy during the Plio-Pleistocene. However, like the previous Darwinian vision of early hominids as dominant hunters (Darwin, 1871) living in forager-like camps, the current paradigm cannot be reconciled with the latest archaeological and taphonomic data.

Many researchers would argue that archaeological knowledge accumulates gradually, following a continuous progression. This Popperian outlook runs counter to the more pragmatic vision of academia as a circle of power. Both Kuhn (1962) and Lakatos (1978) have argued that academia is often a slave to its paradigms and thus reluctant to modify them. Archaeology is no different. The European tradition, which pioneered the study of early humans, upheld the hunting hypothesis without any real empirical support until the North

American processual tradition introduced new methodologies that questioned this interpretation. Largely through the control of Citation Index journals, this new paradigm was globally incorporated in a relatively short period of time. As mentioned, this volume will both challenge and build upon the ideas espoused by the processual critique of the hunting hypothesis.

It was clear from the beginning that the research conducted on the Bed I archaeofaunas needed to be presented in book form. Given that each site provided a unique taphonomic problem, the analyses are presented independently. When studying multiple sites, it has become customary to combine all sites together in the final analysis. The philosophy behind this approach is that taphonomic processes may affect sites similarly, and if they do not, a uniform analytical approach (which necessarily limits the number of variables used) can detect the differences. However, we advocate a different approach. Each site is the result of a unique depositional history. The collective analysis of a group of sites may use the same set of variables and even the same referential frameworks; however, the uniqueness of any given site is only understood if it is studied as an individual entity. Only in this way can all its defining characteristics (e.g., lithology, postdepositional winnowing, biotic impacts, exposure) be studied together. This is specifically true of the Olduvai sites: despite the similarities among them, each site is distinct from the others in terms of taphonomic history. Previous analyses, which lumped these sites together, produced ambiguous or incorrect interpretations of these sites as natural death and/or carnivore-made assemblages with specific signs of carnivore-hominid interaction (Binford, 1981) or, alternatively, hominid-created stone caches (Potts, 1988). The individual treatment of each site in the chapters which follow has helped differentiate the taphonomic history of one site from another, which in turn has yielded individual site interpretations.

We emphasize the individual nature of site analyses also because each of them is very useful to address specific and different hypotheses: for instance, interdependent or independent multiple agents in site formation (Chapter 8, Chapter 12, Chapter 15), scavenging from leopard kills (Chapter 9), natural death arenas (Chapter 7, Chapter 15), carnivore dens (Chapter 13), and linking carnivore ravaging to paleoecology (Chapter 11). Given the diverse nature of these problems, each chapter could not be homogenized to the exact same format. However, the structure is the same throughout: introduction (describing the problem to be analyzed), results (divided by several sections including, in this order, skeletal part representation, bone breakage, and bone surface modifications), discussion, and conclusion. That is, a similar structure to a paper but without the methods, which are described in Chapter 3.

Chapters 1 and 2 provide the history of research on site formation and hominid behavior during the Plio-Pleistocene in order to place the current analyses into a theoretical and methodological context. This summary is organized around two general issues: (i) the socioeconomic function of early archaeological sites (the home base debate) and (ii) the carcass acquisition strategies employed by hominids (the hunting-versus-scavenging debate). Chapter 3 summarizes various approaches to tackling taphonomic problems, and introduces the methodological underpinnings of our own research, which we refer to as the "physical attribute" approach to taphonomic analysis. The following chapters apply this approach site by site. As the most anthropogenic of the Bed I sites, an in-depth treatment of FLK Zinj is provided. We conclude by discussing the implications of our findings for characterizing the ecological role of hominids within the broader context of Plio-Pleistocene carnivore guild dynamics, at both the site and landscape level.

This project began after a reanalysis of the FLK Zinj fauna (carried out by MDR and RB in November, 2004) revealed significant incongruities with previous research. Advances in taphonomic methods subsequent to the last comprehensive treatment of many Bed I sites demanded analysis of all the Bed I assemblages, and CPE joined the project during the summer of 2005. Each site was analyzed separately under the direction of one of the three authors using specific techniques appropriate to each assemblage. Nevertheless, each analysis is united by the same overarching methodology, and the authors consulted each other and exchanged (sometimes opposing!) opinions at all stages of analysis. It is noteworthy that much of the Bed I equid material reported by previous research was not available to us during our stay at the National Museums of Kenya in 2004–2006, as it is currently on international loan.

There are a number of individuals and institutions we wish to thank for their support and input throughout this research. We thank COSTECH (Tanzanian Commission for Science and Technology) for permission to study the Olduvai materials. Special thanks are due to Emma Mbua (Head of Paleontology, National Museums of Kenya) and all her staff for their invaluable help during our research at the museum. We are particularly thankful to Elia Organista for her contribution to two chapters of the book and to Rafael Mora and Ignacio de la Torre for their contribution in one chapter. We greatly appreciate the comments made on the early drafts of this manuscript by T. Pickering, T. Plummer, P. Andrews, Sixto Fernández, L. Alcalá, R. Meadow, Manuel Santonja, Gerardo Vega and Nines Querol, and one anonymous reviewer. T. R. Pickering's and H. T. Bunn's works have been a continuous source of support and inspiration.

1. The "home-base" debate

M. DOMÍNGUEZ-RODRIGO, C.P. EGELAND, AND R. BARBA

The dense concentrations of faunal remains and stone tools at some Plio-Pleistocene sites at Olduvai Gorge (Tanzania) and Koobi Fora (Kenya) have traditionally been interpreted as favored locations to which hominids repeatedly transported carcass parts for processing (Leakey, 1971; Isaac, 1978, 1983, 1984; Bunn, 1982, 1991; Potts, 1982, 1988; Bunn and Kroll, 1986; Schick, 1987; Blumenschine, 1988, 1991, 1995; Bunn and Ezzo, 1993; Schick and Toth, 1993; Oliver, 1994; Rose and Marshall, 1996; O'Connell, 1997). These co-occurrences, which typically contain bones from several individuals within a vertically discrete horizon (referred to by Isaac [1978] as "Type C" sites), often preserve high densities of archaeological material in spatially restricted concentrations. For many researchers, these archaeological sites represent places where hominids may have stayed for extended periods, very likely performing activities beyond stone tool manufacture and carcass manipulation (Leakey, 1971; Isaac, 1978, 1984; Bunn, 1982; Stanley, 1992; Domínguez-Rodrigo, 1994a; Oliver, 1994). The seemingly frequent processing of nutrient-dense large mammal tissue by hominids between 2 and 1.5 Ma led many evolutionary anthropologists to suggest meat eating as the critical adaptation for understanding the emergence of stone tool use. These debates necessarily gravitated towards the evidence from Olduvai, and more specifically, from the single site of FLK Level 22 (the *Zinjanthropus* Floor). Despite its importance for reconstructing hominid behavior, an almost exclusive focus on the archaeological evidence from FLK *Zinj* has limited archaeologists' views in two important ways. First, it forced researchers to neglect regional variability in hominid adaptive patterns (clearly stressed by Potts [1994]). Second, it hindered the development of explanatory frameworks that could deal with a diversity of site formation scenarios. Although the debates over these "Type C" sites have clearly generated fruitful discussions over the past two decades, it is also evident that other types of archaeological occurrences representing stone tool–using activities, not necessarily linked to carcass manipulation, have yet to be fully appreciated.

When Mary Leakey (1971) published the results of her excavations in volume three of the Olduvai Gorge monograph series, she identified four types of sites: (1) living floors; (2) butchering or kill sites; (3) channel sites; and (4) vertically dispersed deposits where hominid behavior could not be accurately characterized. The blueprint provided by the "Man the Hunter" paradigm of the 1960s clearly influenced Leakey's (1971) interpretations of the Olduvai sites, which centered on home bases and cooperative hunting activities. Isaac (1978) explored this issue further, defining "humanness" around a series of features that included symbolic behavior, articulated language and, especially, archaeologically detectable home bases where intentional food sharing took place. This "Home-Base" or "Food-Sharing" model interpreted early sites as

M. Domínguez-Rodrigo, R. Barba, and C.P. Egeland, Deconstructing Olduvai: A Taphonomic Study of the Bed I Sites, 1–10.

resulting from the delayed consumption of transported carcass parts. Delayed consumption implied intentional food sharing, which in turn triggered cooperation, sexual division of labor, and the emergence of home bases, which were themselves seen as *loci* for social activities.

Subsequent taphonomic research began to question both the methodological and theoretical framework of the "Home-Base/Food-Sharing" model. Some researchers even contended that nonhuman agents were largely responsible for the formation of many Plio-Pleistocene sites, giving rise to radically different interpretations of early hominid behavior (e.g., Binford, 1981). Because they did not require essentially modern hunter-gatherer behavior to accompany the emergence of the earliest archaeological record (as the initial formulations of the "Home-Base/Food-Sharing" hypothesis did), such interpretations appeared to better fit within a gradualistic Darwinian framework. Early archaeological sites were thus interpreted as derived hydraulic jumbles, carnivore kill sites, or hyena dens where hominids played little or no role in site formation and, when present, were limited only to eating scraps of flesh and marrow (Binford, 1981, 1985). This in turn implied that although biological agents may have participated in site formation, the faunal remains themselves were not actively accumulated but rather occurred more or less at the original death site. However, subsequent hands-on taphonomic analyses initiated in the late 1970s suggested that these early sites in fact represented special locations on the landscape where carcasses were actively accumulated, many of them presumably by hominids. This argument, summarized usefully by Potts (1988), was based on the following:

1. The density of faunal remains at early sites seemed extremely high when compared to natural bone scatters observed in modern savanna landscapes. That is, attritional mortality simply did not generate the spatial clumping of carcasses seen at early sites. In fact, it was observed that as animals die, their bones become scattered if they are not consumed or otherwise destroyed (Hill, 1975, 1979a, 1979b, 1980; Behrensmeyer and Boaz, 1980; Behrensmeyer, 1983). Other potential accumulating agents (such as water flow) also did not produce the density or patterning of faunal remains seen at early sites (Hill, 1975, 1979b; Shipman, 1975; Potts, 1982, 1988).
2. Under normal circumstances, remains from several individuals were rarely encountered in modern savannas (Hill, 1975, 1979b; Behrensmeyer, 1983). For example, Behrensmeyer (1983) found an average of only one individual and a maximum of three in her 1,500 m^2 sampling units in Amboseli National Park in Kenya. A similar study conducted in a smaller area near a pond, north of Lake Ndutu in Tanzania, yielded a maximum of five individuals (Domínguez-Rodrigo, 1993). In the archaeological levels from Olduvai Gorge, however, 34–36 individuals were represented in areas smaller than the sampling units used in modern landscape studies (Potts, 1982, 1988).
3. The taxonomic and ecological diversity of the faunal assemblages was relatively high (Bunn, 1982; Potts, 1982, 1988). This would not be expected if early sites represented serial predation by carnivores, as the range of taxa represented at such sites tends to be relatively narrow (Kruuk, 1972). In addition, animals are often preyed upon and consumed within their ecological niche.
4. Because axial bones were observed to be more abundant at death sites (Hill, 1975, 1979a), the high representation of limb bones at many early sites suggested postmortem transport instead of *in situ* carcass processing (Potts, 1982, 1988).
5. Finally, the spatial and functional (based on butchered bone specimens) association

of stone tools and fauna implicated hominids as primary accumulating agents at "Type C" sites.

Consequently, it became broadly accepted that the sites from Bed I at Olduvai Gorge – and possibly other "Type C" sites in East Africa – were the result of active carcass transport. Because wind and water action were largely excluded as major accumulating agents (Hay, 1976; Potts, 1988), it seemed that hominids or other savanna carnivores were largely, though not solely, responsible for transporting carcass parts (Potts, 1988; see review in Domínguez-Rodrigo, 1994a, 1994b). The goal then became to determine the relative roles of hominids and carnivores in the accumulation of Plio-Pleistocene faunas.

The observation that felids accumulate bones led several researchers to stress the role of these carnivores in site formation (Brain, 1969, 1981; Tappen, 1992). However, felids usually do not create large bone accumulations because they consume their prey near the site of acquisition, only moving the carcass if they are in danger of losing it to other predators (Domínguez-Rodrigo, 1994a). However, large bone accumulations can be created by felids at serial predation sites. This site formation scenario was seen as unlikely given that the establishment of such hunting arenas, particularly by lions, requires proximity to open-air water sources such as ponds and is viable only at the end of the dry season. Such settings are relatively rare, and, when present, create bone assemblages that are spatially diffuse and that only contain prey from the ecological context in which the site is situated (Domínguez-Rodrigo, 1993). This, as mentioned earlier, ran counter to the patterning observed at early sites, which preserve extremely dense accumulations represented by a diversity of ecological niches.

Leopards are exceptional among felids in that they are known to repeatedly transport carcasses into caves, rock shelters, and trees (Brain, 1981; Bunn, 1982; Cavallo, 1998). However, leopards prey mainly on smaller mammals (Kruuk and Turner, 1967), whereas a majority of individuals represented at some Plio-Pleistocene sites were medium-sized (Bunn, 1982; Potts, 1982, 1988). This was inconsistent with the idea that leopards accumulated bones at early sites.

Hyenas, which accumulate substantial numbers of bones at dens, were also seriously considered as agents of site formation. Based largely on skeletal part frequencies, Binford (1981, 1985) in particular attributed many Plio-Pleistocene faunas to the activities of hyenas. Although hyenas do preferentially transport limb bones (Hill, 1975; Brain, 1981; Domínguez-Rodrigo, 1994a) and consume their epiphyses, subsequent taphonomic research revealed that such patterns are not necessarily indicative of primary and sole access by hyenas to carcasses. It was shown that an underrepresentation of epiphyses and axial bones can also result from hyenas ravaging hominid food refuse (Potts, 1982, 1988; Marean *et al.*, 1992; Capaldo, 1995, 1998b).

Both hominids and hyenas were confirmed as taphonomic agents by the presence of cut marks and tooth marks on the surfaces of several bones from Olduvai Gorge and Koobi Fora (Bunn, 1981, 1982, 1983a; Potts and Shipman, 1981; Bunn and Kroll, 1986; Shipman, 1986). The issue then was to determine if early sites were hyena accumulations in which hominids participated only secondarily (Binford, 1981, 1985) or whether they represented hyena scavenging of hominid food refuse (Bunn, 1982; Isaac, 1983; Bunn and Kroll, 1986; Blumenschine, 1988; Marean *et al.*, 1992).

Bunn (1982; Bunn and Kroll, 1986, 1988) argued that the presence of cut marks on meat-bearing bones indicated primary access by hominids to carcasses. It was thus logical to conclude that hominids were the primary and hyenas the secondary agent in site formation. This "standard argument" (as Binford [1985] termed it) became increasingly supported by several middle-range studies (Blumenchine, 1988; Marean *et al.*, 1992; Selvaggio, 1994; Capaldo, 1995, 1997, 1998b). When combined, several lines of data clearly indicated that early sites could not have been the result of hominids scavenging from hyenas:

1. The high frequency of meat-bearing elements with cut marks could only be satisfactorily explained if hominids had defleshed them first (Bunn, 1981; Potts and Shipman, 1981; Isaac, 1984; Bunn and Kroll, 1986), as large carnivores rarely leave appreciable amounts of meat on bones (Domínguez-Rodrigo, 1999a) (see later).
2. About 25% of bone specimens from Plio-Pleistocene sites were tooth-marked, which was much less than that observed in assemblages accumulated and consumed by hyenas (about 80%) (Bunn, 1982; Potts, 1982, 1988; Blumenschine, 1988; Domínguez-Rodrigo, 1994a). Moreover, the frequency of tooth-marked specimens at early sites was consistent with hyena ravaging of a hominid-generated assemblage (Blumenschine, 1988; Capaldo, 1995).
3. Habitat reconstructions suggested that archaeological sites were found in riparian woodlands (lacustrine and gallery forests). This was not consistent with the habitat preferences of hyenas, which are commonly found in open plains away from forested areas (Blumenschine, 1986; Domínguez-Rodrigo, 1994a).
4. Hyenas at least partially excavate their dens, and no stratigraphic evidence had been detected to suggest that burrows or other aspects of den structure existed at Plio-Pleistocene sites.
5. If hominids had scavenged bones from hyenas, it was presumably to extract the marrow they contained (Binford, 1981). Notwithstanding the fact that little marrow would have been left by bone-crushing hyenas, it seemed logical to expect marrow harvesting to be reflected in the lithic assemblages. However, although nodular artifacts like hammerstones and anvils would have been the most useful tools for such activities, many Plio-Pleistocene lithic assemblages were dominated by flakes and debitage (Leakey, 1971; Isaac, 1976; Merrick and Merrick, 1976; Toth, 1982, 1985). Therefore, hominids were creating cutting tools to fulfill functions such as stripping meat from carcasses. Keeley and Toth's (1981) use-wear analysis of several artifacts from Koobi Fora demonstrated that only flakes preserved wear polish and that some polishes clearly indicated contact with meat.

The wide range of studies undertaken to address this issue had also produced an extremely useful methodological framework for disentangling the complex web of agents contributing to the formation of Early Stone Age sites. After two decades of such research, archaeologists at the millennium felt confident attributing the formation of early faunal assemblages largely to the actions of hominids (Blumenschine and Bunn, 1987; Bunn, 1991; Rose and Marshall, 1996). However, the behavioral implications of these artifact and bone concentrations were not yet resolved. Two socioeconomic models of early site formation were proposed that fit what was known of the hominid contribution to early site formation: (1) the "Stone Cache" model (Potts, 1982, 1984, 1988) (later renamed the "Resource Transport" model [Potts, 1991]) and (2) the "Central-place Foraging" model (Isaac, 1983).

Potts' (1988) model rested on the simple fact that early sites unambiguously reflected the manufacture of stone tools and the processing of carcasses. For Potts, any other interpretation of site function became secondary inferences. While he never rejected food sharing or cooperative behavior; he simply said that these behaviors could not be clearly demonstrated by his analysis of the archeological evidence. The original formulation of Isaac's (1978) model argued that Plio-Pleistocene sites represented home bases where intentional food sharing occurred, all of which resulted from a cooperative social system based on a sexual division of labor. This implied that home bases were safe places on the landscape where individuals could carry out social activities and information exchange. This conception of site function was rejected by the "Stone Cache" model, as several lines of evidence seemed to indicate that sites were not conducive to long-term hominid occupation. The abundance of tooth-marked bones demonstrated that both large carnivores and hominids frequented the same localities. As inferior competitors, hominids would have minimized the time spent at sites. The high frequencies of complete bones found at sites, which suggested incomplete processing and thus short-term occupation, also supported this claim. Finally, bone weathering studies suggested that site formation could have taken up to 10 years, which exceeded the time during which many hunter-gatherers reuse campsites.

According to Potts (1988), the co-occurrence of lithic material and bones on the landscape could be due to: (1) the transport of carcasses to raw material sources; (2) the movement of stones to carcass acquisition points; or (3) the transport of both stones and carcasses to referential places. Arguing from an optimal foraging standpoint, Potts concluded that the third option would have been the most energetically efficient. If encounter rates with carcasses increased, hominids were expected to establish several referential places in order to reduce transport distances. This stood in contrast to Isaac's "single-place" strategy where hominids would transport carcasses to a single home base regardless of transport distance.

It was therefore suggested that early sites represented "stone caches" – collections of stored lithic material – where carcasses would have been transported for quick processing. Such a strategy, like that employed by, for example, leopards, which bring prey to particular spots for consumption, was meant to avoid competition with other carnivores and to cut down on energetic costs in faunal processing. Thus, these caches were subsistence-based *foci* only. According to Potts, this model was appealing because hominids could be understood wholly within an ape behavioral framework. They were not required to possess premeditation or a fundamentally different type of social organization that fostered extensive food sharing. Therefore, the model predicted an intermediate range of behaviors, somewhat different from extant apes, but not quite fully "human."

When scrutinized, the "Stone Cache" model did in fact require relatively complex behavior. For example, although chimpanzees were known to use, reuse, and transport stone tools in the wild, stone usage was sporadic; reuse was simply a by-product of seasonal exploitation of nearby tree resources; and transport was limited to very short distances (Boesch and Boesch, 1984). Hominids, on the other hand, were, according to the "Stone Cache" model, repeatedly supplying stones to particular spots where they were intentionally flaked for carcass processing. The fact that hominids were transporting lithic material over several kilometers (Hay, 1976; Toth, 1982) also suggested a reliance on stone tools well beyond that of apes.

Stone caching as envisioned by Potts also implied forethought and strategic anticipation, as hominids were selecting particular spots on the landscape to accumulate stones prior to their actual utilization. Such planning depth lies outside the range of modern ape behavior, and Binford (1985) even argued that this type of behavior was known only among behaviorally modern hunter-gatherers. To counter this argument, Potts (1993) argued that stone cache production was a biological strategy and that cache production may have been an unintentional consequence of hominid discard behavior. The *a priori* rejection of cooperative behavior and food sharing was also problematic. Given the risk of predation on the open savanna, it is probable that several individuals at a time participated in stone transport, and hominids likely would have used the stones collected by other individuals. If large lithic assemblages from sites like FLK *Zinj*, where over 20 kg of quartz debitage (whose original source is several kilometers away) were recovered, were the result of cooperative transport, it seemed likely that shared use of the resources obtained from using the material (animal tissues) would be expected. Although innovative in many ways, the "Stone Cache" model implied many of the hominid behaviors it tried not to assume (strategic planning of future events, cooperative behavior, resource sharing).

The "Stone Cache" model was also at odds with some of the archaeological data. For instance, if stones were cached for future usage, unutilized cores or manuports were expected to dominate the lithic assemblages as they do at some Olduvai sites (Potts, 1982, 1988). However, at Plio-Pleistocene sites all along the African Rift (e.g., Omo, Hadar, Koobi Fora, West Turkana) flaked material appeared in high frequency relative to manuports. This forced Potts to suggest that some Koobi Fora sites represented an archaeological occurrence distinct from the stone caches at Olduvai. As Potts clearly recognized, it was also expected that if carcasses were carried to the nearest processing spot, the taxonomic representation of the fauna should reflect the habitat in which the site appeared. Instead, as noted earlier, the Olduvai Bed I sites contained taxa from a variety of habitats. This indicated that their transport could not be explained solely by short distance transport and carnivore avoidance. The paleoecological context of many Plio-Pleistocene sites also ran counter to the idea that they were dangerous places. Most sites were situated at the proximal ends of alluvial plains, areas associated with closed-vegetation habitats such as lacustrine and gallery forest. Among the range of savanna habitats, these areas are associated with the lowest predation risk (Blumenschine, 1986; Domínguez-Rodrigo, 2001). Therefore, early sites would have been conducive to prolonged stays where activities other than just carcass processing may have been carried out. For example, microwear analysis suggested that in addition to carcass butchery, some stone tools were used for cutting soft plant material and scraping and sawing wood (Keeley and Toth, 1981). Furthermore, the presence of tooth-marked bones did not necessarily indicate high levels of interference competition and predation risk. More likely, this indicated serial use of the site by both hominids and carnivores. Potts' contention that early sites may have taken up to 10 years to form was based largely on data from subaerial weathering. Although several factors, including skeletal element, taxon, and depositional environment condition weathering rates (Lyman and Fox, 1989), Bunn and Kroll (1986) used similar data to argue that some sites may have formed within 2 years. (As will be seen, our own analyses of the Bed I sites from Olduvai suggest relatively quick site formation.) Finally, a recent reanalysis of the Bed I manuports suggested that their small size and poor quality (being porous basalts), together with the fact that most detached pieces are in a different raw material (quartz), identified these stones as ecofacts (de la Torre and Mora, 2005).

For these reasons, Isaac's (1983) "Central-place Foraging" model was more appealing to many researchers. According to Isaac, early sites represented referential places established by hominids to act as points of dispersal and fusion. Individuals would temporarily split apart to procure resources and reunite again to share food and potentially carry out other social activities. After two decades of taphonomic work, this model appeared to adequately account for the formation of Plio-Pleistocene sites (Rose and Marshall, 1996), and implied innovative hominid behaviors such as

1. Relatively sophisticated planning abilities expressed in the procurement and intentional modification of lithic raw material. Hominids also established referential places (which later became archaeological sites) before food and stones were actually transported to them.
2. The delayed consumption of animal tissues, which were transported and perhaps shared at the referential place (vegetal foods were also processed at [Keeley and Toth, 1981], and presumably transported to, sites as well).

Central-place foraging is currently a widely accepted interpretive framework for characterizing early archaeological sites, particularly those from Olduvai and Koobi Fora. A refined version of this model can be found in Schick's (1987) "Favored Place" model, where hominids are presented as being dependent on raw materials for stone tool production and discard, which would have conditioned the *loci* on the landscape where stone tools and bones were more likely to accumulate. According to Schick (1987), repeatedly visited sites with nearby lithic sources would have a net import of artifacts (and potentially food residue, like carcass parts) greater than on other parts of the landscape, because the need to transport stone was less immediate and artifact discard rates were increased, due both to the proximity of raw material sources and the use of tools on these *loci*. In this model, carcass acquisition strategies are not discussed as they are in Isaac's central-place foraging model.

However, by emphasizing hunting/confrontational scavenging and male provisioning, many researchers over the last 20 years have argued that this model is male-biased. Alternative models have downplayed the role of meat and male provisioning while stressing the importance of female gathering activities. O'Connell *et al.*'s (2002) "Male-display" model is the most recent version of this critique. They argue that male hominids displayed in front of the social group by confronting carnivores at kills. Faunal accumulations were created very near the site of carcass acquisition and did not fulfill any of the socioeconomic functions of central-place foraging, including male provisioning. This model is provocative because it draws upon a wealth of archaeological and ethnographic data and correctly recognizes the importance of female subsistence activities. However, there are several issues that need to be worked out before this model will adequately explain Plio-Pleistocene site formation.

O'Connell *et al.*'s (2002) model is structured around (1) studies of one hunter-gatherer group (the Hadza in Tanzania), (2) the replacement of central places with "near-kill" locations, and (3) skeletal part profiles and bone surface modifications. O'Connell *et al.* (2002) claim that the consumption of meat among the Hadza is relatively infrequent and that hunting is not a reliable source of energy. Further, they document that most of the meat from a hunt is distributed not to the successful hunter's offspring, but to other members of the social group. Despite this, Hadza men continue to hunt because successful hunters gain social status and more frequently participate in extramarital sex. Linking these ideas together, O'Connell *et al.* (2002) suggest that meat acquisition by hominid males, carried out by usurping carcasses at carnivore kills, was meant to procure mating opportunities rather than provide a regular source of energy to other group members. Females would then selectively mate with the males most successful at "power scavenging."

Such a uniformitarian leap requires that modern, observable processes have specific and unique effects that can be detected archaeologically (Gifford-Gonzalez, 1991; Lyman, 1994). However, O'Connell *et al.*'s (2002) ethnographic analogy leaves many of their points open for alternative explanations. Most importantly, a similar pattern of hunting prestige needs to be documented systematically in other hunting and gathering groups. Another major objection concerns the hunting tactics employed by the Hadza. In addition to increasing hunting returns, an important secondary effect of adopting sophisticated bow and arrow technology like that used by the Hadza was to enhance the success of individual hunters. Other forms of hunting, including thrusting spears, traps, and stone throwing, which would have been used throughout the Pleistocene, often require the participation of a number of individuals. Such collective efforts provide less opportunities for display and also call for what Potts (1988: 290) has termed an "expectancy of sharing" among the group. Therefore, if group hunting was the primary strategy throughout the Pleistocene, prestige like that gained by Hadza hunters (perhaps only made possible through bow and arrow technology) may be a relatively recent phenomenon.

The suggestion that Plio-Pleistocene hominids could actively displace carnivores larger in size than modern ones with stones and branches also remains untested. As O'Connell *et al.* (2002) report, the ability of the Hadza to displace carnivores depends largely on their use of heavy bows, which are effective at 40 m. The lighter and smaller bows of the San, however, make them much less formidable in confrontations with large carnivores. Recent reports from Uganda of human deaths also attest to the danger associated with scavenging from predators, even with modern technology (Treves and Naughton-Treves, 1999).

O'Connell *et al.* (2002) also argue that family provisioning is an unlikely goal for hunting given the irregularity of meat acquisition. However, meat may be an unreliable food source for the Hadza because, like many modern hunter-gatherers, they live in marginal environments (O'Connell *et al.* [2002] readily admit that the modern Eyasi Basin ecosystem is not a perfect analog for Plio-Pleistocene environments) and regularly acquire agricultural products. Nevertheless, other foraging groups that depend more heavily on meat than do the Hadza use it as an important aspect of family provisioning (Damas, 1972; Nooter, 1975; Wenzel, 1995; Bodenhorn, 2000; Cordain *et al.*, 2000). It has also been suggested that infant provisioning is the main reason for even Hadza hunting (Marlowe, 2005). There are physiological reasons to suspect that meat was an important component of hominid diets during the early Pleistocene. Hominids throughout the Pleistocene were much more robust and muscular, and it has been estimated that Neanderthals for example required double the calories of modern humans (Sorenson and Leonard, 2001). The bone architecture of *Homo ergaster*, which is characterized by thick cross sections, suggests a heavily muscled species with relatively high caloric requirements (Wolpoff, 1999). Therefore, the pressure for Pleistocene hominids to acquire high-quality, easily digestible foods like meat may have been accentuated relative to more gracile modern humans like the Hadza.

O'Connell *et al.* (2002) suggest that early sites represent "near-kill" locations reflecting short-distance transport from the site of carcass acquisition. This runs counter to the "Central-place Foraging" model, which posits repeated long-distance carcass transport to favored locations. As a comparison, O'Connell *et al.* (2002) present a Hadza near-kill locale with a total of 400 specimens representing at least 300 elements from 11 individuals. Although these frequencies are comparable to some of the smaller Olduvai sites, FLK *Zinj,* which is the quintessential Plio-Pleistocene central place, has around 60,000 bone specimens that represent over

50 individuals. Clearly, other taphonomic and ethnoarchaeological data will be required to justify such a comparison. For example, at Hadza near-kill locations, bones from a single individual are usually clustered together, sometimes in nearly or fully articulated sets (O'Connell *et al.*, 1992). To date, a similar pattern has not been documented in spatial analyses of the Olduvai sites, although more research on this aspect of site formation is clearly warranted.

The lithic data are also inconsistent with the idea that Plio-Pleistocene sites are simply near-kill butchery locations as O'Connell *et al.* (2002) argue. Butchery locations are usually characterized by short-term occupations where only a limited number of activities are carried out. If this were the case for early sites, we would expect relatively low frequencies of lithic artifacts relative to the number of carcasses present (after all, it only takes a few good flakes to butcher a medium-sized carcass). However, the sites from Olduvai preserve some of the highest densities of stone artifacts for sites between 2.0 and 1.5 million years old. Some of these artifacts were even brought on-site for purposes clearly beyond just butchery. Furthermore, if hominids were maximizing the efficiency of their stone raw material for butchery during brief occupational episodes, the reduction of cores and/or the number of discarded pieces should covary with the distance from the material's source. This pattern is in fact documented at Pliocene archaeological sites like Gona and Bouri in Ethiopia and Lokalalei in Kenya (Roche *et al.*, 1999) and at Peninj, Tanzania at 1.5 million years ago (de la Torre *et al.*, 2003). However, the pattern does not hold at many of the classic early Pleistocene "Type C" occurrences from Olduvai.

Drawing on data from body part representation and bone surface modifications, O'Connell *et al.* (2002) contend that the early Pleistocene occurrences from especially Olduvai and Koobi Fora are consistent with aggressive, but opportunistic, hominid scavenging of large carnivore kills. Therefore, although O'Connell *et al.* (2002) concede that hominids enjoyed at least occasional early access to carcasses, such opportunities were not frequent or consistent enough to generate returns worth transporting back to the social group on a regular basis. However, body part profiles are ambiguous indicators of hominid behavior given the bias introduced by ravaging carnivores. Furthermore, the cut mark data do not support aggressive scavenging any more than they support active hunting, as signature criteria for distinguishing these scenarios have yet to be fully developed. As O'Connell *et al.* (2002) point out, experimental controls modeling hominid scavenging of carcasses in a variety of ravaging states are clearly needed.

Given that the current cut mark data from Olduvai, Koobi Fora, Peninj, and, most recently, Swartkrans Member 3 indicate the defleshing of fully fleshed carcasses (Bunn and Kroll, 1986; Domínguez-Rodrigo, 2002; Domínguez-Rodrigo *et al.*, 2002; Pickering et al., 2004), the critical issue revolves around how often hominids encountered carcasses in this state. Estimates of Plio-Pleistocene site formation times vary from 1 year to a millennium, which means carcass encounter rates could have been either quite frequent or "meaninglessly low" (O'Connell *et al.*, 2002: 853). Detailed zooarchaeological and taphonomic analyses of both the fossil and actualistic assemblages themselves may help resolve this issue on a site-by-site basis. For example, decreasing encounter rates should lead to more intensive carcass exploitation, which itself may be reflected by fragmentation, surface mark frequencies, and the range of carcass resources and carcass parts processed. Systematic patterns of carcass manipulation (e.g., patterned cut mark and percussion mark locations) may indicate that carcass butchery was a regular feature of a particular hominid group's behavioral repertoire, in turn suggesting relatively frequent carcass encounters. Again, these issues must be addressed on a site-by-site basis using a myriad of zooarchaeological and taphonomic data (some of which are presented in this volume).

Overall, O'Connell *et al.* (2002) bring together a vast quantity of data to provide an interesting model for early site formation. However, the reliance on one modern hunter-gatherer group (the Hadza) living in a marginal environment with access to bow and iron technology and agricultural products is problematic. In addition, more zooarchaeological and taphonomic data must be brought to bear on important issues such as carcass encounter rates and the timing of hominid access to carcasses. In fact, African zooarchaeologists are still engaged in the debate surrounding carcass access and the regularity of meat consumption by hominids. This debate, probably more than any other, has directly influenced many of the pioneering studies in archaeological taphonomy over the past 25 years.

2. The Hunting-versus-scavenging debate

M. DOMÍNGUEZ-RODRIGO, C.P. EGELAND, AND R. BARBA

Introduction

The hunting-versus-scavenging debate has focused primarily on ascertaining the order of access by hominids to carcass resources. Primary access implies either confrontational scavenging (Bunn, 1995, 1996) or hunting, while secondary access suggests passive scavenging. As the previous chapter indicated, after 25 years the meaning of the earliest archaeological sites is still controversial. However, most Africanists would now agree that some sites were created by hominids repeatedly carrying carcass parts and stones to particular places on the landscape (Bunn, 1982, 1983a, 1991; Potts, 1982, 1988; Isaac, 1983; Bunn and Kroll, 1986; Blumenschine and Bunn, 1987; Blumenschine, 1988, 1991, 1995; Blumenschine and Marean, 1993; Bunn and Ezzo, 1993; Schick and Toth, 1993; Blumenschine *et al.*, 1994; Domínguez-Rodrigo, 1994a; Oliver, 1994; Selvaggio, 1994; Capaldo, 1995, 1997; Rose and Marshall, 1996; Cavallo, 1998). This consensus stands as one of the most important achievements of Plio-Pleistocene archaeological taphonomy. However, the socioeconomic function of these sites (Isaac, 1978, 1983; Potts, 1982, 1988; Stanley, 1992; Schick and Toth, 1993; Blumenschine *et al.*, 1994; Domínguez-Rodrigo, 1994a; Oliver, 1994; Rose and Marshall, 1996) as well as the nature of hominid involvement with carcasses (Bunn and Kroll, 1986; Blumenschine, 1991, 1995; Blumenschine and Madrigal, 1993; Bunn and Ezzo, 1993; Selvaggio, 1994; Capaldo, 1995, 1997; Shipman and Philipps, 1976; Domínguez-Rodrigo, 1997a; Cavallo, 1998); remain disputed. Both issues revolve around the central question of the long standing hunting-versus-scavenging debate: the extent of meat-eating by early hominids. Regular meat-eating implies early access to carcasses and a food surplus that may have promoted food-sharing and other cooperative behaviors as detailed in Isaac's (1978, 1983) well-known models. In this scenario, sites are social *loci* that reflect the beginnings of a "human-like" behavior similar to that observed in modern hunter-gatherer base camps. If, instead, carcass access was late, as would be the case for a passive scavenging strategy, meat would have been unimportant in early hominid diets (Blumenschine, 1986). In addition, low-yielding carcasses provide little or no surplus resources for sharing and the potential for other complex cooperative behaviors to develop in early hominids is thus diminished. This chapter provides a selected summary and critique of the various methodological and theoretical approaches to this hotly debated issue.

An extended version of this chapter appeared in *Journal of World Prehistory* (Domínguez-Rodrigo, 2002). Reproduced with permission from Kluwer.

M. Domínguez-Rodrigo, R. Barba, and C.P. Egeland, Deconstructing Olduvai: A Taphonomic Study of the Bed I Sites, 11–22.

The Use and Misuse of Skeletal Part Profiles

Because skeletal part frequencies and taxonomic identification have long been the primary duties of zooarchaeologists, discussions naturally focused on these data in the early phases of the hunting-versus-scavenging debate. Researchers have written extensively on developing methods for quantifying the various anatomical parts and individuals represented in fossil bone assemblages (e.g., Binford, 1978, 1981, 1986, 1988a; Bunn, 1982; Klein and Cruz-Uribe, 1984; Grayson, 1984; Bunn and Kroll, 1986, 1988; Gamble, 1986, 1999; Stiner, 1990, 1991, 1994). That is not to say, of course, that other forms of taphonomic data such as bone breakage patterns and surface modifications were ignored. However, identifying and interpreting the perimortem and postmortem damage undergone by carcasses parts (butchery, burning, carnivore ravaging, trampling, weathering, etc.) was a secondary analytical procedure carried out largely to reinforce inferences drawn from primary analyses.

The implications of skeletal part profiles for hominid behavior have often been controversial. For instance, the skeletal part analyses of Binford (1981, 1984) and Binford and Ho (1985) at Zhoukoudian, Olduvai, or Klasies River Mouth have been contested, respectively, by the alternative explanations of Bunn and Kroll (1986) and Klein (1982a, 1982b). The African Plio-Pleistocene also provides clear examples of disparate interpretations based on skeletal part analyses. Some researchers suggested, as we have seen, that hominids were marginal scavengers (Binford, 1981, 1984, 1985, 1988a, 1988b; Shipman, 1983, 1986; Binford *et al*., 1988), whereas others depicted them as successful hunter/scavengers targeting flesh rather than marrow (Bunn, 1981, 1982, 1983a, 1983b; Isaac, 1983, 1984; Bunn and Kroll, 1986). Using the same skeletal part profiles, some researchers argued that hominids were transporting high-yielding meat sections from carcasses (Potts, 1988; Bunn and Ezzo, 1993), while others contended that high-yielding marrow bones already devoid of flesh were being selected (Blumenschine, 1991, 1995; Blumenschine and Marean, 1993), or that hominids were opportunistic scavengers conditioned by processing costs rather than calorie yields alone (Lupo, 1998). This controversy convincingly demonstrates the limited interpretive power of skeletal part profiles (at least given current analytical procedures) for inferring hominid behavior. Two issues in particular are highlighted below.

USING MODERN HUMANS AS ANALOGS FOR EARLY HOMINIDS

The analysis of skeletal part profiles is partly based on the assumption that a diagnostic pattern exists in the way humans and other agents transport and accumulate bones. Several well-known ethnoarchaeological studies on carcass transport among modern foragers have been carried out to provide a framework for understanding these processes (Binford, 1978, 1981; Bunn *et al*., 1988, 1991; O'Connell *et al*., 1988, 1990, 1992). These data are believed by some researchers to indicate that humans selectively transport certain anatomical units. The most influential model, based on the work of White (1952) and Perkins and Daly (1968), posits that humans preferentially transport limbs and leave axial units at the site of carcass acquisition (the so-called "schlep effect"). Therefore, assemblages dominated by limb bones are interpreted as the result of transport (Binford, 1981; Bunn, 1982, 1991; Bunn and Kroll, 1986; Bunn *et al*., 1988, 1991).

Such a simplistic model, however, fails to fully account for many of the factors that condition how foragers transport and accumulate carcasses. Transport costs are probably the most important issue, which are determined by a number of interrelated variables including the distance between the carcass and the base camp, the number of carriers, the time of day, and the size of the carcass (O'Connell *et al.*, 1988, 1990; Metcalfe, 1989; Metcalfe and Barlow, 1992). Another factor that must be taken into account is the strategy of field butchery employed before transport. Some human groups prepare carcasses for transport by first disarticulating them and discarding some bones at the kill site (O'Connell *et al.*, 1992) whereas others do not (Domínguez-Rodrigo and Martí Lezana, 1996). Cultural preferences for particular anatomical sections (O'Connell *et al.*, 1990, 1992) and social obligations that determine the way carcasses are shared (Marshall, 1994) will also determine transport and accumulation patterns. Humans are also known to distort their own bone accumulations when clearing living areas of refuse (O'Connell *et al.*, 1991). Unfortunately, it appears that at least some of these behaviors will not show a clear archaeological signature (Bartram *et al.*, 1991; Gargett and Hayden, 1991; Marshall, 1994).

A main objection that can be levelled against skeletal part profiles for zooarchaeological purposes is that there simply is not a unique "human" pattern of bone transport and accumulation. The way that the Nunamiut transport carcasses (Binford, 1978, 1981) is not the same as, for instance, the Hadza (O'Connell *et al.*, 1990) or the Kua (Bartram, 1993). Among the Hadza themselves there is significant variation in the parts transported to base camps. The Hadza data show that the Whitean proposition of preferential appendicular transport is wrong in many cases (Monahan, 1998; O'Connell *et al.*, 1990). When dealing with wildebeest and hartebeest carcasses, the Hadza seem to preferentially transport vertebrae, pelves, and ribs, followed by the head, scapulae, and limbs. For impala, the transport pattern is similar. For eland, the Hadza transport vertebrae and pelves followed (in order) by the head, ribs, and limbs, while for buffalo (despite their similarity in size to eland) the limbs are the most likely to be transported. With zebras and warthogs, a high proportion of axial elements are transported (O'Connell *et al.*, 1990).

Overall, the Hadza data demonstrate two things: (1) preferential appendicular transport by humans has been exaggerated in ethnoarchaeological models (*contra* Binford, 1978, 1981; but see Bunn *et al.*, 1988, 1991); and (2) that although carcass transport can be patterned, it varies by carcass size and even species. If such variation is present in a single foraging group, it is not surprising that such differences become exaggerated when comparing several human groups (Bartram, 1993; Domínguez-Rodrigo and Martí Lezana, 1996).

MULTI-AGENT PATTERNING IN SITE FORMATION

Most ethnoarchaeological research has focused on the differential transport of carcass parts from kill sites to base camps while largely ignoring the processes that affect transported parts after consumption (but see Bartram *et al.*, 1991). Therefore, the archaeological models built from these observations focused on the variables conditioning carcass part *selection* instead of *destruction*. However, bone assemblages are the result of both processes: humans making decisions at kill sites, transporting selected bones, modifying and destroying part of the assemblage at camps due to consumption and abandonment, and finally, the intervention of other physical (such as flowing water) and/or biological (such as carnivore ravaging) agents. The recognition that archaeological sites are palimpsests (i.e., resulting from the intervention of several agents) weakens single-patterned interpretations (e.g., carnivore den versus human-transported bone assemblage) and has led some researchers to stress the importance of multi-patterned models (Blumenschine, 1988, 1995; Marean *et al.*, 1992; Blumenschine and Marean, 1993; Selvaggio, 1994; Capaldo, 1995).

Given that hominid and carnivore (in particular, hyena) signals are unambiguously reflected at Plio-Pleistocene sites, much research has been devoted to documenting the effects of ravaging carnivores on human-made bone accumulations. Two general conclusions emerge from this important work (Marean *et al.*, 1992; Blumenschine and Marean, 1993; Capaldo, 1995; Marean and Cleghorn, 2003; Cleghorn and Marean, 2004). First, axial bones (ribs, vertebrae, and pelves) and small compact bones are largely deleted from ravaged assemblages; this creates skeletal part profiles dominated by limbs and head parts. Second, limb bone epiphyseal fragments are more likely to be deleted from ravaged assemblages than midshaft specimens. The selective removal of these parts can probably be explained by their low structural density and high grease content (Lyman, 1994; Lam *et al.*, 1999). The main problem with such distorted skeletal element patterning is that it is difficult to discern selective from complete carcass transport. This is unfortunate as both types of transport behavior clearly have implications for hominid carcass exploitation. That many researchers did not fully appreciate multi-patterning in the formation of bone assemblages goes far in explaining the ambiguity surrounding interpretations of Plio-Pleistocene hominid behavior in the 1980s. This problem has been compounded by the fact that many zooarchaeologists traditionally rely almost exclusively on epiphyseal specimens to calculate the minimum number of elements (MNE). This analytical bias has been argued to result in element profiles artificially dominated by heads and distal feet (Marean and Frey, 1997; Bartram and Marean, 1999). Therefore, skeletal element profiles derived without the systematic inclusion of denser midshaft fragments probably do not accurately reflect the relative frequencies of bones originally present at a site.

The Ecology of Scavenging

Although the work of Binford (1981, 1985, 1988a) and Shipman (1983, 1986; Shipman and Rose, 1983, 1984) made important contributions to this issue, it is the research of Blumenschine (1986, 1987, 1988, 1989, 1991, 1995) on the ecology of scavenging that has shaped current ideas about passive scavenging among Plio-Pleistocene hominids. These seminal studies have provided important data on savanna ecology, carnivore behavior, and, most importantly, the taphonomic signals of trophic dynamics and competition, resource availability, and multi-patterning in assemblage formation. Blumenschine demonstrated that regular scavenging is feasible only at felid kills, as hyaenids and canids are able to thoroughly consume much of the meat, marrow and grease from small- and medium-sized carcasses. Although small scraps of flesh are available at felid kills, the primary resources available to hominids practicing such a passive scavenging strategy are the head contents and marrow (Blumenschine, 1986). Scavenging from tree-stored leopard kills is an exception to this, provided carcass access occurs before its total consumption (Cavallo and Blumenschine, 1989). Blumenschine also showed how the availability of scavengable resources varied by habitat (more resources are available in lower competition riparian woodlands) and season (more carcasses are encountered during the dry season). Although similar studies in different ecosystems have documented variation in the availability of carcasses by habitat and season (Tappen, 1992, 1995), what is actually available for scavenging does not seem to vary. Nevertheless, a strategy focusing on low-yielding carcasses does not necessarily require cooperative behavior and thus precludes delayed consumption and food distribution (Blumenschine, 1986; Domínguez-Rodrigo, 1994b).

Further actualistic work by Blumenschine (1988, 1995), Selvaggio (1994) and Capaldo

(1995, 1997) led to a proposed triple-stage model for early site formation in which felids initially defleshed carcasses followed by hominid consumption of marrow and brain contents and finally hyena processing of grease from axial bones and limb bone epiphyses (see below). The main conclusion from these actualistic studies is that passive scavenging does not provide significant amounts of meat to secondary consumers. Given this, it is difficult to reconcile arguments that insist scavenging at carnivore kills could have enabled hominids to obtain high meat yields or that scavenging from hyena dens could have provided them any food at all (Binford, 1988a; Bar Yosef, 1989, 1994; Auguste, 1991, 1995; Stiner, 1991, 1994; Fosse, 1995, 1996; Brugal *et al.*, 1997). Scavenging meat from small- and medium-sized carcasses is only possible through confrontational means (O'Connell *et al.*, 1988, 1990, 1992), or, more rarely, after catastrophic events such as mass drownings (Capaldo and Peters, 1995).

Testing the Passive Scavenging Hypothesis

Researchers initially focused on differentiating early from late carcass access through the study of skeletal part profiles (e.g., Binford, 1978, 1981, 1988a, 1988b; Bunn, 1982; Grayson, 1984; Bunn and Kroll, 1986, 1988; Stiner, 1991, 1994). As highlighted above, many of these studies did not fully consider the bias introduced by postdepositional destruction (particularly carnivore ravaging). The systematic inclusion of midshaft fragments into MNE estimates stems some of this bias (Bunn, 1986; Bunn and Kroll, 1986; Todd and Rapson, 1988; Marean, 1998; Marean and Kim, 1998); however, it remains difficult to provide reliable MNEs for less-dense, grease-rich bones and bone portions. For that reason, bone surface modifications, although often used only as ancillary support for inferences drawn from skeletal part analyses, provide a critical source of data for ascertaining the order of hominid and carnivore access to carcasses. Bones processed by humans and carnivores leave distinctive surface modifications in the form of butchery marks (cut marks and percussion marks) and tooth marks, respectively. Given that diagnostic criteria that reliably distinguish these marks have been published (Bunn, 1981; Potts and Shipman, 1981; Shipman, 1983; Shipman and Rose, 1983; Shipman *et al.*, 1984; Blumenschine, 1988; Blumenschine and Selvaggio, 1988; Fisher, 1995; Blumenschine *et al.*, 1996), they provide unambiguous evidence for the exploitation of carcass resources by humans and carnivores.

Blumenschine (1988) demonstrated that the frequencies of tooth-marked specimens by limb bone section (epiphyseal, near-epiphyseal, midshaft) could be a sensitive indicator of the order of carnivore access to carcass resources. Specifically, his experiments found that primary carnivore access results in a high percentage of midshaft specimens bearing tooth marks (usually >75%); this can result from defleshing only (felids) or defleshing and/or demarrowing (hyaenids and canids). Secondary access by carnivores to defleshed and demarrowed bones (e.g., ravaging of human food refuse) results in low percentages of tooth-marked midshaft specimens (5–15%). This makes sense as these bone sections retain little or no nutritious tissue after human processing (Blumenschine, 1988; Capaldo, 1995, 1997) (Figure 1). Tooth mark percentages on epiphyseal and near-epiphyseal specimens remain relatively high regardless of the order of carnivore access, as they retain grease both before and after human processing (assuming that boiling technology is not used to extract it).

Human carcass processing leaves both cut marks and percussion marks on bones that reflect, respectively, the extraction of outer tissues (skin, tendons, meat) and within-bone nutrients (marrow). Cut marks are typically long, parallel striations with internal striae that

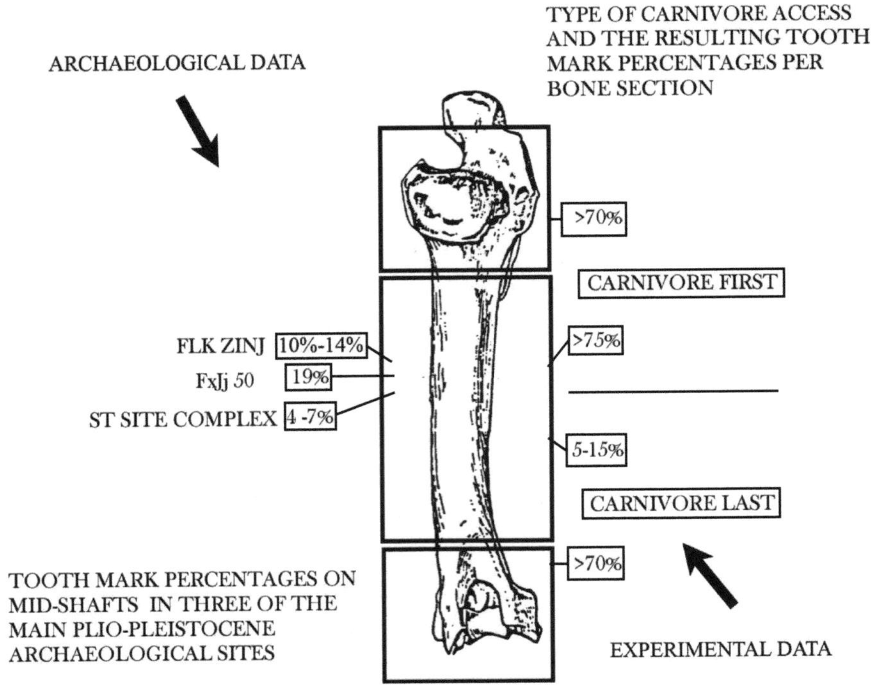

Figure 1. Tooth mark distribution according to long limb bone section. Epiphyseal sections are highly tooth-marked irrespective of the type of carnivore access to ungulate carcasses. Midshaft sections show a high contrast according to whether carnivores have primary or secondary access to carcasses. Tooth-mark distributions in the archeofaunas indicate that carnivores were predominately secondary consumers of carcass parts at these early sites. (Data from Blumenschine [1995] for the experimental samples, and Domínguez-Rodrigo [2002] for FxJj 50, Domínguez-Rodrigo and Barba [2006] for FLK *Zinj*, and Domínguez-Rodrigo *et al.* [2002] for the ST site complex [Figure modified from Domínguez-Rodrigo and Pickering, 2003, *Evolutionary Anthropology*]).

are produced by the contact of sharp-edged stone tools with bone surfaces. Cut mark occurrence on limb bones depends largely on the butchering activity. When defleshing, cut marks appear on midshaft and near-epiphyseal sections (Domínguez-Rodrigo, 1997a, 1997b). Dismembering leaves cut marks on the epiphyseal and near-epiphyseal sections (Binford, 1981). The removal of small scraps of flesh and periosteum can be done by scraping rather than cutting. This activity leaves scrape marks in the form of multiple shallow, parallel striations. The difficulty of cutting through tendons during dismemberment sometimes requires the use of unifacial or bifacial choppers, handaxes, or cleavers for chopping, an activity that results in deep chop marks. Ethnoarchaeological and experimental studies indicate that tool-assisted defleshing results in 15–30% of long limb specimens bearing cut marks (Bunn, 1982; Domínguez-Rodrigo, 1997a, 1997b, 1999a; Lupo and O'Connell, 2002). Furthermore, it has been shown that the frequency and anatomical distribution of cut marks can usefully inform on the order of hominid access to carcass resources (Domínguez-Rodrigo, 1997a, 1997b). Percussion marks are pits often associated with emanating microstriae (Blumenschine and Selvaggio, 1988). Blumenschine and Selvaggio (1988) and Pickering and Egeland (2006) report that in experiments where humans process all limb bones for marrow, 20–30% of the total specimens bear at least one percussion mark. The refinement of these actualistic datasets, particularly in the early and mid-1990s, led to fruitful applications to the archaeological record.

APPLICATION TO THE ARCHAEOLOGICAL RECORD

The target for many researchers interested in Plio-Pleistocene hominid behavior was the large and well-preserved fauna from FLK *Zinj* at Olduvai Gorge. Blumenschine (1995), Selvaggio (1994), and Capaldo (1995, 1997) concluded that the archeofauna from FLK *Zinj* reflected hominid scavenging of abandoned felid kills that were subsequently transported on-site to be demarrowed. The final stage of site formation involved the deletion of grease-bearing bone and bone portions by ravaging carnivores (namely hyenas). Carnivore ravaging would have selectively removed limb bone epiphyses, axial elements and compact bones, ultimately skewing the original frequencies of elements transported by hominids (Blumenschine, 1988, 1995; Marean *et al.*, 1992; Blumenschine and Marean, 1993; Capaldo, 1995, 1997). Support for this triple-stage sequence was found in tooth mark and percussion mark frequencies and their distribution according to limb bone section. Hominid involvement in marrow removal was clearly indicated by the high percentage of specimens bearing percussion marks, which was similar to experiments where humans demarrow limb bones before carnivore intervention (Blumenschine, 1995). High frequencies of tooth marks on epiphyseal and near-epiphyseal specimens suggested that scavenging carnivores likely played a major role in modifying and destroying carcass parts subsequent to their abandonment by hominids. Interestingly, tooth mark percentages on midshaft specimens were also relatively high, near to those obtained in experiments modeling primary carnivore access to bone marrow (Blumenschine, 1988, 1995; Selvaggio, 1994; Capaldo, 1995, 1997) (Figure 1). This anomalous situation could only be explained if felids tooth-marked bones before hominid marrow removal, thereby increasing tooth mark frequencies above those expected in a simple two-stage hominid-to-carnivore scenario.

Largely ignored in these studies is the significant presence of cut marks on many specimens from FLK *Zinj* (Figure 2). This stems not from ignorance but from a perceived potential for equifinality[1] in their interpretation. For example, cut marks on bones could either be the result of hominids manipulating fully fleshed carcasses (Bunn and Kroll, 1986) or simply removing marginal scraps of flesh that commonly survive carnivore consumption (Blumenschine, 1991, 1995). Binford (1986) suggested that some cut marks from FLK *Zinj* may reflect the removal of meat from desiccated carcasses (although Lupo [1994] demonstrates that similar cut mark patterns may arise due to the butchery of relatively fresh but rigored carcasses). According to Blumenschine (1995) and Capaldo (1995), the lack of controlled experiments that compare these processes and their associated cut mark distributions and frequencies render this class of bone surface modification inadequate for inferring hominid behavior. Therefore, by excluding the cut mark data and focusing largely on the frequencies of tooth-marked midshaft fragments, Blumenschine (1995) and Capaldo (1995) concluded that hominids were passively scavenging from felid kills, a model first subjected to experimentation by Selvaggio (1994). In contrast with this, Oliver (1994) argued that some mid-shaft tooth-marking followed hominid processing. According to him, a high frequency of tooth marks does not necessarily mean a primary access to carcass remains by carnivores.

[1] Equifinality is commonly understood as the phenomenon of different processes producing the same results. It can also be understood as diverse processes producing different results that are not distinguishable because of limitations of currently available analytical methods (see Lyman, 2004).

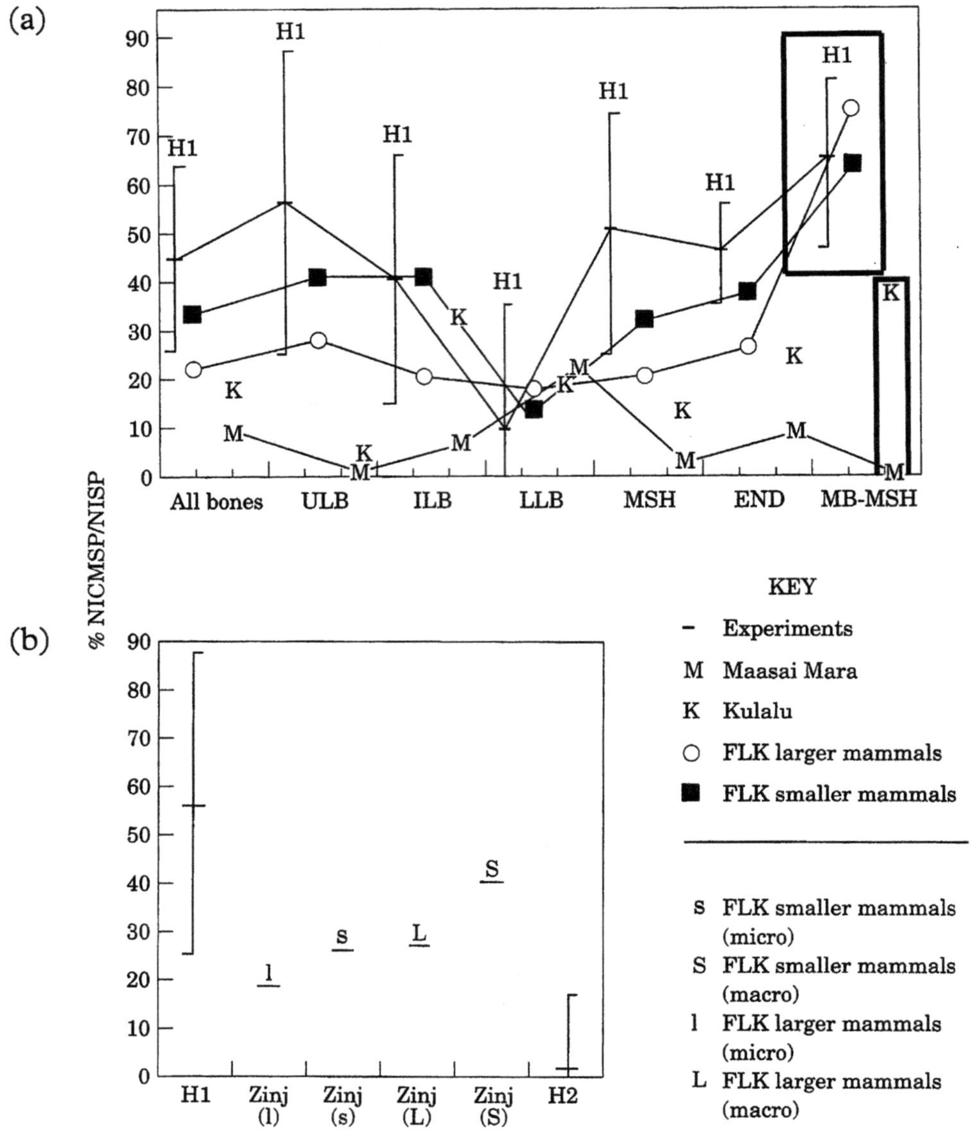

Figure 2. Frequencies of cut marks according to element type and bone section. (A) Range of variation of Hypothesis 1 experiments (primary access to fleshed carcasses), the mean for Hypothesis 2 experiments (passive scavenging from felid kills at Kulalu and Maasai Mara) (K, M) and the data from smaller and larger animals from the FLK *Zinj* site are shown. 95% C.I. analyses were not made for the experiments simulating secondary access to carcasses because the sample is too small and also because they were made in two different ecosystems, under different processes; one in which lions deflesh carcasses thoroughly (MM) and another one in which lions were interrupted from consuming carcasses by humans and therefore carcasses are more fleshed. Note the clustering of the FLK *Zinj* data together with the Hypothesis 1 mean (primary access to fleshed carcasses), with respect to meaty bone midshafts (MB-MSH). (B) Despite sample size, the range of variation of upper limb bone specimens showing cut marks for both experimental samples does not overlap. Note the differences in representation of cut-marked specimens from meaty long bones (humerus, femur, radius, and tibia) when comparing access to fleshed carcasses with access to defleshed carcasses from felid kills, especially to the most commonly documented pattern of felid consumption (Maasai Mara; see Figure 3). Boxes show the difference in the ranges of variation of cut-marked midshafts from meaty bones documented in primary access to fleshed carcasses and secondary access to felid-consumed carcasses. Frequencies show the number of identified cut-marked specimens (NICMSP) according to total number of identified specimens (NISP) (Modified from Domínguez-Rodrigo, 1997a, *Journal of Human Evolution*).

ON THE UTILITY OF CUT MARKS FOR INFERRING HOMINID CARCASS EXPLOITATION

As we have seen, Blumenschine (1986) demonstrated that passive scavenging would only be feasible at felid kills, as these carnivores do not possess the dental battery to fully exploit within-bone nutrients. Therefore, within-bone nutrients and marginal flesh scraps would have been the only carcass resources available to passively scavenging hominids (see Figure 3). Given this, the key issue became whether or not the frequency and distribution of cut marks observed at Plio-Pleistocene sites could result from the removal of small flesh scraps at felid kills. To test this specifically, and, more generally, to address the issue of equifinality in cut mark interpretation, observations on the exact anatomical distribution of flesh scraps from felid-modified carcasses and the global distribution of cut marks produced in scrap removal were required. These data could then be compared to experiments modelling meat-stripping from fully fleshed carcasses.

These experiments were carried out by Domínguez-Rodrigo (1999a; see also Chapter 6), who documented a patterned distribution of flesh scraps after the consumption of middle-sized bovid carcasses by lions in an African savanna ecosystem. These parameters provide a good proxy for Plio-Pleistocene dynamics because (i) middle-sized bovids dominate many Plio-Pleistocene faunas; (ii) lions are the only modern felid that focus on medium-sized prey; (iii) although specific variables such as humidity, temperature, vegetation and carnivore and herbivore densities no doubt differ between modern and Plio-Pleistocene African savannas, the overall ecological structure is broadly similar (Domínguez-Rodrigo, 1994a, 2001); and (iv) the morpho-functionally dictated carcass processing abilities of modern carnivores and their Plio-Pleistocene counterparts are similar.

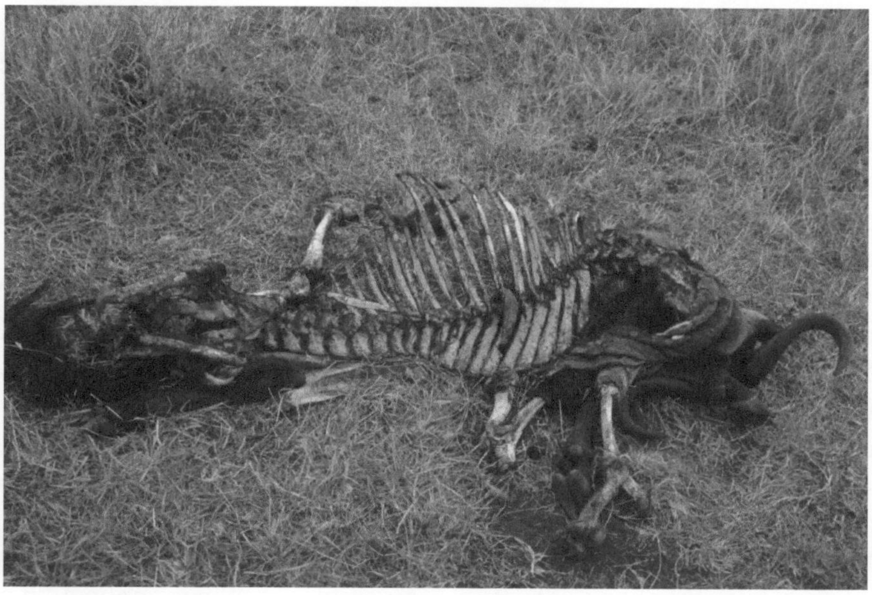

Figure 3. Defleshed wildebeest carcass from a lion kill after its consumption by a pride of seven lions. It is one of the carcasses belonging to the sample reported in Domínguez-Rodrigo (1999a). No flesh has survived and very few scraps of flesh were documented.

As flesh specialists, lions (and other felids) leave very few edible scraps on most of the skeleton, with only the cranium possessing significant amounts of flesh after consumption. This also applies to cheetahs and leopards when feeding on the ground (MDR personal observation; Brain, 1981). When documented, flesh scraps were small and observed most frequently on the neck. Almost 90% of the flesh scraps observed on limb bones occur on the proximal or distal portions, and upper limb bones (humerus and femur) in particular preserved no scraps at all on their midshaft portions (Domínguez-Rodrigo, 1999a). This pattern differs markedly from the frequency of flesh scraps after human butchery, where they tend to be much more common (especially on vertebrae and limb bone midshafts). This is probably explained by the fact that whereas lions and other felids tear entire muscle masses from bones, humans cut through the meat so that muscles are not removed as complete units. Domínguez-Rodrigo (1999a) also observed that flesh scraps survived only on carcasses located on the open plains. This may be due to the fact that lions could spend more time at kills in lower competition riparian woodlands (from several hours to even days) than at kills on the open plains (1–3 h). Given these results, the simple prediction emanating from this study is that if hominids were passively scavenging from felid kills, cut marks on the bones they processed should appear where modern felids leave flesh scraps. Conversely, cut marks on bone sections where felids do not commonly leave flesh scraps cannot be linked to passive scavenging. More specifically, cut-marked midshafts from upper limb bones and the middle and distal sections of ribs clearly indicate hominid access to fully fleshed carcasses.

This hypothesis was further tested through the stone-tool-assisted butchery of fully fleshed carcasses from natural deaths and human settlements and defleshed felid kills. Comparing cut mark frequencies and distributions from these scenarios allowed primary and secondary access by hominids to be reliably inferred (for a complete statistical dataset see Domínguez-Rodrigo, 1997a, 1999a and Chapter 6). Arguments about the meaning of cut marks from FLK *Zinj* (Capaldo, 1998a; Selvaggio, 1998) ignored these data and that of others (e.g., Marshall, 1986; Perez-Ripoll, 1992; Blasco Sancho, 1995; Martinez, 1998; Pumarejo and Cabrera, 1992), all of which indicated a relationship between cut mark representation and the quantity of meat on a carcass. However, a clear picture of the order of hominid intervention with carcasses emerges when the analytical framework outlined above is applied to FLK *Zinj*.

At FLK *Zinj*, upper and intermediate limb bones make up a majority of the cut-marked limb specimens (88%), and upper limb bone specimens are more frequently cut-marked (40%) than intermediate specimens (30%) (Oliver, 1994; Domínguez-Rodrigo, 1997a). These percentages match closely the experiments simulating early access to fully fleshed carcasses, clearly contradicting the conclusions of passive scavenging based on tooth-marked midshaft specimens. Because such fundamental contradictions have important implications for the veracity of current actualistic frameworks, we examine in-depth the utility of tooth marks for directly inferring hominid behavior.

ASSUMPTIONS OF MULTI-PATTERN ACTUALISTIC FRAMEWORKS AND THE UTILITY OF TOOTH MARKS

Although the experimental frameworks of Blumenschine, Selvaggio, and Capaldo have correctly stressed the importance of multi-agent interactions in the formation of bone assemblages, their application requires three critical assumptions. First, other agents such as hominids (Pickering and Wallis, 1997) or other primates (e.g., baboons; Domínguez-Rodrigo,

1999b) are excluded as potential tooth-marking agents. Second, these models assume that hominids exploit each and every carcass thoroughly; that is, all bones are completely defleshed and/or demarrowed. Finally, these multi-part models assume that hominids and carnivores always acted interdependently in carcass modification (i.e., every carcass accumulated on-site was exploited by both hominids and carnivores).

Experimental studies of both captive and wild baboons have shown that they can produce tooth mark frequencies and distributions that mimic those in experiments where carnivores ravage humanly produced bone assemblages (Domínguez-Rodrigo, 1999b). Even if primates are excluded as significant tooth-marking agents, violating the latter two assumptions still poses serious problems for tooth mark-based passive scavenging scenarios.

Complete limb bones make up a relatively large portion (9–10% of the total limb bone MNE) of many Bed I sites at Olduvai (Potts, 1988). The abandonment of complete limb bones by hominids suggests that carcass yields and/or carcass encounter rates were high enough to merit a less than complete carcass processing strategy (Egeland and Byerly, 2005). An inference of high carcass yields at least indirectly implies the presence of significant amounts of flesh on some carcasses (Domínguez-Rodrigo, 1997a; Cavallo, 1998). At Koobi Fora, Bunn (1994) documented several complete bones bearing cut marks, again suggesting incomplete carcass processing focused on flesh removal. These data are especially significant in that Capaldo (1995) demonstrated convincingly that even in lower-competition riparian habitats hyenas will destroy all grease- and marrow-bearing bones. Thus, the fact that almost 10% of limb bones from Bed I Olduvai sites are complete, and that hyena ravaging has been documented at many of the same sites (Blumenschine, 1988; Blumenschine and Marean, 1993), suggests that the original number of complete limb bones was significantly higher (Capaldo, 1995).

Previously reported tooth mark frequencies on midshaft specimens at FLK *Zinj* (see Chapter 5 for updated numbers) can also be more clearly understood if hominids were not exploiting all limb bones for marrow. Actualistic studies show that an average of 75% of midshaft fragments from medium-sized carcasses preserve tooth marks in "carnivore having primary access" feeding experiments (Blumenschine, 1988; Capaldo, 1995). However, tooth mark frequencies on midshafts from FLK *Zinj* (57%) fall outside the 95% confidence intervals of both these experiments and those that model secondary carnivore access to defleshed and demarrowed limb bones (Blumenschine, 1995; Capaldo, 1995). As we have seen, a stage before both hominid demarrowing and carnivore ravaging of hominid food refuse (felid defleshing) was invoked to account for the anomalous tooth mark frequencies. However, Domínguez-Rodrigo (1999b), assuming high carcass yields would have led to the abandonment by hominids of some marrow-bearing bones, provided tooth mark data on partially demarrowed carcasses. Although epiphyseal and near-epiphyseal tooth mark frequencies were similar to those obtained in "hominid-to-carnivore" experiments (Blumenschine, 1988, 1995; Capaldo, 1995), midshaft frequencies were similar to those documented in the FLK *Zinj* sample (Figure 1). (Capaldo's [1995] "whole bone-to-carnivore" experiments also produced similar midshaft tooth mark frequencies to FLK Level 22; however, the lack of hominid bone breakage in these experiments, an activity clearly documented at FLK *Zinj* by the presence of percussion marks and bone breakage pattern [Oliver, 1994], precludes the use of this scenario in interpretations of site formation.)

Coupled with cut mark data summarized above, this all suggests that hominids were exploiting carcasses not solely for marrow, but also for flesh. Further support for this contention is found in patterns of axial bone representation. At many Bed I Olduvai sites, axial bones make up a significant portion of the total MNE and number of identified specimens (NISP) (between 20% and 40%; Capaldo, 1995). Again, Capaldo's (1995) experiments show that axial bones survive hyena ravaging at rates of about 5% (see also Marean *et al.*, 1992). Their high relative representation at many Bed I sites suggests that they were much more abundant before carnivore ravaging, further suggesting complete carcass transport. Heavily fleshed but devoid of marrow, Capaldo (1995: 350) acknowledges that "[a] flesh surplus might help to explain the presence of so many axial elements, pelves and scapulae at FLK *Zinj*; that is, why would hominids transport these items to the site if they were defleshed?" The abundance of cut marks on scapulae, pelves, and ribs at FLK *Zinj* (Bunn and Kroll, 1986) also demonstrates that hominids were defleshing these skeletal elements. Overall, the fact that hominids left many limb bones unbroken, especially in the Bed I sites from Olduvai, contradicts the marrow-based foraging strategy envisioned by passive scavenging models.

Another important assumption of passive scavenging scenarios and the experimental samples they are based on concerns the level of hominid-carnivore interdependence in assemblage accumulation and modification (terminology follows Egeland *et al.*, 2004). Actualistic samples reflect tooth mark frequencies in situations where carnivores and hominids always modify the same carcasses. However, many archeofaunas are palimpsests in the extended sense (Binford, 1980), where multiple agents contribute both interdependently *and* independently to site formation (Domínguez-Rodrigo *et al.*, in press; Egeland *et al.*, 2004). Domínguez-Rodrigo *et al.* (in press) demonstrate that if carnivores and hominids are accumulating and modifying carcasses independently of each other (i.e., they are not transporting or consuming the same carcasses), tooth mark frequencies may mimic "carnivore-first" scenarios, *despite the fact that this says nothing about hominid access to carcasses.*

In summary, tooth mark frequencies can be susceptible to equifinality, and the actualistic frameworks based largely on this class of surface modifications must be applied carefully. High frequencies of tooth marks can be obtained in experiments modelling both primary carnivore access and in scenarios modelling secondary carnivore access to carcasses abandoned by hominids with some marrow bones unexploited. In addition, if hominids and carnivores contribute independently to site formation, at least to some extent, tooth mark frequencies can be misleading. This is not to say that tooth marks cannot be used to make inferences of hominid behavior. However, they must be used in conjunction with cut mark frequencies and distribution, which inform directly on hominid carcass exploitation. The FLK *Zinj* assemblage provides an excellent example. Cut mark data suggest primary hominid access to carcasses and a focus on flesh removal. High carcass yields may have resulted in hominids discarding some marrow-bearing bones intact, a suggestion confirmed by the intermediate (and initially anomalous) frequencies of tooth-marked midshaft fragments.

The next chapter describes the taphonomic methodology employed in the analysis of the Bed I faunas from Olduvai Gorge.

3. The "physical attribute" taphonomic approach

M. DOMÍNGUEZ-RODRIGO, C.P. EGELAND, AND R. BARBA

Introduction

In this book, we distinguish between two complementary approaches to taphonomic investigation. The first approach is more traditional in nature, focusing on the differential representation of skeletal parts and the construction and interpretation of mortality profiles and taxonomic lists. Because this approach developed first among paleontologists, we will refer to it here as the "paleontological approach." This general methodology dominated the early taphonomic debates over early site formation (see Chapter 2), and Binford's (1981) well-known critique in particular was based on the premises of such a "paleontological approach." Potts' (1988) work on the Bed I faunas is another example. Ecological in nature, Potts' seminal research quantified patterns of bone frequency and representation created by natural processes in modern ecosystems as a comparative framework for interpreting fossil assemblages. Based largely on such comparisons, Potts (1988) argued for hominid authorship of many Bed I sites.

The other, which we consider a "physical approach," concentrates on changes in the physical attributes of bones throughout their taphonomic history. The physical alterations undergone by bones often leave diagnostic signatures. The action of nonbiotic processes is reflected in physical attributes such as weathering, abrasion, polishing, size sorting, and chemical modification. Biotic processes like butchery and carnivore ravaging also leave distinctive alterations in the form of butchery marks, tooth marks, and patterns of fragmentation.

Although many traditional paleontological approaches, including Potts' (1988), take into account variables of a physical approach, the data generated (e.g., bone surface modifications, fragmentation, etc.) are often used only as ancillary support for conclusions drawn from more traditional analyses. For instance, it has recently been argued that skeletal part abundances, species representation, and mortality profiles reflect behavioral differences between Pleistocene Eurasian hyaenids and their modern African counterparts (Brugal *et al.*, 1997; Fosse *et al.*, 1998). Such differences are potentially important, given the common usage of modern carnivores as proxies for extinct species. Bone assemblages accumulated by Eurasian hyaenids show a higher relative proportion of carnivore remains than do modern African dens. As Brugal *et al.* (1997) argue, this could indicate more intensive occupation of dens by cave hyenas. On the other hand, a high representation of carnivore remains may simply reflect the fact that caves are focal points of activity that attract carnivores constantly, whereas modern African open-air hyena dens do not. Arguing again from taxonomic lists, Brugal *et al.* (1997) suggest that Eurasian hyenas appear to select larger prey compared to modern African hyenas. Pastoralists in modern African ecosystems probably provide these hyenas with ample opportunities to capture and

M. Domínguez-Rodrigo, R. Barba, and C.P. Egeland, Deconstructing Olduvai: A Taphonomic Study of the Bed I Sites, 23–32.

accumulate bones from smaller goat and sheep. In contexts where pastoralists live in very low densities or are absent, as in the Maasai Mara (Kenya), hyenas accumulate bones from prey as large as their Eurasian Pleistocene counterparts (Kerbis, 1990).

Arguing from a more physical approach, Marra *et al.* (2004) and Villa *et al.* (2004) point out that skeletal part abundances, taxonomic representation, and mortality profiles do not differ between Eurasian Pleistocene and modern African hyenas, and, in fact, neither differs substantially from human bone accumulations ravaged postdepositionally by modern hyenas. The differences are only apparent when taking into account the physical attributes of the assemblages themselves, including fragment sizes, levels of fragmentation, and frequencies of digested bone.

We therefore feel that a true physical approach to the interpretation of faunal assemblages has yet to be fully developed. This chapter attempts to outline and defend such a methodology, which will contradict previous interpretations of site formation during Bed I times at Olduvai Gorge. In formally presenting this methodology, we do not propose that traditional data sets such as skeletal part abundances and species lists be abandoned; these data are found throughout this volume. However, we do argue that physical attributes are more reliable taphonomic indicators and should play a primary role in understanding site formation.

Analytical Parameters of the Physical Attribute Taphonomic Approach

NONBIOTIC AGENTS OF SITE FORMATION

Like many taphonomic methodologies, the physical approach requires that the contribution of natural, nonbiotic agents such as water transport and soil movement to site formation be understood and accounted for. Data on specimen size distributions and polishing and abrasion provide evidence for the presence and intensity of water disturbance and transport, respectively. However, water can also affect bone surfaces without transporting them to or from sites. For example, soil humidity and pH can produce biochemical modifications (exfoliation, cracking) that mimic subaerial weathering. Other times, bones lying on humid soils show discoloration, and, in some cases, circular pitting. Biochemical processes are often discernable from subaerial weathering in that they occur in isolated areas (i.e., those areas in contact with the ground surface). Bones exposed to specific conditions can also show modifications caused by bacteria and fungi (Domínguez-Rodrigo and Barba, 2006; see also Chapter 5). All of these modifications provide valuable information on the microenvironmental context of site burial.

CALCULATING SKELETAL PART ABUNDANCES

This book includes information on the number of identified specimens (NISP), minimum number of elements (MNE), and minimum number of individuals (MNI). Long limb bones are classified as upper (stylopodials), intermediate (zygopodials), and lower (metapodials). There has been heated debate over the most appropriate methodology for estimating the MNE, especially for limb bones. Traditional approaches to limb bone MNE estimation rely largely on readily identifiable epiphyseal portions. It has been argued that this analytical bias is responsible for the head- and foot-dominated skeletal part patterns (referred to as "Type II" [Marean *et al.*, 2004]) documented at many archaeological sites (Marean and Frey, 1997; Marean, 1998; Marean and Kim, 1998; Marean *et al.*, 2000; Pickering *et al.*, 2003). Modern observations clearly show that limb bone shafts survive carnivore ravaging at higher rates than epiphyseal fragments (Pickering *et al.*, 2003). These observations can be explained by the fact that shaft fragments are both denser and lack the nutritional tissues (e.g., grease) found in epiphyseal fragments. In the analyses that follow, MNEs were calculated using both epiphyseal fragments and the systematic inclusion of shaft fragments (Barba and Domínguez-Rodrigo, 2005). Epiphyseal fragments are included not because they provide reliable MNE estimates for particular limb bones (*contra* Stiner, 2002, 2004), but because epiphyseal representation (in terms of MNE or NISP) provides a good measure of the intensity of carnivore ravaging (Brain, 1967, 1981; Marean *et al.*, 1992; Blumenschine and Marean, 1993; Capaldo, 1995).

SHAFT CIRCUMFERENCE

Limb breakage by humans and carnivores typically produces numerous shaft fragments and some articular ends, with the latter often consumed by carnivores (e.g., Bunn, 1986, 1991; Binford *et al.*, 1988; Blumenschine, 1988; Marean and Spencer, 1991). However, Bunn (1982) noted that bone breakage (marrow extraction by humans in this case) produced a specific pattern of shaft fragment representation in terms of the preserved cross section. Generally, the articular ends of broken limb bones preserve an attached section of shaft that is complete in circumference (Bunn's [1982] "Type 3"), whereas the isolated shaft fragments preserve either more than half (Type 2) or less than half (Type 1) the original circumference. Bunn (1982) found that in broken assemblages the ratio of Types 3 and 2 (complete and more than half circumference, respectively) to Type 1 (less than half circumference) ranges from 0.44 to 0.10; that is, shaft specimens preserving less than half the circumference outnumber the other types. Importantly, carnivore-broken assemblages tend towards the high end of this range (more specimens with complete and more than half the circumference), relative to the human-broken assemblages. Therefore, the relative distribution of these three circumference types can help gauge the contribution of hominids and carnivores to bone breakage.

Marean *et al.* (2004) recently proposed that Bunn's (1982) circumference types can also be used to measure bias in a faunal assemblage. Specifically, they argue that assemblages not subjected to the selective retention of readily identifiable fragments (often at the expense of less diagnostic shaft fragments) will be dominated by Type 1 shaft fragments.

BONE SURFACE MODIFICATIONS

A systematic search for cut marks, percussion marks, and tooth marks in addition to naturally derived biochemical and abrasion marks was conducted. All surface modifications were identified by more than one researcher under strong light with the aid of 10X–20X hand lenses. The diagnostic criteria of Bunn (1981, 1982) and Potts and Shipman (1981) guided our identification of cut marks. We have found that secure cut-mark identifications only rarely require SEM images[1]; we therefore agree with Blumenschine's (1995) suggestion that most cut marks can be reliably identified with a hand lens.

---

[1] Two of us (MDR and RB) have performed blind tests on cut-marked and trampled specimens and have been able to interpret most of the marks with less than 20X. The use of SEM is necessary when doubts arise after the microscopic examination with hand lenses has failed to identify any given type of mark. The application of SEM to hundreds of specimens like the assemblages analyzed in this book would be costly, time consuming, and usually does not significantly increase the number of identifiable marks, compared with using hand lenses (Bunn and Kroll, 1986). Close-up SEM photographs of marks such as tooth pits and trampling have been included in several parts of the book to show details that are easily observable when the specimen is microscopically inspected with hand lenses.

Tooth marks were identified based on criteria in Binford (1981), Shipman (1983), Blumenschine and Marean (1993), Blumenschine (1988, 1995), and Fisher (1995). Blumenschine (1995: 29) describes tooth marks as follows:

> Carnivore tooth marks contain bowl-shaped interiors (pits) or U-shaped cross-sections (scores; see also Bunn, 1981) that commonly show crushing that is conspicuous under the hand lens, and which, macroscopically, gives the mark a different patina than the adjacent bone surface

Experimental work suggests that the dimensions of tooth pits can aid in distinguishing the type of carnivore responsible for their creation (Andrews and Fernández-Jalvo, 1997; Domínguez-Rodrigo and Piqueras, 2003). All tooth pits with clearly delineated borders were molded in high-resolution silicone. Pit dimensions (length and breadth maxima) were measured to the nearest 0.01 mm with high-precision digital calipers following Domínguez-Rodrigo and Piqueras (2003). Pits appearing on midshaft sections were analyzed separately from those appearing on epiphyseal sections, as carnivores create larger pits on cancellous bone (Andrews and Fernández-Jalvo, 1997; Selvaggio and Wilder, 2001; Domínguez-Rodrigo and Piqueras, 2003).

Percussion marks were identified following criteria in Blumenschine and Selvaggio (1988, 1991), Fisher (1995), and Pickering and Egeland (2006). Blumenschine (1995: 29) also provides an excellent description of classic percussion marks, which appear as

> pits, grooves or isolated patches of microstriations. Pits and grooves are usually associated with densely packed and shallow patches of microstriations oriented approximately transverse to the long axis of the bone... these patches of microstriations occur within and/or emanate from the depression. Microstriations also occur as isolated patches.

Although Blumenschine (1995) claims that percussion marks lack the internal crushing associated with carnivore tooth marks, our unpublished experiments suggest that chopping tools can produce marks with internal crushing. However, the angular contour of such percussion marks makes them readily distinguishable from tooth marks. We have also observed that with unmodified hammerstones a substantial proportion of percussion marks – as high as 30% – will lack associated microstriations. Nevertheless, the present analysis considers only those marks with all the diagnostic features described by Blumenschine and Selvaggio (1988) as true percussion marks.

For comparative purposes, surface modifications were recorded following Blumenschine's (1988, 1995) epiphyseal, near-epiphyseal, and midshaft bone portion system. Modifications were also quantified by element and bone section as described by Domínguez-Rodrigo (1997a, 1999b). The bone section methodology takes into consideration the actual location of surface marks. For example, in Blumenschine's (1988, 1995) bone portion system, an epiphyseal fragment may be represented not only by the epiphysis itself, but also by attached near-epiphyseal and shaft sections, any of which may preserve a surface mark. However, because surface marks should only be created where nutrients (flesh, marrow, grease) are present, the actual location of a surface mark is very sensitive to the nature of hominid and carnivore involvement with carcasses. The aspect (e.g., cranial, caudal) on which each surface modification appeared was also recorded (see especially Chapter 6).

PATTERNS OF BONE BREAKAGE

Breakage patterns were analyzed from a multifaceted analytical approach and included the analysis of notches and breakage planes. Notches appear as semicircular outlines along the otherwise rectilinear edge of a fracture surface and are associated with a negative flake scar on the medullary surface. The experimental work of Capaldo and Blumenschine (1994) demonstrated quantitatively that the dynamic force of hammerstone percussion produces

Table 1. Frequencies of main notch types in carnivore-made assemblage (Maasai Mara hyena den) and human-butchered carcasses

	Maasai Mara hyena den	Human-butchered small carcasses	Human-butchered large carcasses	Human-butchered carcasses, all sizes
% complete notches/all notches	45/132 (34)	42/70 (60)	59/77 (76.6)	101/146 (69)
% overlapping notches/all notches	46/132 (34.8)	9/70 (12.8)	10/77 (12.9)	19/146 (13)
% double-opposing notches/all notches	21/132 (15.9)	4/70 (5.7)	7/77 (9)	11/146 (7.5)

notches that are broader and shallower in cortical view compared to the notches created by the static loading of carnivore teeth. Therefore, following Capaldo and Blumenschine (1994), notch shape was measured to the nearest 0.1 mm using two ratios:

1. Notch breadth: notch depth (in cortical view)
2. Flake scar breadth: notch depth

The platform angles of bone flakes removed from percussion notches also tend to be more obtuse than those removed from carnivore tooth notches. Using a goniometer, the platform angle was taken at the loading point on the negative scar of the detached flake. Finally, each notch was classified according to notch type (modified from Capaldo and Blumenschine [1994]):

1. Complete notches have two inflection points on the cortical surface and a nonoverlapping negative flake scar.
2. Opposing complete notches are two complete notches that appear on opposite sides of a fragment and result from two opposing loading points.
3. Incomplete Type A notches are missing one of the inflection points.
4. Incomplete Type B notches have a collapsed loading point, resulting in an incomplete negative flake scar on the medullary surface.
5. Incomplete Type C notches have negative flake scars that overlap with an adjacent (incomplete Type C) notch.
6. Bifacial notches have double scars; one emanating from the cortical surface and the other from the medullary surface.
7. Pseudonotches.

Supplementing the work of Capaldo and Blumenschine (1994), we have found that the relative distribution of notch types differs in assemblages broken by humans and carnivores.[2] Our experimental samples (Egeland *et al.*, in preparation) include limb bones from eight carcasses butchered by humans and limb bones from a spotted hyena den in the Maasai Mara (Table 1). As the data from Table 1 demonstrate, incomplete Type C and opposing notches are twice as abundant in the carnivore-broken sample (Figure 4). These data provide important supplementary information on the agent of breakage in archeofaunal samples, although they are still preliminary and need further confirmation.

Breakage patterns were also analyzed by measuring fracture angles on the breakage planes of limb bones (Villa and Mahieu, 1991). The way that a bone breaks follows basic physical principles (as in the creation of notches).

[2] The study was carried out by Egeland and Domínguez-Rodrigo in the National Museums of Kenya, Nairobi, in 2005 with kind permission from J. Kerbis.

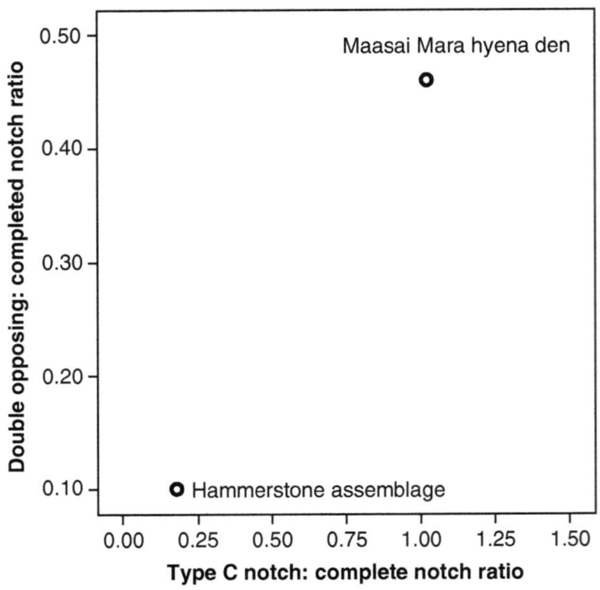

Figure 4. Ratio of double opposing: complete notches and double overlapping: complete notches in the Maasai Mara shaft sample and in experimental human-butchered bone assemblages.

Dynamic loading through hammerstone percussion creates an impact on bone that expands, following physical principles – according to the density of the bone and the force of the impact – detaching a fragment with an angle value that tends to be either acute or obtuse. The same physical principle applies to the obtainment of stone flakes through percussion. The resulting flake, being the product of raw material type, point of impact, and applied force can even be reconstructed, if broken, only by taking some basic measurements, in which the angle of the flaking platform is essential. The same principle applies to bones. Experimental work (Alcántara *et al.*, 2006) shows that the dynamic loading of percussion breakage produces fracture surfaces with more acute and obtuse angles, which occur especially on oblique breakage planes, and are similarly acute/obtuse to those obtained on the flaking platform of stone flakes. Static loading from carnivore breakage, on the other hand, creates more right-angled fracture planes, just as flakes obtained though pressure show platform angles closer to right angles. This approach was experimentally modeled on bones from small- and middle-sized carcasses with varied density (Alcántara *et al.*, 2006) and successfully applied to archaeological and paleontological contexts (Pickering *et al.*, 2005a).

Following the previous studies, all oblique, longitudinal, and transverse (relative to the long axis of the specimen) planes >4 cm in length were measured to the nearest degree with a goniometer.

Referential Frameworks of the Physical Attribute Taphonomic Approach

Sixto Fernández (2000) has emphasized the importance of distinguishing between "deposited entities," "preserved entities," and "recorded entities" in a taphosystem. Deposited entities are those that are deposited on the ground and subjected to deposition and redeposition and sedimentation and resedimentation. Preserved entities are fully interred in the sediment and subject to fossil diagenesis. Recorded entities are those that are discovered and finally analyzed by researchers. Importantly, many zooarchaeologists that rely on observational ethnographic data rarely document deposited entities (e.g., O'Connell *et al.*, 1990, 1991, 1992). This focus on predepositional assemblages creates a disjunction between the reference ethnographic sample and the taphonomically biased samples available to zooarchaeologists. In addition, much of this research centered on identifying the accumulation agent using skeletal part abundances, while the physical properties that directly inform on the modifying agent were largely ignored.

The taphonomic approach in this study takes as its starting point the physical modifications on bones in order to unambiguously identify the modifying agents in a faunal assemblage. Only then can the nature of carcass accumulation and acquisition be assessed.

Therefore, bone surface modifications and fracture patterns are the primary sources of data, while skeletal part abundances, species lists, and mortality profiles serve only as ancillary and taphonomically biased reflections of the behavior of the modifying agent(s).

Given the need to eliminate equifinality to the greatest extent possible, a rigorous grounding in experimental archaeology is a major theme of a physical approach. Notches and fracture angles created through both dynamic and static loading have been studied experimentally (Capaldo and Blumenschine, 1994; Alcántara *et al.*, 2006; Pickering and Egeland, 2006). For tooth marks, Blumenschine and his colleagues have documented that frequencies of marked midshaft fragments are sensitive indicators of the order of carnivore access to carcasses (Blumenschine, 1988, 1995; Blumenschine and Marean, 1993; Selvaggio, 1994; Capaldo, 1995). When carnivores have access to complete bones, they tooth-mark midshafts in the process of breaking them for marrow, which results in >80% of midshaft fragments bearing tooth marks. However, when carnivores are limited to ravaging humanly demarrowed limb bones, only 5–15% of midshaft fragments display tooth marks (Blumenschine, 1988; Capaldo, 1995).

This is not to say, of course, that experimental work is infallible. After all, interpretations based on experimentation are limited by the scope of the experiment itself. For example, the invocation of felids as primary carcass consumers in interpretations of the FLK Level 22 fauna was based not on experimental observations of felid tooth-marking but on experiments in which hyenas served as carcass modifiers (Blumenschine, 1988) or on experiments in which results from felid bone modifications were lumped together with those created by canids and hyaenids to produce a mixed "carnivore-first" pattern (Selvaggio, 1994). Nevertheless, further experimental work has documented differences in tooth-mark frequencies between felids and hyenas, suggesting that the previously proposed felid–hominid–hyena sequence of carcass modification at FLK Level 22 is unlikely (Selvaggio, 1994; Domínguez-Rodrigo *et al.*, in press).

Similar issues surround the use of cut marks. Recently, Lyman (2006) has argued that "cut marks are merely epiphenomena with only variously fortuitous relationships to hominid behavior." This assertion is in fact upheld by some experimental work, where, for example, it has been shown that cut-mark frequencies have little relationship to the number of tool strokes (Egeland, 2003). However, other experiments clearly indicate the usefulness of cut marks for addressing certain aspects of hominid behavior. Based on controlled experiments, Domínguez-Rodrigo (1997a) suggested three types of cut-mark data that could distinguish primary from secondary access to flesh by hominids: overall frequency, the differential distribution of cut marks on meat-bearing and non-meat-bearing limb bones, and the presence of cut marks on midshaft sections of meat-bearing limb bones. These suggestions are also supported by ethnographic data from the Hadza (Lupo and O'Connell, 2002). Experimental work dictates the questions that can be answered using cut marks, and such studies indicate that a combination of frequency and distribution is required to fully appreciate the utility of cut marks.[3]

When hominids crack bones for marrow, hammerstones leave a distinctive percussion mark on the bone surface. Blumenschine and Selvaggio (1991) calculated percussion-mark frequencies in assemblages first processed by

[3] However, this is not completely understood by all researchers. Recently, DeGusta *et al.* (2003) have provided a good example of a flawed interpretation of the behavioral meaning of cut marks by using incomplete information regarding experimental sample sizes, disregarding important statistical tests, and neglecting fundamental principles of actualistic research. See Domínguez-Rodrigo and Barba (2005) for a critique.

humans and subsequently modified by hyenas. Percussion-mark frequencies on limb bone specimens (irrespective of bone portion) varied by carcass body size: size class 1 and 2 limb bone assemblages showed percussion-mark frequencies of ~30%; size class 3 had frequencies of 21.5%; size class 4 had frequencies of 53.8%. Further, Blumenschine and Selvaggio (1991) established percussion-mark to tooth-mark ratios in these simulated hammerstone-to-carnivore assemblages that range from 1.5:1 (in size class 3 carcasses) to 2:1 (in size class 2 carcasses). In other words, percussion-marked specimens are nearly twice as frequent as tooth-marked specimens in cases in which hyenas act as secondary modifiers of limb bones first demarrowed and abandoned by humans.

Reflections on the Validity of Actualistic Frameworks for Interpreting Early Hominid Behavior

The potential of actualistic studies to inform on hominid and carnivore interactions in the formation of bone assemblages remains unappreciated by some researchers, and it seems that many of those critical to such an approach are non-Africanists in specialization. However, given the complex formational histories of early archaeological sites and the paramount importance of demonstrating rather than assuming hominid involvement in site formation, actualism must be an integral part of taphonomists' toolkits. Controversy over the application of modern referential frameworks to the past and the use of analogies can be properly addressed if the very basic premise of actualism is understood: the fact that the past is unrelated to our temporal experience with reference to the present makes it challenging to reconstruct historic processes. The only way to assume we can "reconstruct" the past is to assume that there are certain regularities in the way the world works which are atemporal, and therefore observable in the present and inferred for the past.

Science was born with such a uniformitarian assumption. Running counter to catastrophism and deistic explanations of the world, uniformitarianism as originally formulated by nineteenth century geologists argued that the world was created through the gradual action of cyclical natural forces (Haneberg, 1983). Simply put, the same laws that govern the modern observable world acted in the past (Rudwick, 1971; Smith, 1977). According to Simpson (1970), uniformitarianism established two basic principles: (1) the history of the earth can be explained by natural forces that can be observed in the modern world and (2) the formation of the earth was a gradual process during which these forces interacted.

The assumption of uniform rates and the implication of slow and gradual change in this "substantive uniformitarianism," to use Gould's (1965) term, have been shown to be incorrect in many instances. The modern conception of uniformitarianism therefore does not assume this rate of change and acknowledges that the agents of change cannot be verified empirically. However, the laws that govern these agents remain permanent. This new uniformitarianism is therefore methodological and vital to scientific procedure. Spatial and temporal invariability in the laws that control processes is absolutely critical if any general conclusion about the past is to be made from observations in the present (Gould, 1965). Methodological uniformitarianism does not directly inform on nature, but provides an approach with which to understand it (Shea, 1982). This approach, in assuming that natural laws are invariant in time and space, does not invoke unknown hypothetical processes if the observed results can be explained through modern processes (Gould, 1965).

Simpson (1970) elaborated further on this concept by separating those aspects of the world that remain unmodified in time and space (immanent properties) and those that are contingent on particular interactions in each moment and place (configurational properties). Immanent properties allow historical processes (or parts of them) to be interpreted precisely because they are not subject to variation in space and time; that is, they are universal. Configurational properties must be approached in a different way. Unlike universal immanent properties, configurational properties are based on *regularities* in the variables that regulate them. It is important to note, however, that there are two types of configurational process: (1) those that are highly variable and therefore difficult to predict and (2) those that those that are highly regular and therefore predictable. Obviously, only the latter can be reliably applied to past dynamics.

We agree with Gould (1979, 1980) in that only those processes whose properties and range of variation can be measured should be used in scientific archaeology. Some researchers believe that only geological/physical processes can be understood from such an approach (e.g., Nairn, Hawkes). However, Simpson (1970) stresses that for any process to be understood it needs only to be uniform; that is, we need to be able to establish regularities. This is possible whether reconstructing the mechanic aspects of the world or biotic behaviors. Much misunderstanding emanates from the misconception that only universal, or immanent, laws are applicable to the past. However, because even universal laws are never absolute (Popper, Watson), we are left with heuristic explanations (Lakatos) that are grounded in the accuracy of their regularities.

Regularity is derived from probability. However, Simpson (1970) notes two difficulties with inferring historical processes: (1) multiple processes may have similar results (that is, equifinality) and (2) configuration makes processes unpredictable. Simpson (1970) himself suggests a solution for the latter. Scientific prediction depends on what is periodical and repetitive. Although historical events are unique and therefore unpredictable on at least some level, there are different degrees of uniqueness, and historical events can be considered predictable to the extent to which we understand their causes and the regularity of their behavior. This is how Simpson (1970) defines historical laws based on probabilities and with similar heuristic power (*sensu* Lakatos) as natural laws. An historical event is determined by the immanent characteristics of the universe that act upon it, but in a configurational way.

This leads us to the use of analogy. To understand past configurational processes, the context and variables that generate modern processes must also be understood. This understanding differentiates formal (descriptive) analogies from relational (dialectical) analogies (Gifford-Gonzalez, 1991). Analogies play an important role because they can discern and document variability in observed regularities. Most importantly, analogies can be observed and replicated.

We can now reexamine modern actualistic frameworks and their applicability to past dynamics. Africanists have carried out a number of studies on the influence of carnivore ravaging on human-made bone assemblages (although non-Africanists have done much work on this and similar topics, we restrict our discussion to the Africanist literature here). We believe that regularity and the range of variability of this process has been established given that similar observations have been made in a variety of contexts during several seasons of the year. The application of these studies to early archaeological sites is also appropriate given the broad similarities in ecological context (African savannas). Simply put, the regular expression of hyaenid bone destruction, based on 15 years of research, makes it (by probability) a good proxy for hyaenid behavior in Plio-Pleistocene African savannas. This inferential process follows Simpson's (1970) guidelines for methodological uniformitarianism.

In sum, we feel that methodological uniformitarianism is an appropriate framework with which to understand and interpret historical processes. In addition, the referential frameworks selected to aid interpretations of hominid and carnivore behavior during Bed I times at Olduvai Gorge are acceptable because they abide by uniformitarian rules and have a solid experimental background. Finally, we suggest that an analytical approach focused on physical attributes will provide a more detailed and rigorous interpretive framework for reconstructing site formation.

4. Geological and paleoecological overview of Olduvai Gorge

C.P. EGELAND, M. DOMÍNGUEZ-RODRIGO, AND R. BARBA

Introduction

The Plio-Pleistocene of Africa (and elsewhere) is marked by a general trend towards cooler and drier climatic regimes. This is superimposed on what has been summarized as "a succession of wet-dry cycles with a long-term shift toward drier conditions, punctuated by step-like shifts in characteristic periodicity and amplitude" (deMenocal, 2004: 8). These punctuations, which have been found to coincide with the onset and amplification of high latitude glacial cycles (Shackleton, 1995; deMenocal, 1995, 2004), occurred at 3.0–2.6 Ma, 1.8–1.6 Ma, and 1.2–0.8 Ma. Profound changes in vegetation structure accompanied these large-scale environmental patterns. Especially significant is the rise and expansion of grasslands and savannas in Africa beginning in the Mio-Pliocene and continuing into the Plio-Pleistocene (Cerling, 1992; Cerling *et al.*, 1993, 1997). Importantly, the establishment of savannas probably signals more-or-less modern patterns of seasonality and rainfall because such C_4 ecosystems (i.e., with tropical grasses and sedges and woody species that do not form continuous cover [Bender, 1971]) are more tolerant of arid, seasonal environments (Jacobs *et al.*, 1999). However, modern levels of C_4 dominance do not develop in East Africa until the Middle Pleistocene (Cerling, 1992), which suggests that Plio-Pleistocene savannas were generally wetter and more closed than today. This theme is confirmed by isotopic data from other studies in the Baringo (Kingston *et al.*, 1994) and Turkana (Cerling *et al.*, 1988; Wynn, 2004) basins of Kenya, and at Gona, Ethiopia (Levin *et al.*, 2004), although subtle variation in the timing and pattern of savanna expansion in these regions is documented as well.

These environmental changes are also reflected in patterns of faunal turnover. In the Omo-Turkana Basin, for example, high turnover rates began occurring among bovids, equids, suids, and primates by 2.5 Ma and peaked after 1.8 Ma, reflecting the gradual expansion of grasslands (Behrensmeyer *et al.*, 1997; Bobe *et al.*, 2002; Bobe and Behrensmeyer, 2004). Increased climatic variability is also indicated in the Omo-Turkana Basin as rapid fluctuations in faunal composition between 2.5 Ma and 1.8 Ma have been documented. Broadly contemporaneous faunal turnover at Olduvai Gorge at about 1.7 Ma (Gentry and Gentry, 1978a, 1978b) and at Konso, Ethiopia between 1.7–1.4 Ma (Suwa *et al.*, 2003) also show a shift towards open-adapted taxa.

Overall, broad-scale analyses suggest three punctuated, step-wise shifts in aridification in Africa during the Plio-Pleistocene. The concomitant expansion of C_4-dominated savannas occurred at slightly different rates in different regions. It is also important to realize that "the record does not support unidirectional shifts to permanently drier conditions" (deMenocal, 2004: 18). Rather, the general trend toward aridification progressed in the context of increased environmental variability. That is, relative to earlier periods, Plio-Pleistocene

M. Domínguez-Rodrigo, R. Barba, and C.P. Egeland, Deconstructing Olduvai: A Taphonomic Study of the Bed I Sites, 33–38.

climates displayed both marked differences and rapid transitions between wet and dry periods. We now turn to a detailed discussion of the geology and paleoecology of the Olduvai Gorge sites.

Geology and Paleoecology

Before proceeding, it is important to define the term "site." In this study, each of the geological and archaeological localities (e.g., FLK) defined by Leakey (1971) and Hay (1976) are considered "sites;" that is, areas with stratified layers of (potentially) varied sedimentology and different archaeological and paleontological material. This is distinguished from a "level," which, as defined by Leakey (1971), is a conglomeration of lithics and/or fauna within a vertically defined unit. Levels can be concentrations of material within thin paleosols (e.g., FLK NN Level 3, FLK Level 22) or more diffuse aggregations scattered throughout almost a meter of sediment (e.g., DK Level 2). An "assemblage" is any sample of lithics and/or fauna and usually this refers to collections from a single level.

GENERAL GEOLOGY AND DATING

Olduvai Gorge rests on the eastern edge of the Serengeti Plain in northern Tanzania (Figure 5). Although the paleontological and archaeological importance of Olduvai Gorge had long been recognized (summarized in Leakey, 1971), it was only after the discovery of OH 5 (*Zinjanthropus boisei* = *Australopithecus boisei*; Leakey, 1959) that large-scale excavations began throughout the gorge. As outlined in

Figure 5. Location of Olduvai Gorge and other significant Plio-Pleistocene localities in Africa.

Chapters 1 and 2, the large and generally well-preserved faunas excavated by Mary Leakey from Olduvai have figured prominently in the ensuing three decades of debate over the behavior of early hominids.

Richard Hay's (1976) seminal monographic treatment continues to lay the foundation for our current understanding of the geology of Olduvai Gorge. The uplift of volcanic highlands to the east and south formed the Olduvai Basin about 2.0 Ma and downcutting stream activity over the past 200 thousand years (Ka) eventually created the modern gorge, which splits into two fingers (the main and side gorges) (Figure 6).The area is underlain by a basement of metamorphic rocks that still outcrop today as isolated inselbergs including Naibor Soit and Kelogi. Hay (1976) defined seven geological formations within the gorge (from oldest to youngest): Beds I, II, III, IV, and the Masek, Ndutu, and Naisiusiu Beds. Because this volume focuses on the Bed I sites, only the Bed I geology is considered in detail.

The lava flows of Bed I are overlain by lake, lake-margin, alluvial fan, and alluvial plain deposits. A number of potassium–argon (^{40}K–^{40}Ar) and argon–argon (^{40}Ar–^{39}Ar) dates from the lavas and marker tuffs of Bed I are available (Figure 7) (Walter *et al.*, 1991, 1992; Manega, 1993; Blumenschine *et al.*, 2003). The bottom of Bed I is marked by underlying lava flows dated to around 2.0 Ma. The first Bed I marker tuff (Tuff IA) lies above these

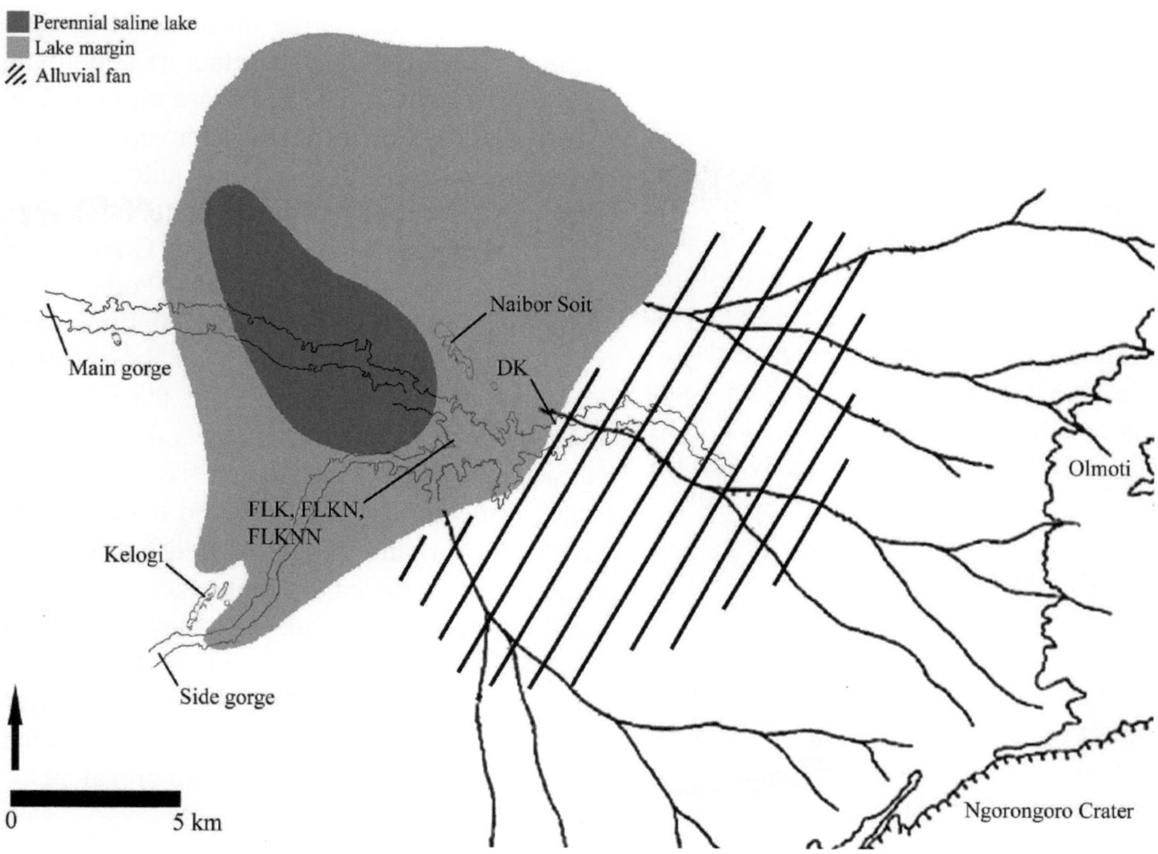

Figure 6. Map showing site locations, significant geographic features and paleogeographic reconstruction of the Olduvai Basin during Bed I times. Extent of lake-margin zone is based on the paleogeography of Tuff IF. Base map adapted from Hay (1976: Figure 29) and Peters and Blumenschine (1995: Figure 4).

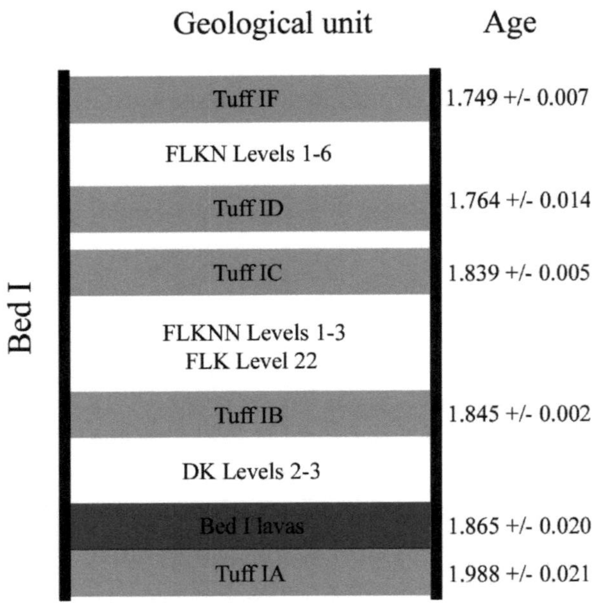

Figure 7. Simplified composite stratigraphy of Bed I showing both the stratigraphic position of the Olduvai sites and the dates of important marker tuffs. Dates from Walter *et al.* (1991, 1992), Manega (1993), and Blumenschine *et al.* (2003).

basal lavas and dates to 1.98 Ma. Capping Tuff IA are the Bed I lavas, which are dated to 1.87 Ma. A series of well-dated marker tuffs are found above the Bed I lavas. The first of these marker tuffs, Tuff IB, overlies the oldest archaeological occurrences in the gorge and is dated to 1.85 Ma while the last, Tuff IF, marks the boundary between Beds I and II and dates to 1.75 Ma. Therefore, Bed I spans approximately 100 Ka.

GENERAL PALEOECOLOGY

Paleogeographic reconstructions indicate that during Bed I times the Olduvai Basin was dominated by a saline and alkaline paleolake of fluctuating size (Hay, 1976; Peters and Blumenschine, 1995; Hay and Kyser, 2001). An alluvial fan and plain were situated on the eastern margin of the basin and intermittent streams drained the volcanic highlands to the south and east of the lake (Figure 6). A majority of the Bed I archaeological occurrences occur in Hay's (1976) Eastern Lake Margin lithofacies, which is likely due to the presence of fresh water along the eastern margin of the basin (Hay, 1976; Peters and Blumenschine, 1995; Deocampo *et al.*, 2002).

Cerling and Hay's (1986) isotopic data suggest a mean annual rainfall of 800 mm and average temperatures of 16°C during Bed I times, which is wetter and cooler than the Olduvai Basin is today (mean annual rainfall = 566 mm; average temperature = 22°C). Wooded grasslands seem to have dominated around the paleolake (Sikes, 1994, 1999). Two major environmental episodes can be distinguished in the Bed I sequence. The first occurred before 1.76 Ma (below Tuff ID). The lake level was high early in this period and fluctuated thereafter (Hay and Kyser, 2001). Urocyclid slugs (Verdcourt, 1963) and closed-habitat rodents (Jaeger, 1976; Fernández-Jalvo *et al.*, 1998) indicate the presence of densely vegetated and humid habitats (Kappelman, 1986). The bovid data are more equivocal in this regard, especially at DK where open habitat alcelaphines and antilopines are quite common; however, closed- and mixed-habitat bovids are more common below Tuff ID (Gentry and Gentry, 1978; Kappelman, 1984; Potts, 1988; Shipman and Harris, 1988; Plummer and Bishop, 1994; Kappelman *et al.*, 1997). Isotopic data from Level 22 at the FLK site suggest a riparian or grassy woodland at this location about 1.8 Ma (Sikes, 1994). The second environmental episode occurred between Tuff ID (1.76 Ma) and the top of Bed I (1.75 Ma). Significant aridification took place during this period and pollen evidence suggests rainfall may have been as low as 350 mm/year (Bonnefille, 1984). The lake level also dropped during this period and open habitat rodents and bovids became more prevalent.

THE BED I SITES

DK is among the oldest of the archaeological sites within the gorge. It lies above the Bed I lavas and is overlain by Tuff IB, dating the deposits to about 1.85 Ma. Leakey (1971) distinguished three levels at the DK complex, Levels 1–3 from youngest to oldest, each varying in thickness from about 30–80 cm. A majority of the Level 3 material lay atop an eroded 9 cm-thick paleosol. The site was situated near the intermittent, northwest-flowing streams that drained the highlands to the east. The deposits consist mainly of claystone and tuffs interspersed with volcanic conglomerates (Hay, 1976). The paleolake was at its greatest extent during the deposition of the DK sediments (Hay and Kyser, 2001) and a variety of indicators, including crocodile remains (Leakey, 1971), urocyclid slugs (Verdcourt, 1963), aquatic and semiaquatic turtles (Auffenberg, 1981), and papyrus rhizomes and flamingo remains (Hay, 1976) suggest nearby permanent shallow water and marshland. The presence of reduncine bovids supports this interpretation, although open-adapted alcelaphines and antilopines are represented as well (Gentry and Gentry, 1978a, 1978b). A mixed habitat, albeit with a significant grassland component, is suggested by more recent taxonomic (Potts, 1982, 1988; Kappelman, 1984; Shipman and Harris, 1988) and ecomorphological (Plummer and Bishop, 1994; Kappelman *et al.*, 1997) studies of the bovids.

FLK North North (FLK NN) has levels (1–4 from youngest to oldest) in the lower and middle parts of Bed I. Levels 1, 2, and 3 occur between Tuffs IB and IC while Level 4 lies below Tuff IB. Level 3 was found within a thin paleosol deposit. The lake appears to have regressed slightly by this time (Hay and Kyser, 2001), which placed the site about a kilometer east of the lake shore (Hay, 1976). Although sedimentologically similar to DK (clays and a tuff), FLK NN lacked the conglomerates found at DK. An extremely closed environment at FLK NN is indicated by several lines of evidence. Taxonomic analyses show an abundance of reduncines (Gentry and Gentry, 1978a, 1978b; Bunn, 1982; Potts, 1982, 1988; Kappelman, 1984; Shipman and Harris, 1988) and ecomorphological studies also suggest a significant presence of closed-vegetation species (Plummer and Bishop, 1994). Closed and wet habitat murid rodents are also common at FLK NN (Jaeger, 1976; Fernández-Jalvo *et al.*, 1998). Fernández-Jalvo *et al.* (1998: 166) offer the following reconstruction:

The FLK NN levels can be reconstructed as having been thickly wooded, dominated by a single tree canopy as in present-day African closed woodlands. By analogy with such habitats today, there would have also been a thick understorey of bushes and small trees and abundant ground vegetation consisting of herbs and grasses.

FLK North is located in upper Bed I between Tuffs ID and IF. Sediments at FLK North also consist mainly of clays although several thin ferruginous bands and tuffs were found scattered throughout. Faunal and lithic material were uncovered in six defined levels (1–6 from youngest to oldest) ranging in thickness from several centimeters to almost a meter. Decreases in lake level during the deposition of the FLK North sequence (Hay and Kyser, 2001) are likely linked to significant aridification during upper Bed I times. Consistent with this trend is the dominance of open-adapted bovids (Gentry and Gentry, 1978; Kappelman, 1984; Shipman and Harris, 1988; Plummer and Bishop, 1994) and rodents (Jaeger, 1976) especially in the upper levels of the FLK North sequence. Land birds also occur at the site (Hay, 1976). It does appear, however, that the relatively high frequencies of open habitat gerbils in Levels 4–6 are an artifact of predator selection (Fernández-Jalvo *et al.*, 1998). Therefore, only the upper levels sample open habitats while the lower levels more likely reflect woodlands that were richer than any modern savanna (Fernández-Jalvo *et al.*, 1998).

The FLK site is best known for the Level 22 (*Zinjanthropus*) assemblage. However, samples of lithic and/or faunal material were uncovered from several levels overlying Level 22. Under consideration in this volume are the assemblages from Levels 10–13 and 15.

The Level 22 material was excavated from an eroded paleosol that lies below Tuff IC, providing an age of about 1.84 Ma. The other levels, which lie above Tuff IC and run to the top of Bed I, were found in clayey or sandy tuffs (some of them reworked) and the thickest deposit (Level 13) was about 60 cm. The rodent (Fernández-Jalvo *et al.*, 1998) and isotope (Sikes, 1994) data suggest a closed canopy forest or wooded grassland, respectively, at Level 22. The bovids indicate a mixed habitat with nearby grasslands (Shipman and Harris, 1988; Plummer and Bishop, 1994).

5. New estimates of tooth-mark and percussion-mark frequencies at the FLK Zinjanthropus level: the carnivore–hominid–carnivore hypothesis falsified (I)

M. DOMÍNGUEZ-RODRIGO AND R. BARBA

Introduction

The abundance of skull and limb bones at the FLK *Zinjanthropus* level provides an exceptional data set for reconstructing early hominid behavior. As discussed in Chapter 2, this assemblage has been variously interpreted as the result of: (1) hominids hunting and selectively transporting those parts from complete carcasses (Isaac, 1978, 1983, 1984; Bunn, 1982, 1983b, 1991; Bunn and Kroll, 1986, 1988; Bunn and Ezzo, 1993; Domínguez-Rodrigo and Pickering, 2003; Oliver, 1994; Rose and Marshall, 1996; Domínguez-Rodrigo, 1997a, 2002); (2) hominids transporting complete skeletons from partially defleshed carcasses (Capaldo, 1995, 1997); or (3) hominids passively scavenging the brain and marrow-bearing long limb bones from defleshed carcasses at felid kills (Blumenschine, 1986, 1991). The last of these scenarios was further explored in a landmark publication (Blumenschine, 1995) in which the high frequency of tooth marks on long bone midshafts at FLK *Zinj* was explained by the triple-stage (carnivore–hominid–carnivore) model described in Chapter 2, which was supported by data from earlier experiments with human- and carnivore-modified bones (Blumenschine, 1988).

According to the triple-stage model, some carnivores (felids) had primary access to carcasses, removing flesh partially (Capaldo, 1998b) or completely (Blumenschine, 1986, 1995). Hominids had secondary access, processing marrow-bearing bones (reflected in the frequency of percussion-marked bones). Finally, other carnivores (hyaenids) consumed the grease-bearing cancellous bone. This model was experimentally replicated only once, by Selvaggio (1994), whose samples were heavily biased towards small carcasses; moreover, several types of carnivores (both flesh-eating felids and bone-crunching hyaenids) were lumped in the first stage of carcass consumption. Further experiments by Capaldo (1995, 1997) provided more robust data for carnivore-only and hominid–carnivore models, since his samples included both small and large carcass sizes.

The models built by Blumenschine *et al.* strongly relied on tooth marks and percussion marks as the main bone surface modifications, virtually ignoring cut marks. Yet the cut-mark data from FLK *Zinj* originally reported by Bunn and Kroll (1986) paint a very different picture from Blumenschine's (1995) model (discussed in Domínguez-Rodrigo, 1997a, 1999b); this contradiction prompted our reanalysis of the FLK

A different version of this chapter was previously published in *Journal of Human Evolution* (Domínguez-Rodrigo and Barba [2006], vol. 50, pp. 170–194).

M. Domínguez-Rodrigo, R. Barba, and C.P. Egeland, Deconstructing Olduvai: A Taphonomic Study of the Bed I Sites, 39–74.

Zinj assemblage (see Chapter 6). The carnivore–hominid–carnivore model at FLK *Zinj* relies on an "anomalous" high frequency of tooth marks on midshafts, presumably resulting from felids having initially defleshed the carcasses before hominids scavenged them, since hyenas, having late access to broken limb bones, would only mark the ends while consuming their grease. The tooth-mark frequency observed by Blumenschine (1995) was clearly outside the range of variation for both the carnivore-only and hominid-to-carnivore models developed by him (Blumenschine, 1988). Therefore, the FLK *Zinj* assemblage stood out as an oddity never before replicated experimentally using just felids for the carnivore–hominid–carnivore model (see Selvaggio, 1994).

In the study presented here, we thoroughly examined surfaces of long bone specimens from FLK *Zinj*, and found that some surface modifications with similar morphology to tooth marks were created by biochemical processes, which are overwhelmingly the most common type of bone surface modification in the assemblage. The revised tooth-mark rates at FLK *Zinj* are far lower than those reported by Blumenschine (1995), and are even less than new, lower estimates obtained in recent experiments with felids (Domínguez-Rodrigo *et al.*, in press); they are much closer to those obtained in hominid–carnivore experiments (Blumenshine, 1988; Capaldo, 1995). Additionally, our analysis of tooth pits, percussion marks, and notches provides data which, when compared with experimentally derived referential frameworks, suggest that hominids had primary access to fleshed carcasses. Our multiple lines of evidence point to a simple conclusion: carnivore access to the bone assemblage at FLK *Zinj* was secondary to butchery and marrow extraction by hominids, as originally argued by Bunn (1982, 1983a), Bunn and Kroll (1986, 1988), and Leakey (1971). This conclusion further buttresses the results of earlier cut-mark studies at FLK *Zinj* (Bunn, 1982, 1983a; Bunn and Kroll, 1986, 1988; and interpretations in Domínguez-Rodrigo, 1997a, 2002).

Sample and Methods

The analysis of the long limb bone sample from FLK *Zinj* was carried out in November–December 2004 at the National Museums of Kenya in Nairobi. Specimens were carefully analyzed for surface modifications, including cut marks (Chapter 6), percussion marks, tooth marks, and natural marks (Tables 2 and 3). The number of bones with good cortical preservation analyzed in this study (NISP = 699) is very similar to the sample (NISP = 731) analyzed by Blumenschine (1995), and smaller than the number of specimens (NISP = 1467) analyzed by Bunn and Kroll (1986), since Bunn and Kroll included bones with poor cortical preservation and specimens smaller than 2 cm. The sample is also smaller than Blumenschine's because those specimens showing ambiguous marks that could not be distinguished as either tooth marks or biochemical marks ($n = 26$) were discarded to prevent bias. The additional difference of six specimens between Blumenschine's sample and ours might be accounted for by different NISP counts, since several specimens with recent breaks were counted as one specimen by us.

Both Blumenschine's sample and ours focused on specimens larger than 2 cm that showed good cortical surfaces. Most of the specimens show very good cortical preservation but surface quality is far from excellent. We define excellent condition as the presence of original cortical surfaces on which only modifications made by humans or other animals are present; this contrasts with Blumenschine's (1995: 28) definition of excellent condition as "preserving fine details of mark morphology." Most of the specimens at FLK *Zinj* have been affected by natural processes: exfoliation and cracking created by soil moisture and pH, similar to those observed at various stages of

Table 2. Sample size for tooth marks (TM) and percussion marks (PM) used by Blumenschine and Domínguez-Rodrigo and Barba

Tooth marks		
Sample size studied by Blumenschine and by Domínguez-Rodrigo and Barba = 364		
TM identified by Domínguez-Rodrigo and Barba = 61 (16.8%)		
TM identified by Blumenschine = 255 (70.1%)		
	TM humerus = 75.9% (44/58)	TM femur = 84% (26/31)
	TM radius = 71.7% (43/60)	TM tibia = 69.7% (53/76)
	TM metacarpal = 80% (24/30)	TM metatarsal = 66.7% (24/36)
Number of specimens with biochemical marks = 190 (74%)		
Percussion marks		
Sample size studied by Blumenschine and by Domínguez-Rodrigo and Barba = 364		
PM/percussion notches identified by Domínguez-Rodrigo and Barba = 109 (29.9%)		
	PM = 75 (20.6%)	
	Percussion notches = 34 (9.3%)	
	PM + percussion notch = 96 (26.4%)	
PM identified by Blumenschine = 118 (32.4%)		
	No. of PM coinciding with Domínguez-Rodrigo and Barba = 48 (13.2%)	
	No. of percussion notches coinciding with Domínguez-Rodrigo and Barba = 12 (3.3%)	
	No. of PM + percussion notches coinciding with Domínguez-Rodrigo and Barba = 68 (18.7%)	
Number of specimens with biochemical marks = 190 (74%)		

subaerial weathering (Behrensmeyer, 1978), and biochemical processes created the modifications described later. Despite this biogenic damage, the surfaces are in sufficient condition to preserve the diagnostic features of percussion marks: the shape of the pit, its location with respect to breakage plane and notches, and microstriations as described by Blumenschine and Selvaggio (1988). Cut marks are also clearly identifiable and distinguished from marks created either by trampling/abrasion or by digging tools.

In our analysis, tooth and percussion marks were first scrutinized under a strong light by naked eye; then the complete specimen – not just the mark – was examined with a 10X–20X hand lens, also under strong light. Each mark was individually examined and its features carefully considered in light of experimentally derived research on the identification of tooth and percussion marks, discussed in Chapter 3. We analyzed five to ten bones per hour and identified marks were double-checked by both authors to ensure accuracy. Following Blumenschine's (1995) procedure and his published list of bones bearing tooth marks, we were able to locate all marks that he reported. It should be stressed that despite the divergent results between the present study and Blumenschine's (1995) study, we all followed the same identification protocol. One of us (MDR) received further training using Blumenschine's actualistic collection of carnivore-ravaged bones in 1996. We all agree on the diagnosis of conspicuous and inconspicuous[1] tooth marks on bones from modern experimental assemblages; our results differ on fossil assemblages, where other natural processes (i.e., biochemical processes) are at work.

[1] Conspicuous marks are those easily detected and identified by naked eye. Inconspicuous marks need the aid of magnification to be either detected or properly identified.

Table 3. Mean percentages of tooth-marked specimens in relation to the total number of specimens in experimental assemblages and in the FLK Zinj assemblage, according to bone portion: epiphysis (EPIPH), near-epiphysis (NEP), and midshafts (MSH)

		EPIPH	NEP	MSH	All
Small-sized carcasses					
Carnivore only*					
	Mean %	100	75	69.1	70.8
	95% CI	(–)	(0–100)	(47.5–90.7)	(33.1–100)
Hammerstone-to-carnivore*					
	Mean %	50	60.7	15.9	21.9
	95% CI	(0–100)	(29.3–92.1)	(2.6–29.2)	(5.1–38.7)
Hammerstone-to-carnivore**					
	Mean %	–	31.4	14.9	19.4
	95% CI	–	(19.9–42.9)	(11.2–18.6)	(15.5–23.3)
FLK *Zinj*					
	Percent	24	23.5	13.8	17.3
	No. marked/total	(12/50)	(4/17)	(17/123)	(33/190)
Large-sized carcasses					
Carnivore only*					
	Mean %	100	90.5	86.5	87.7
	95% CI	(–)	(76.5–100)	(75.7–97.3)	(78.5–96.9)
Hammerstone-to-carnivore*					
	Mean %	85.7	39.5	5.1	16.4
	95% CI	(35.6–100)	(8.4–70.6)	(0.2–10)	(10.5–22.3)
Hammerstone-to-carnivore**					
	Mean %	–	48.4	16.2	27
	95% CI	–	(40.5–56.3)	(10.4–22.8)	(22.5–31.5)
FLK *Zinj*					
	Percent	28.3	27.7	9.7	14
	No. marked/total	(17/60)	(15/54)	(39/392)	(71/506)

Analyses of 95% CI for sets of experiments are included. They were calculated using the t distribution, where 0.025 is the critical value of t with $n-1$ degrees of freedom
* Blumenschine (1988, 1995)
** Capaldo (1997)

Blumenschine (1995) claims that percussion marks lack the internal crushing associated with carnivore tooth marks. However, we have experimentally observed that percussion marks obtained using modified hammerstones (i.e., chopper or chopping tools) frequently display marks with internal crushing on bone surfaces, which can often be distinguished from tooth marks because of their angular contours (work in progress). These types of marks are observed on bone specimens from FLK *Zinj* (Figure 8). Unmodified hammerstones also generate a substantial proportion of marks – which may be as high as 30% – without striations, which may be mistaken for tooth pits lacking internal crushing (work in progress). In the present study, only marks that showed the diagnostic features as described by Blumenschine and Selvaggio (1988) were identified as percussion marks.

The locations of surface modifications were recorded by bone portion, bone section, and element. Specimens were classified as

Figure 8. Rectangular percussion marks on a specimen from FLK *Zinj*, probably made with a chopper. None of the marks were associated with microstriations when analyzed using SEM. Note the angular contour of the mark made by the irregular edge of the hammerstone, which (in the absence of microstriations) is the best criteria for differentiating rectangular percussion pits from those created by teeth (scale = 1 cm).

small carcasses (sizes 1 and 2) and large carcasses (sizes 3 and 4) as defined by Bunn (1982). Most carcasses in the large size category were size 3. Additionally, tooth pits were also measured to identify the type of carnivore. They were molded and measured following Domínguez-Rodrigo and Piqueras's (2003) protocol. Breakage patterns were analyzed, since notch type can point to the agent responsible for marrow extraction. We also tallied breakage planes and quantified the angle range on all longitudinal and oblique breakage planes. For more detail on these methods, see Chapter 3.

Throughout our work, we use experimentally derived referential frameworks (discussed in Chapter 2) to interpret the new FLK *Zinj* data and to compare them with those of previous researchers. Frequencies of biochemical marks, tooth marks, percussion marks, and notches in both the experimental and fossil assemblages were described by using means and 95% confidence intervals (CIs). Although 95% CIs can be problematic when comparing many small samples (which have a larger CI than large samples), they have been used by previous researchers and therefore these statistics enable comparisons between past and present work.

Biochemical Marks in Modern and Fossil Bone Assemblages

Biochemical marks on bone are mainly created by fungi and bacteria, which form colonies on bone surfaces (Figures 9 and 10A), altering the underlying bone. This process begins with a change in the bone color; subsequently, cortical layers begin to exfoliate, generating marks that resemble scores or pits. These marks may be shallow (affecting one or two cortical layers) or deep (affecting several cortical layers). In Domínguez-Rodrigo and Barba (2006), it was not fully explained how fungi can create such a diversity of marks. Studies of biochemical modification of bone are still in their infancy, given the difficulty in modeling a process which involves several intervening factors and long time periods. Usually bacteria prepare the way for fungi to act on bone surfaces (Dube, 2005). However fungi, given their small size and tendency to form colonies, may also adhere to organic host materials, such as plant roots, and preserve the shape of their host when they modify bone surfaces. Fungi act in symbiosis with roots, obtaining carbohydrates in exchange for decomposed minerals (Dube, 2005). Decomposition creates strong metabolic acids that modify bone surfaces (Dube, 2005).

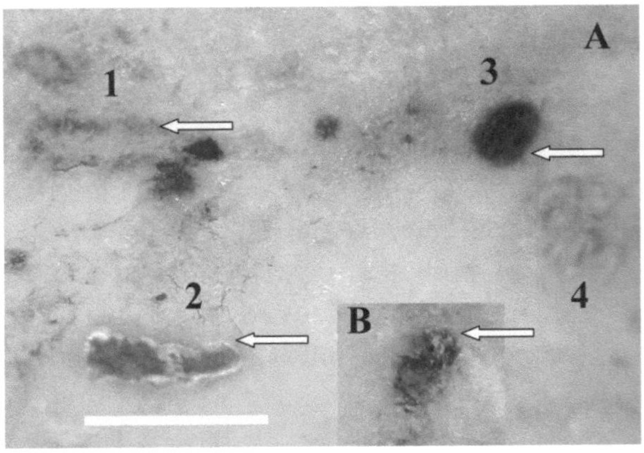

Figure 9. (A) Examples of experimentally created biochemical surface modifications on modern bone. These marks appear as pits and scores with different degrees of modification (reflected in different color tones). A small part of Mark 1 has begun exfoliation. Mark 3 shows a very dark coloring of the cortical surface and Mark 4 shows a powdery texture. Arrows show the diagnostic color outline, which contrasts with the interior of the mark. (B) Fungi found on the cortical surface earlier in the experiment; this colony likely created one of the marks depicted in Figure 9A. Arrow shows the dark outline underlying the fungi. The link between the presence of fungi and the biochemical modification of the cortex can thus be reliably established (scale = 1 cm).

When bones are buried, carbonate often precipitates in the sediment, adhering to both fungi and their organic hosts and preserving their shapes (Figure 10B). These distinctive shapes aid in the identification of biochemical marks. A large portion of the elongated marks observed in the FLK *Zinj* assemblage (Figures 10–12) are probably biochemical modifications of bone surfaces created by fungi which were attached to plant roots.

Experimental work provides more detail on fungal and microbial microscopic modifications of bone. Sharmin *et al.* (2003) undertook a controlled experiment to determine how fungi affect bone structure, exposing murine bones for between 1 and 12 weeks to the pathogenic dimorphic fungi *Coccidioides immitis, Blastomyces dermatitidis, Histoplasma* spp., and *Paracoccidioides brasiliensis*, and to some strains of dematiaceous fungi (*Exophiala* spp., and *Foncecaea pedrosoi*). The authors (2003: 19–20) found that "the fungal isolates exhausted the architecture of compact cortical bones by creating irregular cratered and grooved surfaces and causing impingement. . .[demonstrating] filamentous forms and porulation in and all around the bone structure." Despite this short incubation period, relative to the long-term processes affecting the archaeological record, most of the fungal strains showed degrees of invasiveness to the bone. Not surprisingly, bones left in natural environments over long periods of time show an even higher incidence of cortical modification. Piepenbrink (1984: 241) made observations on buried bone which attest to microbial attack. Microbes invade bone cavities, excrete enzymes and acids in order to release usable energy resources from the bone tissue, and then consume this tissue. The result is "specific osteolysis (microscopical focal destruction)" (Piepenbrink, 1984: 241). Biochemical damage affects both histological structure and the cortical surface (Marchiafava *et al.*, 1974; Hackett, 1981; Child, 1995; Greenlee, 1996).

These results (previously documented to a microscopic scale) must be treated with caution when applied to macroscopic bone modification: experimentally derived models for biochemical damage remain incomplete, and we do not fully understand how the damage which takes place over month- or year-long timescales may be applied to the fossil record. However, we have outlined a preliminary model here, based on an experimentally derived referential framework, for the progression of biochemical bone modification. For a doctoral research project (in progress, under the supervision of MDR), 20 defleshed and demarrowed equid and bovid bones were placed in a cardboard box in a dark storage room for almost 2 years (beginning February 2003) to assess bone decay in the absence of

Figure 10. Several examples of biochemical marks in Stages 1–3 on specimens from FLK *Zinj*. Black bars illustrate cross section of the score. Where they are completely rectangular, no score exists. (A) Brown lines on the bone surface created by fungi (Stage 1). Their orientation is similar to that of tooth scores. (B) Several elongated carbonate formations, under which the cortex is about to exfoliate (Stage 1). (C) Score-shaped lines with a brown outline and cortical deterioration, which sometimes appear exfoliated where a proto-score is forming (Stages 1–2). (D) Score-shaped marks and pits, contrasting in color from the bone surface. The original cortical layer is preserved, though the left-hand mark has exfoliated through more than two cortical layers (where it says north) (Stages 2–3). (E) Detail of a score shown in 2C, showing staining of the score and its contrasting brown outline (Stage 2). (F) Another biochemical score showing original cortex in its proximal and distal ends, but exfoliation in the middle (Stage 3) (scale = 1 cm).

the bone-modifying agents usually present in soils (e.g., pH, microbes, insects). When placed in the box, bones were mostly clean with some miniscule fresh scraps of tendon, periosteum, and flesh remaining; they had been broken, washed, partially dried, and were checked to ensure that they bore no biochemical marks. After 3 months, spongy structures identified as fungi were observed around the patches of periosteum, as well as on some bare bone patches (Figure 9B). In subsequent months, these structures decayed and disappeared by the eighth month, leaving behind dark stains easily distinguishable from the surrounding surface, as are observed at FLK *Zinj*. In most cases, surface color but not cortical structure had been affected; however, a few marks, particularly the more elongated ones, began flaking the outermost cortical layer.

The biochemical marks shown in Figure 9, together with eight more marks created by

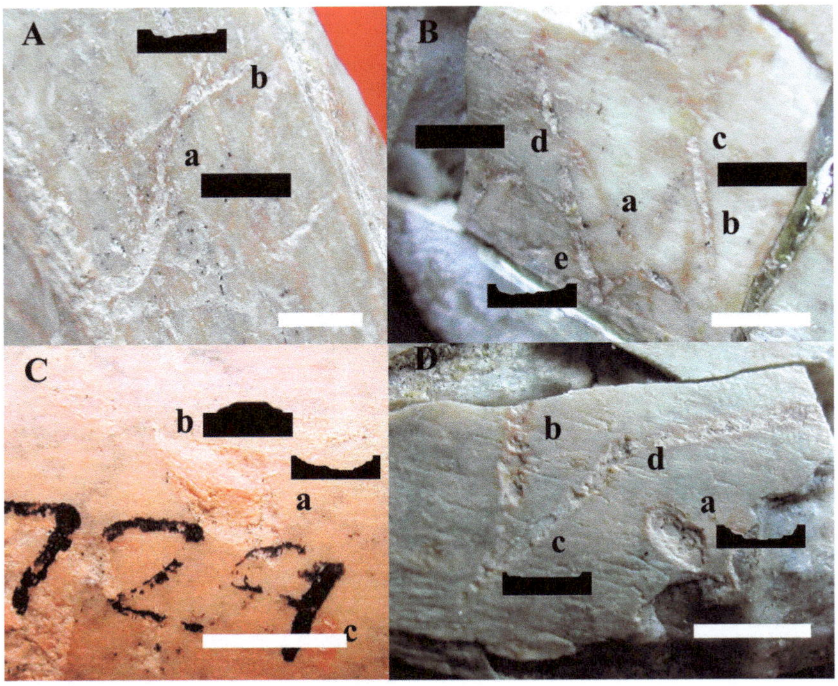

Figure 11. Biochemical marks in Stages 2–4 on specimens from FLK *Zinj*. Black bars illustrate cross section of the score; where they are completely rectangular, no score exists. (A) Bone modified by several biochemical marks; the main score is shown with cross sections at both ends (points "a" and "b"), with the original cortex preserved at one end but exfoliated at the other (Stage 3). (B) Point "a" shows a brown line, the beginning of cortex deterioration (Stage 1); "b" and "c" show two ends of a proto-score, with brown outline and the original cortex present (Stage 2); Points "d" and "e" show an incomplete biochemical score which preserves original cortex at one end ("d"), and has already formed a groove at the other ("e") (Stage 3). (C) A pit which still preserves some of the carbonate which originally created the mark ("b"), and the mark itself ("a"). (D) Point "a" shows a pit created by successive exfoliation, with its diagnostic brown outline (Stage 4); "b," "c," and "d" show scores with irregular outlines, created by independent episodes of exfoliation (Stages 2–3) (scale = 1 cm).

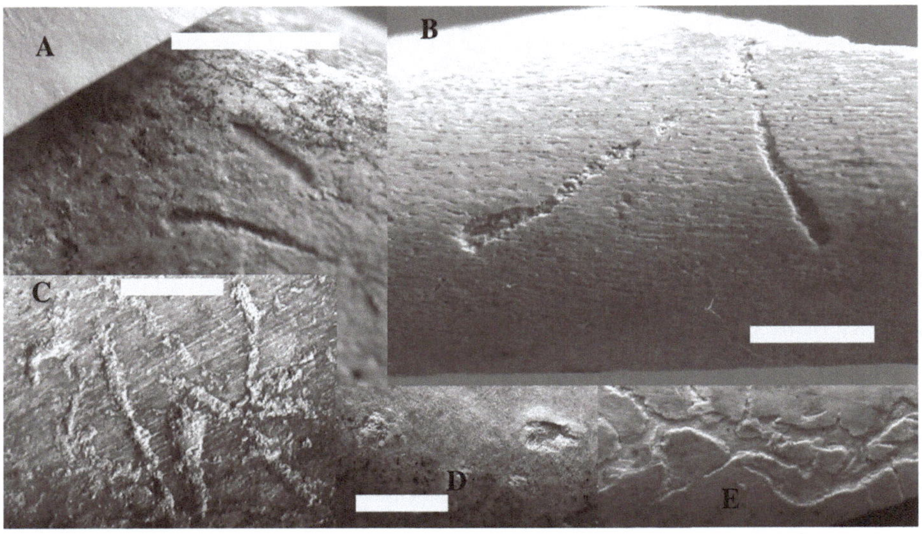

Figure 12. Biochemical marks in Stages 3–4 on specimens from FLK *Zinj*. Scores show irregular, increasing and decreasing width (A, B), and follow the shape of colonies with irregular pits (C). (D) Biochemical pits which can be differentiated from tooth marks based on contrasting color outline and interior exfoliation. (E) Extreme example of multiple biochemical scores with a dendritic appearance (scale = 1 cm).

bone discoloration on the same bone specimen, were biochemically analyzed.[2] The bone specimen had several dark oval-shaped marks. When observed under the microscope, these marks showed a dark central nucleus whose color became lighter towards the periphery. A total of ten microsamples were obtained from ten different marks with a probe containing a sterile solution. Samples were obtained only from within the marks analyzed, and also from the mark-free surface as a control. The mark samples were placed in isolated Petri plaques where fungi and bacteria could develop. The plaques were incubated for a few days so that those microbial components of the samples which were compatible with 6–45°C temperatures could grow.

After 48 h, all plaques showed rapid growth of mycelial fungi (Figures 13–22). Bacteria were also documented, though at much lower frequencies, in all samples. The fungi documented belong to a diversity of taxa (Table 4). This proves that the dark staining of the bone surface corresponds to the action of diverse fungi. The dark pigmentation of some marks indicates prolonged exposure of the area to the acidic action of fungi. The degree of contamination is expressed in colony forming units (CFU)/cm^2, where CFU corresponds to the average number of units forming colonies of fungi or bacteria. The values shown in Figures 13–22 indicate the degree of contamination observed both in agar-based (nutrient-rich) plaques as well as laminated plaques.

It can be concluded that the bone surface is affected by dark circular marks caused by mycelial fungi (Table 4). Bacterial contaminants were also detected but in smaller numbers. Both fungi and bacteria produce metabolites (listed in Table 5) during their growth which are excreted on the bone. These metabolic products include: organic and inorganic acids, enzymes, pigments, and toxins. The marks are the result of the actions of these metabolites altering biochemically the surface

[2] Biochemical study carried out by Andres Sanchez Ledesma, Marcos del Mazo, Ismael Gonzalez, M. Jesus Gomez and Manuel Valiente from Arte-Lab. SL for M. Domínguez-Rodrigo.

Sample N°1

Fungi CFU/cm^2	*Bacteria CFU/cm^2*	*Culture 1 CFU/cm^2*	*Culture 2 CFU/cm^2*	*Culture 3 CFU/cm^2*
14,2	1,6	5,6	1,4	2,5

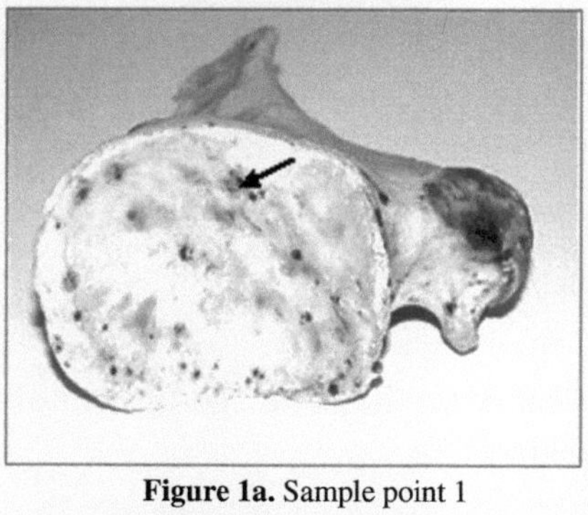

Figure 1a. Sample point 1

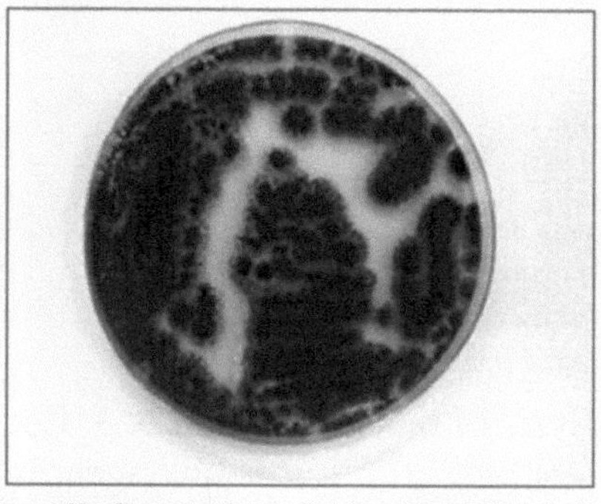

Figure 1b. Growth of mycelial fungi from sample 1

Taxonomic identification: *Clodosporium sp.*

Figure 13. Analysis of biochemical mark 1. Location of mark on the bone and growth of biochemical agent in Petri plaque.

Sample N°2

Fungi CFU/cm²	*Bacteria CFU/cm²*	*Culture 1 CFU/cm²*	*Culture 2 CFU/cm²*	*Culture 3 CFU/cm²*
13.8	18.3	4.2	–	1.7

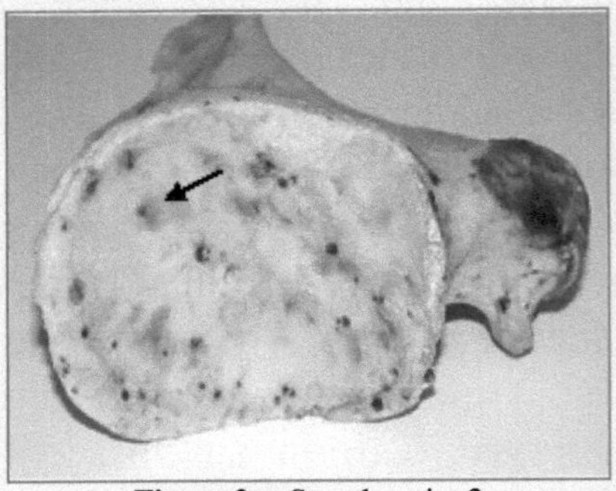

Figure 2 a. Sample point 2

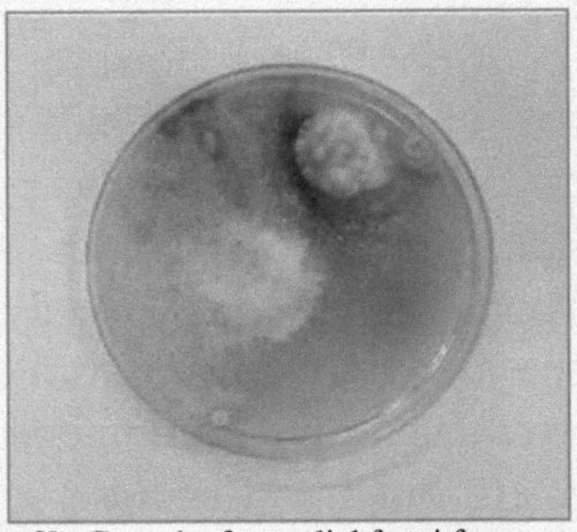

Figure 2b. Growth of mycelial fungi from sample 2

Taxonomic identification: *Actinomyces sp., Penicillion sp., Micrococcus sop.*

Figure 14. Analysis of biochemical mark 2. Location of mark on the bone and growth of biochemical agent in Petri plaque.

Sample N°3

Fungi CFU/cm²	*Bacteria CFU/cm²*	*Culture 1 CFU/cm²*	*Culture 2 CFU/cm²*	*Culture 3 CFU/cm²*
12,5	6,4	2,4	1,4	5,2

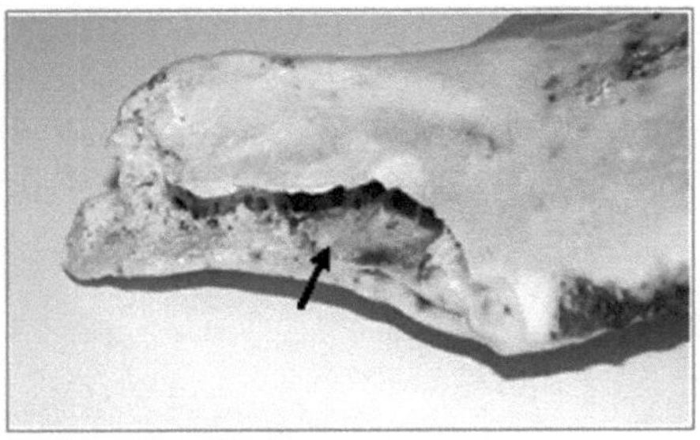

Figure 3 a. Sample point 3

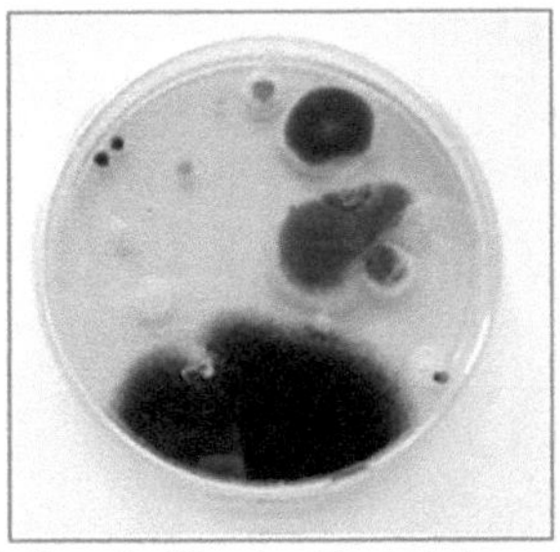

Figure 3 b. Growth of fungi and bacteria from sample 3

Taxonomic identification: *Aspergillus niger, Penicillium sp., Mucor sp., Bacillus cereus, Bacillus subtilis, Micrococcus sp.*

Figure 15. Analysis of biochemical mark 3. Location of mark on the bone and growth of biochemical agent in Petri plaque.

Sample N°4

Fungi CFU/cm²	*Bacteria CFU/cm²*	*Culture 1 CFU/cm²*	*Culture 2 CFU/cm²*	*Culture 3 CFU/cm²*
8,7	16,2	1,3	2,9	3,3

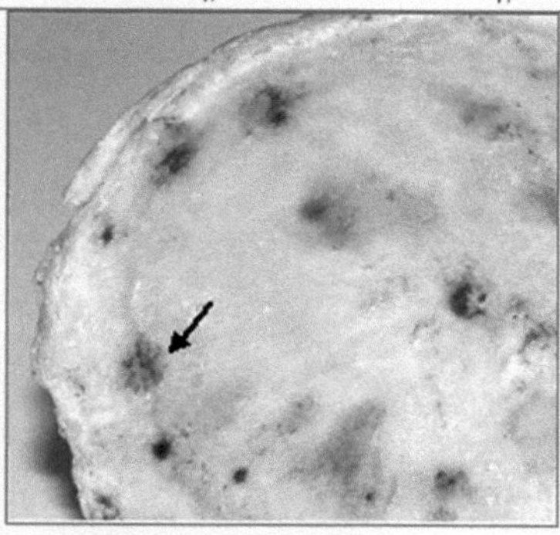

Figure 4 a. Sample point 4

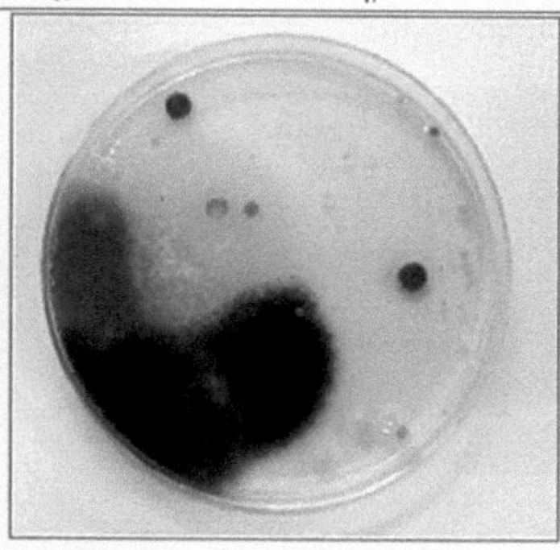

Figure 4 b. Growth of fungi and bacteria from sample 4

Taxonomic identification: *Actinomyces sp., Aspergillus niger, Penicillium sp., Cladosporium sp,. Bacillus cereus, Micrococcus roseus.*

Figure 16. Analysis of biochemical mark 4. Location of mark on the bone and growth of biochemical agent in Petri plaque.

Sample N°5

Fungi CFU/cm²	*Bacteria CFU/cm²*	*Culture 1 CFU/cm²*	*Culture 2 CFU/cm²*	*Culture 3 CFU/cm²*
22,5	2,6	2,5	-	2,4

Figure 5 a. Sample point 5

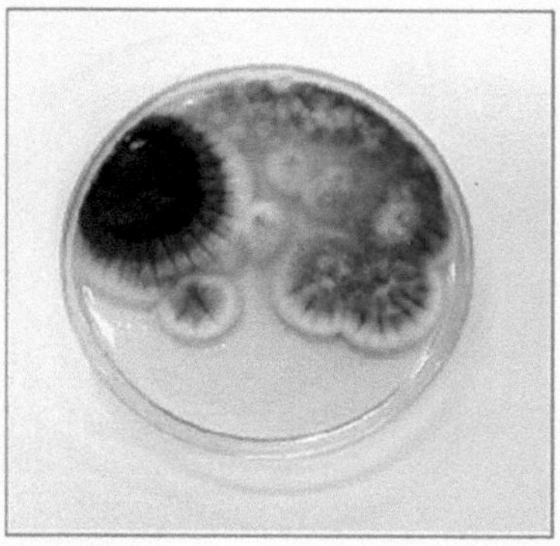

Figure 5 b. Growth of fungi from sample 5

Taxonomic identification: *Penicillium chrysogenum*

Figure 17. Analysis of biochemical mark 5. Location of mark on the bone and growth of biochemical agent in Petri plaque.

Sample N°6

Fungi *CFU/cm²*	*Bacteria* *CFU/cm²*	*Culture 1* *CFU/cm²*	*Culture 2* *CFU/cm²*	*Culture 3* *CFU/cm²*
13,7	4,3	2,4	3,1	-

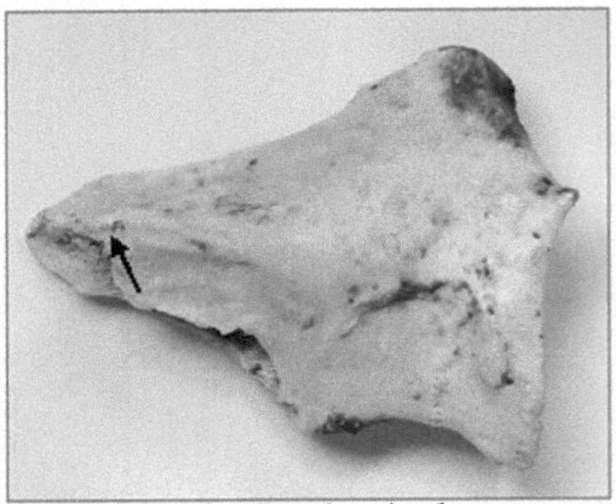

Figure 6 a. Sample point 6

Figure 6 b. Growth of fungi from sample 6

Taxonomic identification: *Penicillium chrysogenum*

Figure 18. Analysis of biochemical mark 6. Location of mark on the bone and growth of biochemical agent in Petri plaque.

Sample N°7

Fungi *CFU/cm²*	*Bacteria* *CFU/cm²*	*Culture 1* *CFU/cm²*	*Culture 2* *CFU/cm²*	*Culture 3* *CFU/cm²*
18,3	22,6	3,2	1,2	6,1

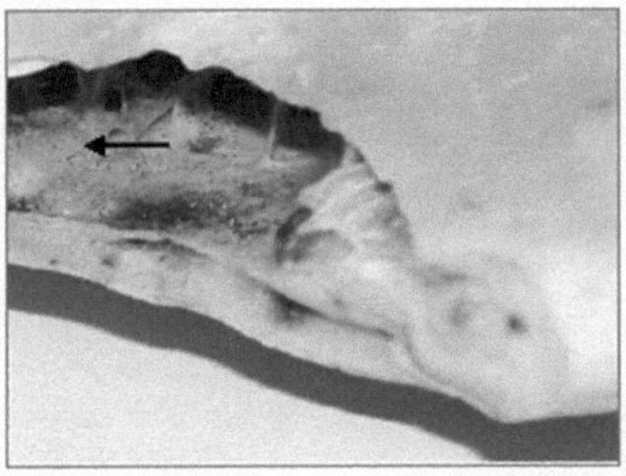

Figure 7 a. Sample point 7

Figure 7 b. Growth of fungi and bacteria from sample 7

Taxonomic identification: *Alternaria sp., Aspergillus sp., Penicillium sp., Cladospoorium sp., Bacillus subtilis, Micrococcus sp.*

Figure 19. Analysis of biochemical mark 7. Location of mark on the bone and growth of biochemical agent in Petri plaque.

Sample N°8				
Fungi* *CFU/cm²	***Bacteria* *CFU/cm²***	***Culture 1* *CFU/cm²***	***Culture 2* *CFU/cm²***	***Culture 3* *CFU/cm²***
4,5	36,5	1,4	2,4	4,1

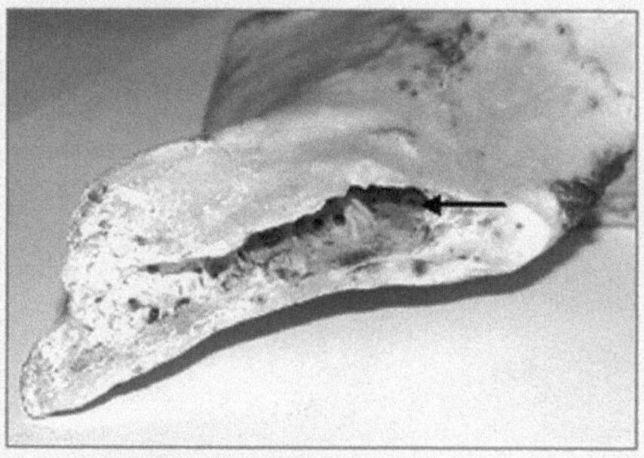

Figure 8 a. Sample point 8

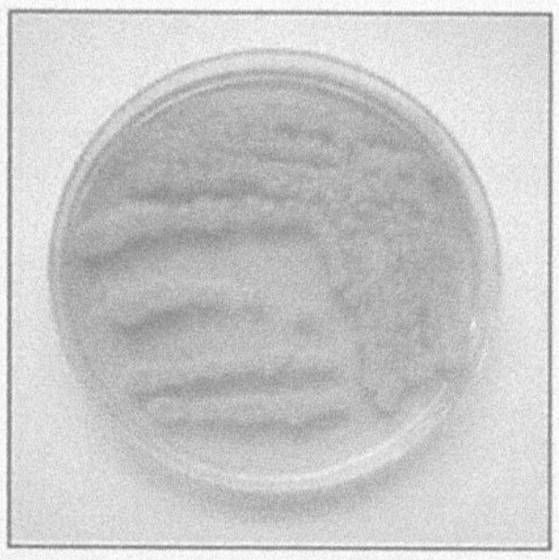

Figure 8 b. Growth of bacteria from sample 8

Taxonomic identification: *Bacillus cereus*

Figure 20. Analysis of biochemical mark 8. Location of mark on the bone and growth of biochemical agent in Petri plaque.

Sample N°9				
Fungi* *CFU/cm²	***Bacteria* *CFU/cm²***	***Culture 1* *CFU/cm²***	***Culture 2* *CFU/cm²***	***Culture 3* *CFU/cm²***
4,2	14,6	12,4	18,6	1,3

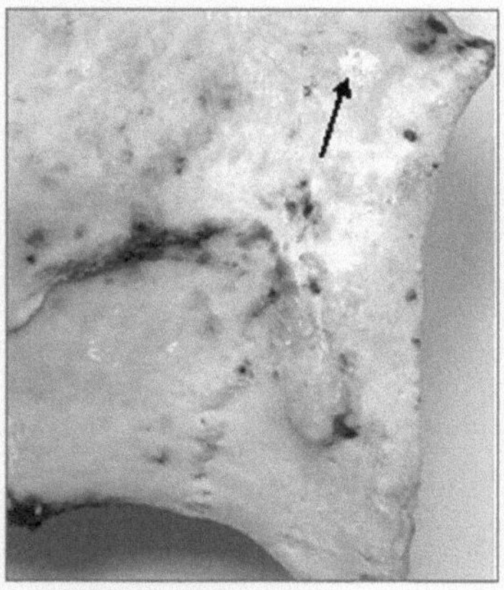

Figure 9 a. Sample point 9

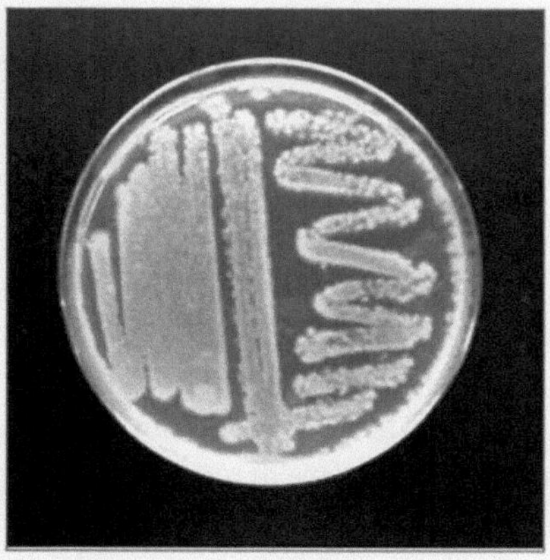

Figure 9 b. Growth of bacteria from sample 9

Taxonomic identification: *Bacillus subtilis*

Figure 21. Analysis of biochemical mark 9. Location of mark on the bone and growth of biochemical agent in Petri plaque.

Sample N°10				
Fungi CFU/cm²	*Bacteria CFU/cm²*	*Culture 1 CFU/cm²*	*Culture 2 CFU/cm²*	*Culture 3 CFU/cm²*
2,6	18,5	7,8	21,5	2,6

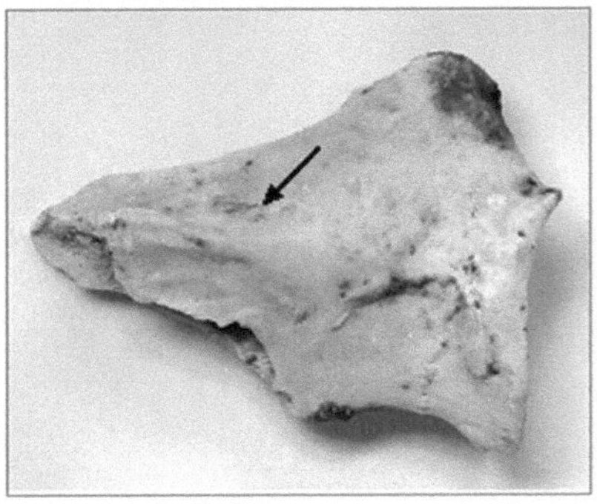

Figure 10 a. Sample point 10

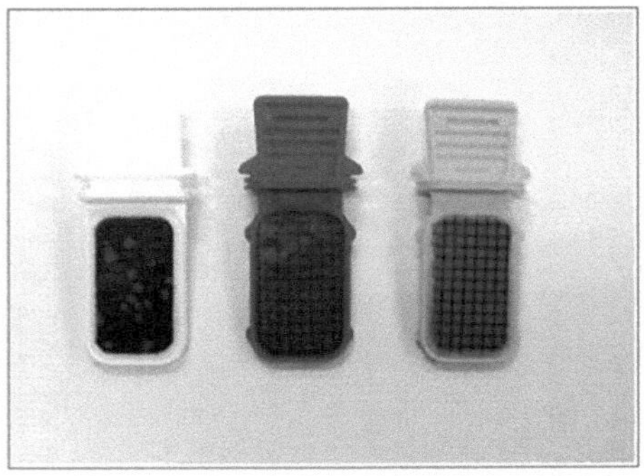

Figure 10 b. Growth of bacteria from sample 10

Taxonomic identification: *Micrococcus sp.*

Figure 22. Analysis of biochemical mark 10. Location of mark on the bone and growth of biochemical agent in Petri plaque.

Figure 23. Radius shaft from FLK *Zinj* showing percussion marks (denoted by arrows on right and left sides) and a "typical" example of a tooth score from FLK *Zinj* (arrow in the middle) as presented by Blumenschine (1995: 35, Figure 1d), which is in fact a biochemical mark (see Figures 7 and 8) (scale = 1 cm).

Table 4. Main fungi and bacteria found on the ten marks analyzed

Microorganisms identified in the ten samples	
Fungi	*Caldosporium* sp. *Penicillium* sp. *Aspergillus niger* *Mucor* sp. *Penicillium chrysogenum* *Alternaria* sp. *Actinomyces* sp. (fungal bacteria) *Micrococcus* sp. *Bacillus cereus*
Bacteria	*Bacillus subtilis* *Micrococcus roseus*

of bone. These metabolites change the pH of the surfaces onto which they are excreted, usually towards more acidic values. The development of metabolites on the analyzed bones was so brief that microfissures and flaking did not develop. However, had they been exposed to metabolic acids for longer, the cortical surface would have begun exfoliation and marking would have been more intense. This study clearly shows that fungi and bacteria can create macroscopic marks on bone surfaces.

This preliminary study clearly illustrates similarities between these marks and those in

Table 5. Main fungi and bacteria documented in the experiment and the associated metabolites excreted by them which are responsible for bone damage

Microorganism	Metabolic substance excreted	Intensity of development in the sample analyzed
Cladosporium	Acetic acid, lactic acid, proteases	High
Penicillium	Citric acid, oxalic acid, lactic acid, lipases	High
Aspergillus	Citric acid, oxalic acid, lactic acid, fumaric acid, malic acid	Moderate
Alternaria	Acetic acid, amylases, proteases	Moderate
Actinomyces	Lactic acid, pyruvic acid, acetic acid	High
Bacillus	Lactic acid, gluconic acid, amylases	High
Micrococcus	Lactic acid, acetic acid, lipases	Moderate

the FLK *Zinj* fossil assemblage (Figure 10). The biochemical-marking process is described in the following four-stage model:

Stage 1. The earliest signs of microbial attack appear as irregular lines of discoloration on the bone surface that may intertwine or appear isolated. These may be accompanied by a greenish/white powdery matrix on the surface responsible for the color change underneath (Figure 10). The initial color change contrasts with the surface color: the lines are dark in bones that have preserved their original color, and light in bones darkened by diagenesis. At this stage the marks cannot be confused with tooth marks, since the external cortical layer of the bone is intact and no groove has formed.

Stage 2. The components of the matrix overlying the bone surface start modifying the uppermost cortical layer. Changes in the surface texture and some microexfoliation appear. Since fungi form colonies, they spread both longitudinally and radially (Figures 10 and 11). Therefore, when they modify the cortical layers they create scores and pits, which at this stage are extremely shallow, and are outlined with a color that contrasts with the surrounding bone. In some cases, the internal surfaces of these marks show microflaking, which is distinctive from the internal crushing of tooth marks. The microflaking may appear as shallow, separate pits. These are clearly distinct from tooth scores, which are formed when a carnivore tooth cusp crushes and drags along the bone surface, evenly lifting and cutting through different cortical layers.

Stage 3. Biochemical marks at this stage may appear incompletely formed. One part of an incipient groove may preserve its original cortex – showing a distinct color that differs from the surrounding surface, but having no depth yet – while another part of the groove may already be deep into the cortical layers, due to flaking and cortical exfoliation (Figures 10 and 11). The difference of this stage from the previous one is that the biochemical degradation of the bone now affects several layers where scoring appears.

Stage 4. By this stage, the uppermost cortical layers have been penetrated, creating pits or scores that could be mistaken for a tooth mark in the absence of a distinct outline coloring (Figure 12). However, some tooth and biochemical scores can be distinguished by their shape, directionality, and degree of symmetry, as described later.

Most biochemical scores show an irregular shape with a winding outline, following the initial formation of the fungi, probably adhering to plant roots (Figures 11A and 12E); tooth scores are almost always straight. Occasionally, the biochemical scores may also be wide, straight grooves. However, the width of biochemical scores varies along its trajectory (Figure 23), which is not the case for tooth scores, where the tooth cusp creates a groove with homogeneous width. The groove may become wider as the tooth cusp crushes deeper into the bone and narrower as the tooth cusp moves upwards and is removed from the bone, but in either case, the width from the longitudinal axis of the mark is symmetrical. Flaking or exfoliation may form on both sides of the tooth score, since carnivore teeth exert uniform pressure on both sides of the mark; in biochemical scores, exfoliations are more random. Finally, tooth scores are usually perpendicular or at oblique angles to the main axis of the bone as the result of carnivores dragging their teeth transversally to the shaft; tooth scores that are parallel to the long axis of the bone are rather exceptional. However, this orientation occurs frequently in biochemical scores.

These traits can be used to differentiate biochemical and tooth marks. Occasionally, however, a biochemical mark loses its distinctive color contrast through diagenesis and cannot be distinguished from a tooth mark in the absence of other diagnostic features. In this case, especially when these marks appear together with other biochemical marks, it is impossible to rule out that the supposed tooth marks are not biochemical marks. For this reason, we use the contextual approach to ensure correct identification of tooth and biochemical marks. The contextual approach to bone surface modification studies was introduced at Swartkrans Member 3 (Pickering *et al.*, 2004, 2005b), where cut marks were safely identified *only* when there were no signals on the bone that would indicate that the marks were natural (microabrasion, random microstriations, or exfoliations by sediment friction). No mark was identified until the whole specimen had been analyzed and the superficial features were taken into account. The contextual approach is essential for the correct identification of bone surface modifications, given the overlapping signatures of different taphonomic processes.

Results

BIOCHEMICAL MARKS IN THE FLK *ZINJ* ASSEMBLAGE

Using the contextual approach, we identified tooth marks only when the rest of the surface did not show any biochemical mark, or showed clear differences between the tooth mark and the biochemical mark. Differences are clearest when the biochemical mark is in its earliest stages (Stages 1–3); biochemical marks in Stage 4 must show clear diagnostic features. Specimens whose marks were ambiguous ($n = 26$) were left out of the sample analyzed ($n = 699$). Biochemical and tooth marks were thus confidently identified. Several figures used here (Figures 10–12 and 23–25) to illustrate biochemical marks are specimens from FLK *Zinj* previously identified as tooth marks by Blumenschine (1995). For example, Figures 23–25 show a specimen that Blumenschine (1995: 35, Figure 1d) presented as a clear example of a tooth mark. A closer inspection reveals the following distinct traits of a biochemical mark:

1. The score is extremely shallow. It was created by removing the top cortical layer of the bone surface. The interior does not show the typical crushing of tooth scores, but rather shows partial flaking and exfoliation, occurring in several separate patches.
2. The outline of the score is irregular and asymmetrical, with flaking occurring randomly on each side of the score.
3. The score width is irregular, alternating increasing and decreasing breadth. This contrasts with the homogeneous width typical of tooth scores.
4. There is a clear outline around the score which is a distinct color from the surrounding bone surface.

Matching the specimens analyzed by Blumenschine and ourselves was not easy,

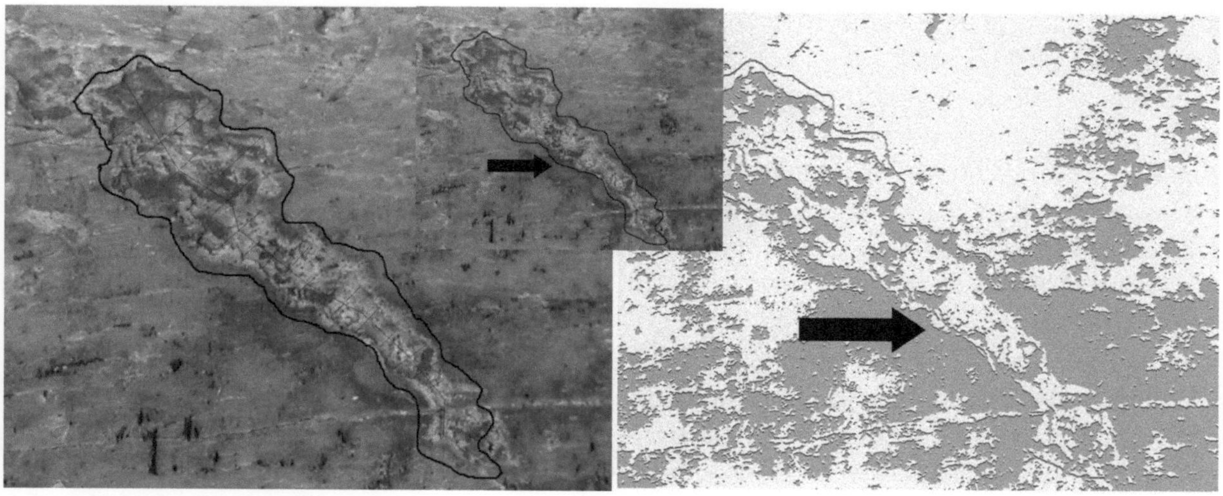

Figure 24. The score depicted in Figure 5 at a higher magnification (*left*) and through a 3D filter (*right*). The interior has been exfoliated but preserves much original cortex and most of the second cortical layer. This shows that the groove was created by exfoliation of the surface layer, and refutes that a tooth created the mark.

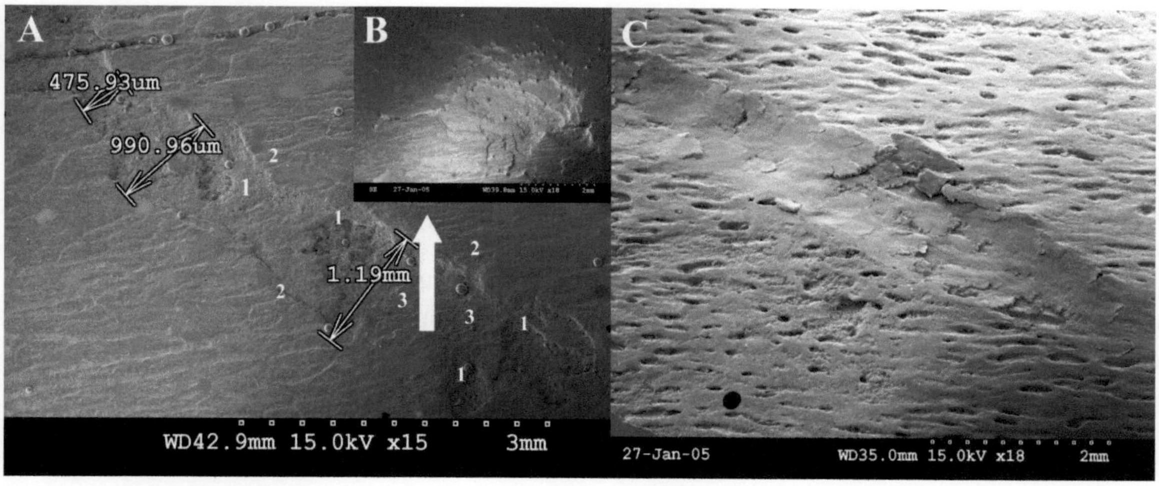

Figure 25. (A) SEM photograph of the score shown in Figures 6 and 7. The score alternates increasing and decreasing width, creating an hourglass shape, indicating that this is not a tooth mark. "1" shows fossae that appear in the score. "2" shows that the score has a step-shaped outline created by the exfoliation of the upper cortical layer. "3" shows that the original surface still occurs in patches in the middle of the score. (B) The same score under higher magnification. The fossae ("1") show flat, unmodified surfaces and stepped outlines caused by the exfoliation of the cortical layer. (C) Tooth-mark score from a modern experimental assemblage, in which a deep, smooth, U-shaped groove can be seen. The score shows a polished surface and homogeneous width along the groove. These traits clearly contrast with the biochemical score.

since we had catalog numbers that do not appear in his list and vice versa. This was also sometimes due to ambiguity in cataloging on the tags or due to double cataloging; additionally, several dozen specimens had the same catalog number. Furthermore, although Blumenschine claimed to use only specimens with ancient breaks, some specimens in his list labeled as "a," "b," and "c" actually belonged to a single, recently broken specimen. Some numbers without letters in our list coincide with numbers with letters in his list, indicating that we both looked at the same specimen. However, to avoid mismatching, we compared only those specimens that were cataloged identically in both lists (n = 364). This sample is sufficiently representative of the identification frequencies and criteria used in both studies; these 364 specimens have good cortical preservation and were analyzed both by Blumenschine and by ourselves.

Of this sample, less than 20% of those specimens identified as "tooth-marked" by Blumenschine were identified the same by us (Table 2). In Blumenschine's "tooth-marked" sample (70% of the sample of 364 specimens), 74% of the specimens also bear biochemical marks in Stages 2–4. There is a correspondence between the presence of natural biochemical marks and Blumenschine's (1995) identification of these as tooth marks. The distribution of tooth marks by element type reveals further inconsistencies between the two studies (Table 2). Tooth marks (as identified by Blumenschine) on metapodials are very abundant, close to the mean for carnivore-first scenarios (Blumenschine, 1988). But a hypothetical felid stage of carcass defleshing would have left almost no marks on these elements, since they are abandoned unmodified by felids most of the time (Brain, 1981; Domínguez-Rodrigo *et al.*, in press). Furthermore, as will be shown later, most of the bones (including metapodials) have been fractured by hominids. The resulting midshaft fragments would have provided no nutritional resources for scavenging carnivores. The high incidence of "tooth-marked" midshaft specimens was therefore alien to our understanding of carnivore modification of bones. The confusion is resolved by the revised tooth-mark frequencies on metapodials (Table 6), which indicate that a substantial amount of biochemical marks were mistaken for tooth marks by previous researchers (Blumenschine, 1995; and interpretations in Selvaggio, 1994; Capaldo, 1995, 1997). The high presence of biochemical marks at FLK *Zinj* is not anomalous. In her analysis of Olduvai Beds I and II, Cushing (2002) remarks that biochemical marks are present on bones from almost half of the occurrences sampled in the Gorge.

The abundance of biochemical modifications caused by fungi in the FLK *Zinj* bone assemblage is also an indirect indicator of paleoenvironmental conditions in the place of

Table 6. Percentage of tooth-marked specimens at FLK Zinj according to element type and bone section

	Humerus	Radius	Metacarpal	Femur	Tibia	Metatarsal
Small-sized carcasses						
Ends	30 (4/13)	16.6 (1/6)	23 (3/13)	14.2 (1/7)	37.5 (3/8)	16.6 (2/12)
Shafts	13.6 (3/22)	18.7 (3/16)	4.7 (1/21)	14.2 (2/14)	22.8 (8/35)	6.2 (2/32)
Large-sized carcasses						
Ends	50 (10/20)	42.1 (8/19)	10 (1/10)	15.3 (2/13)	20 (1/5)	14.2 (1/7)
Shafts	14.6 (11/75)	12.8 (5/39)	14.2 (2/14)	5.7 (3/52)	15.5 (12/77)	5.2 (1/19)

Numbers in parentheses are: number of tooth-marked fragments (numerator) and total number of specimens of each category (denominator)

bone deposition and accumulation. Most fungi do not grow below 12°C (Child *et al.*, 1993). Very dry habitats and permanently waterlogged sites also show very little microbial attack; fungi thrive in shaded environments with intermediate moisture – that is, in proximity of a source of humidity – and temperatures ranging from 20°C to 37°C, preferably at the lower end of this range (Marchiafava *et al.*, 1974; Hackett, 1981; Child *et al.*, 1993; Child, 1995; Hedges *et al.*, 1995; Jackes *et al.*, 2001). These biochemical modifications at the *Zinj* assemblage can, therefore, be taken as a likely evidence of shaded environment, probably because of the presence of trees, given that the savanna environments where fungi thrive are riverine forests and in the vicinity of some wooded swamps and springs (McClanahan and Young, 1996).

TOOTH-MARK ANALYSIS BY BONE PORTION

By restricting the identification of tooth marks to specimens with no ambiguous or confounding biochemical marks, our analysis of tooth marks using Blumenschine's (1988) bone portion system reveals a completely different picture from previous analyses. In the following analysis, we compare our data from FLK *Zinj* with experimentally derived referential frameworks developed by Blumenschine (1988, 1995) and Capaldo (1995) for hominid–carnivore and carnivore-only models. Table 3 and Figure 26 show the distribution of tooth-marked specimens in the *Zinj* assemblage by bone portion. In small-sized carcasses, the FLK *Zinj* limb bones are tooth-marked at similar rates (17.3%) to those in human-to-carnivore experiments. The frequency of midshaft fragments bearing at least one tooth mark is very low (13.8%), also closely matching the percentages obtained in human-to-carnivore experiments. Near-epiphyseal specimens are tooth-marked at very low rates, between half of the mean frequency reported by Blumenschine (1988, 1995), and about 25% less than the mean frequency reported by Capaldo (1995), both for human-to-carnivore experiments, but within the range of variation reported by both authors. Epiphyseal fragments are tooth-marked 50% less than in human-to-carnivore experiments. In sum, global distribution of tooth marks at FLK *Zinj* is distinct from those observed in carnivore-only models.

Long bones from large-sized carcasses are also tooth-marked slightly less (14%) than in human-to-carnivore experiments. Only 9.7% of midshaft specimens are tooth-marked; this is very close to the percentage obtained for human-to-carnivore experiments and is far from the expected percentage (86.5%) for a carnivore-first scenario required by the carnivore–hominid–carnivore hypothesis (Blumenschine, 1995). Near-epiphyseal fragments are tooth-marked about 30% less than the mean frequency for Blumenschine's human-to-carnivore experimental set, and are only one point outside his range of variation. As was the case with small carcasses, epiphyseal fragments are tooth-marked less often than in any experimental assemblage.

The low frequency of tooth-marked epiphyseal and near-epiphyseal specimens is not surprising. Intensive ravaging of bone by hyaenids, as reported by Blumenschine (1988), Capaldo (1995, 1998), and Domínguez-Rodrigo (in Pickering *et al.*, 2003), is usually reflected in a substantial amount of moderate to intensive furrowing in the epiphyseal fragments. At FLK *Zinj* epiphyses are fairly complete, and only 3 out of 110 epiphyseal fragments (2.7%) show clear signs of furrowing. Furthermore, furrowing was observed in only 3 of the sample of 59 complete epiphyses that we studied for biometric purposes (work in progress). This suggests very mild hyaenid bone modification compared to experimental assemblages. The reason may lie in the way researchers construct experimental models.

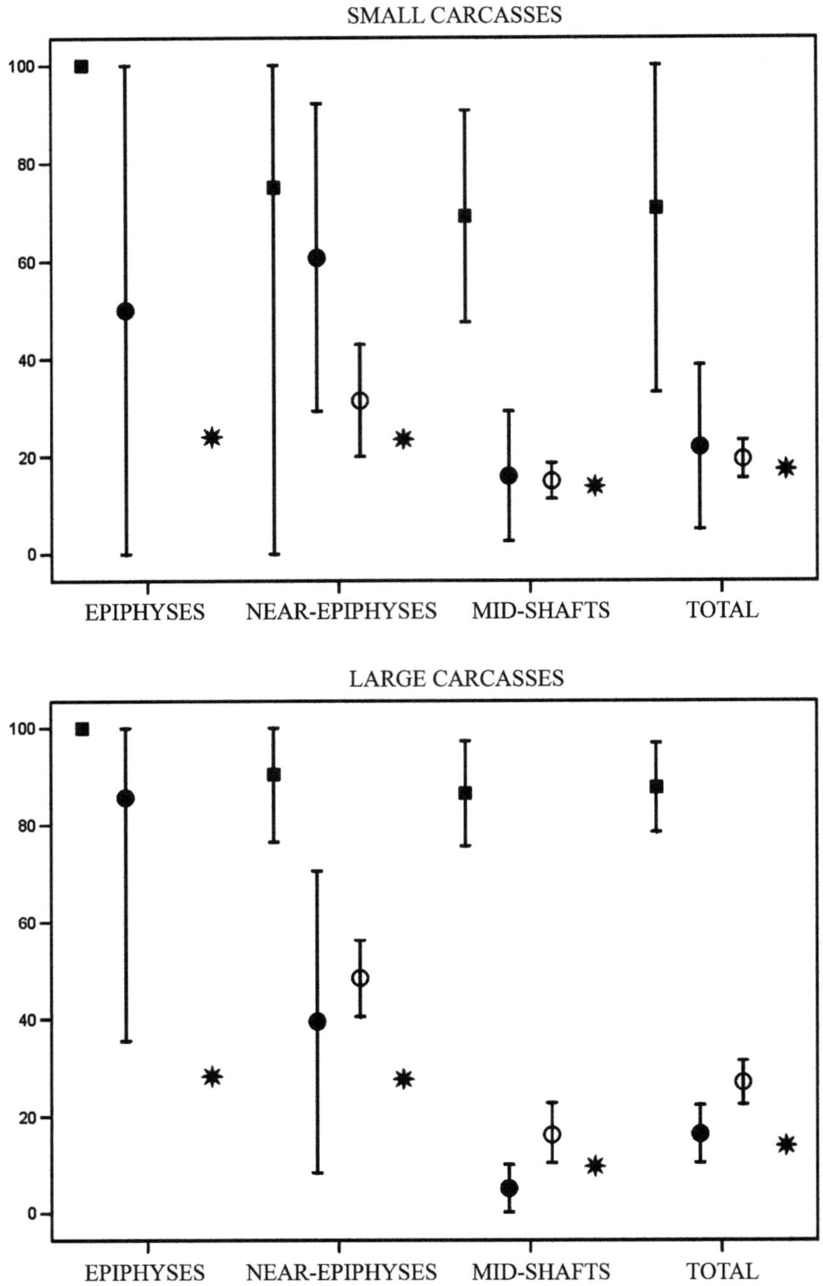

Figure 26. Distribution of the 95% CI (confidence intervals) for the frequency of tooth-marked specimens for each bone portion in experimental assemblages and in the FLK *Zinj* assemblage. ■, Carnivore only; ●, human (hammerstone)-to-carnivore (Blumenschine, 1995); ○, human (hammerstone)-to-carnivore (Capaldo, 1997); ✷, FLK *Zinj*.

Experiments with few bones, often from a single carcass, create a paucity of food that may result in far more intensive damage than when plentiful resources are available for scavenging, as was the case at FLK *Zinj* where bones from several carcasses (including many axial elements) were available. Less intense ravaging would have resulted in a lower rate of tooth-marked epiphyseal and near-epiphyseal fragments.

The frequencies of tooth marks at FLK *Zinj* are dissimilar from those generated by carnivore-first experiments, and thus refute the carnivore–hominid–carnivore hypothesis. Instead, the frequencies are very similar (especially for midshaft specimens) to those obtained in human-to-carnivore experiments, whereas the frequencies reported by Blumenschine (1995) fell outside the ranges of variation for all experimental assemblages.

TOOTH-MARK ANALYSIS BY BONE ELEMENT AND SECTION

A total of 496 bone specimens (513 if including ulnae) were identified to one of the six marrow-bearing long bones (humerus, radius, metacarpal, femur, tibia, metatarsal). Some bones were represented by more than one section (e.g., a shaft that had both near-epiphyseal and midshaft sections) and thus a total of 549 sections (excluding ulnae) are represented in this analysis. Sections were classified into ends (containing either epiphyseal section and/or near-epiphyseal section) and shafts. Data in Table 6 show that ends appear more highly tooth-marked than midshafts for both carcass sizes. In most cases, the shaft sections of small-sized carcasses are more highly tooth-marked than those from larger animals. Tibia shafts in both carcass sizes are the most commonly tooth-marked, while metapodials are the least affected. Humeri (for both carcass sizes) and radii (for large carcasses) show a tooth-mark distribution that matches human-to-carnivore experiments, in which epiphyses are highly tooth-marked and midshafts show a low number of tooth marks (Blumenschine, 1995; Capaldo, 1998b). No midshaft from any element shows tooth-mark frequencies different from those reported for the same bone section in human-to-carnivore experiments (Blumenschine, 1988, 1995; Capaldo, 1995). This further reinforces the inferences drawn from the analysis by bone portion presented earlier.

The precise locations of tooth marks on each section and element are also significant. Figures 27 and 28 show that most tooth marks cluster around long bone ends. Most tooth-marked midshafts are also marked more towards their ends; if we divide the midshaft into three parts, the middle segment is nearly devoid of tooth marks. This applies to every meat-bearing bone. This distribution would not be expected if a carnivore had previously defleshed these bones. The concentration of marks towards the ends of bones corroborates the hypothesis that carnivores (very likely hyenas or canids) intervened only in the grease extraction phase, that is, after hominids had discarded the bones.

ANALYSIS OF TOOTH PIT SIZES AND ATTRIBUTION TO CARNIVORE TYPE

Tooth pit size (length and breadth), considered together with other lines of evidence, can help distinguish the types of carnivores involved in the modification of carcasses. Appendix C shows the specimens that were used for the study of tooth pit size. Although most tooth marks in our total sample were scores, referential frameworks established by Domínguez-Rodrigo and Piqueras (2003) suggest that tooth pits are better than scores for discriminating among diverse carnivore groups. Therefore, only pits were measured here.

Small-sized carnivores and intermediate-sized felids (leopard and cheetah) can be differentiated from larger carnivores (hyena and lion) by tooth pit sizes in both cancellous and cortical bone. A tooth pit size range is used to further differentiate among carnivore taxa. This range equals the mean plus one standard deviation (SD) of tooth pit sizes. Ranges have been measured for specific carnivore taxa in bone modification experiments conducted by Domínguez-Rodrigo and Piqueras (2003), which will be used as a referential framework for the FLK *Zinj* data. Given the ambiguity of

LARGE CARCASSES

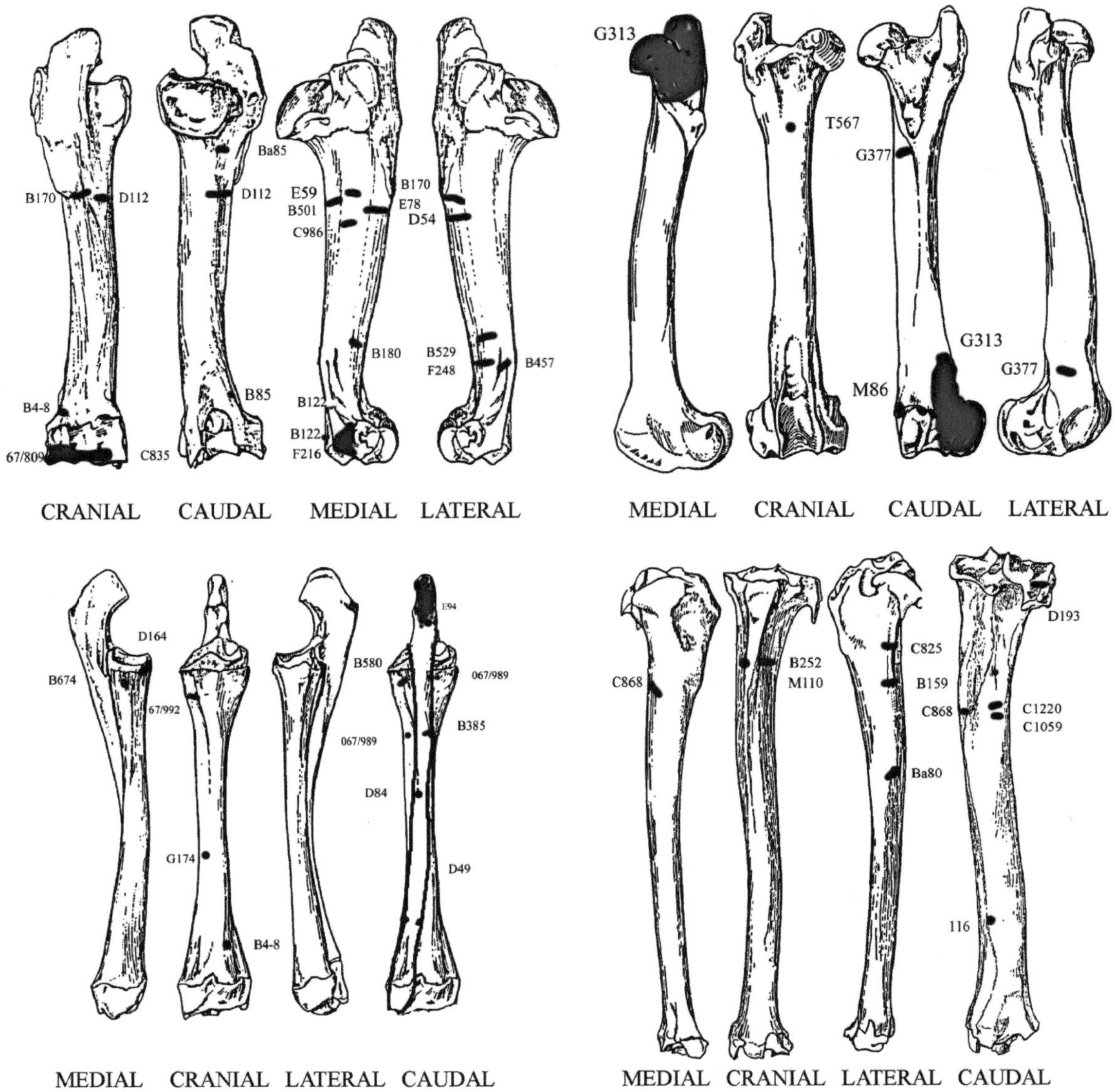

Figure 27. Anatomical distribution of tooth marks on humerus, femur, radio-ulna, and tibia from large carcasses. Letters and numbers indicate specimen number from the FLK *Zinj* assemblage. Specimens-bearing marks which could not be anatomically located were not incorporated into this figure. (Bones are redrawn from Pales and Lambert [1971].)

some overlapping size ranges, studying both tooth pit sizes and bone furrowing intensity in tandem provides a more secure identification of the carnivores involved.

Tooth pit sizes on cortical surfaces of bone shafts from FLK *Zinj* (Figure 29) seem to display the size range exhibited by hyaenids. The mean value is closer to that obtained for hyaenids than to any other type of carnivore, and the range is both too large for intermediate felids (leopard and cheetah) and too small for lions. The distinction between hyenas and lions is not very clear, but it is obvious that the tooth marks on shafts were made by a large carnivore. This contrasts with the data obtained from cancellous tissue, where the tooth pit size range

SMALL CARCASSES

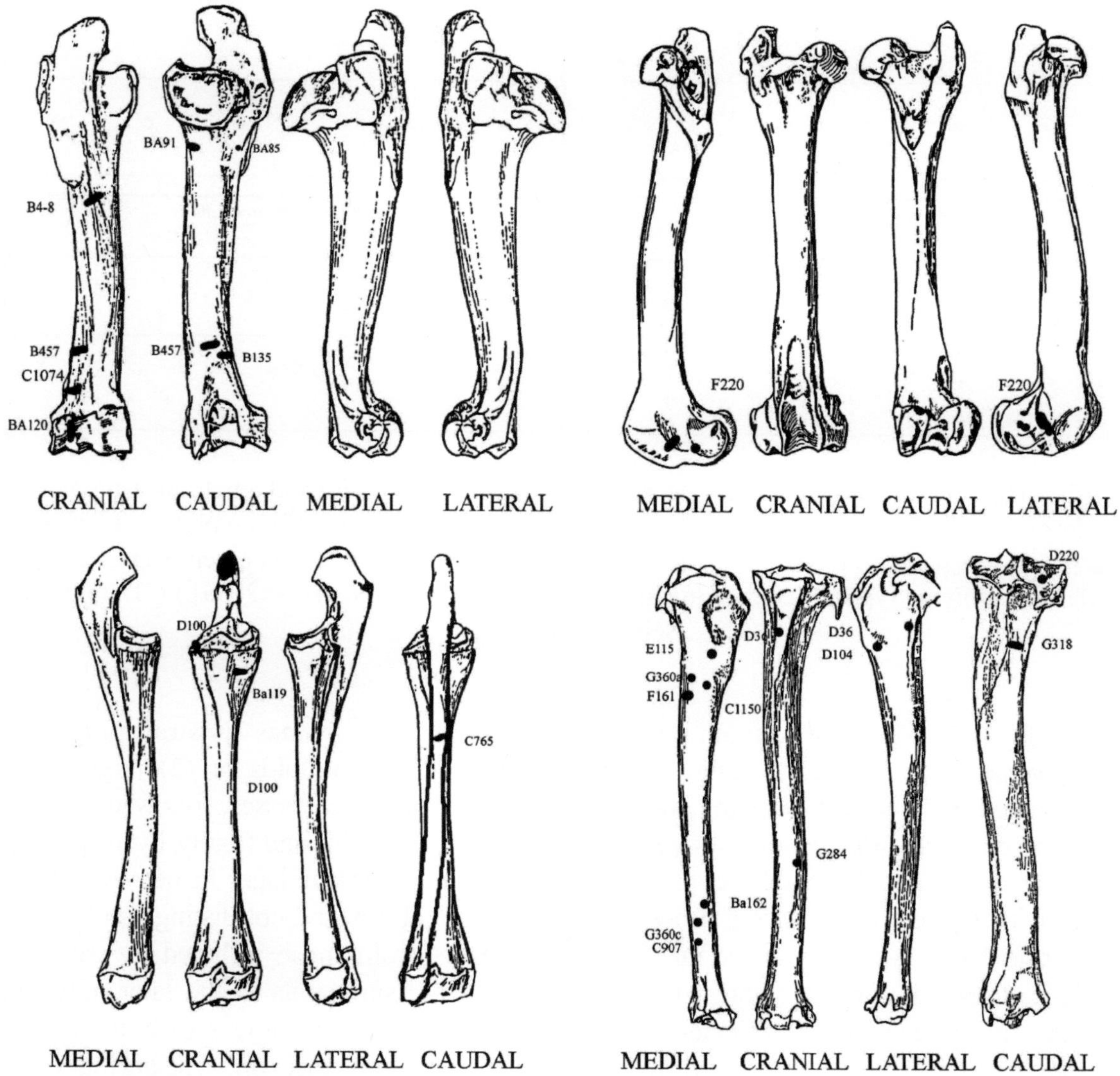

Figure 28. Anatomical distribution of tooth marks on humerus, femur, radio-ulna, and tibia from small carcasses. Letters and numbers indicate specimen number from the FLK *Zinj* assemblage. Specimens-bearing marks which could not be anatomically located were not incorporated into this figure. (Bones are redrawn from Pales and Lambert [1971].)

overlaps with several carnivore types. Combining all carcass sizes, the mean value and the range are larger than those reported for leopard and cheetah, and they overlap with those reported for large carnivores (hyena and lion) and some intermediate carnivores (other than leopard and cheetah). When separated by carcass size, the pit sizes for cancellous tissue from large carcasses are within the range reported for both large and intermediate carnivores. Marks on cancellous tissue from small carcasses exhibit a much wider range of variation, which suggests that they are the result of hyena intervention or a combination of hyaenids and intermediate carnivores.

These interpretations must be taken cautiously, since experimentally derived referential frameworks are imperfect. The experiments

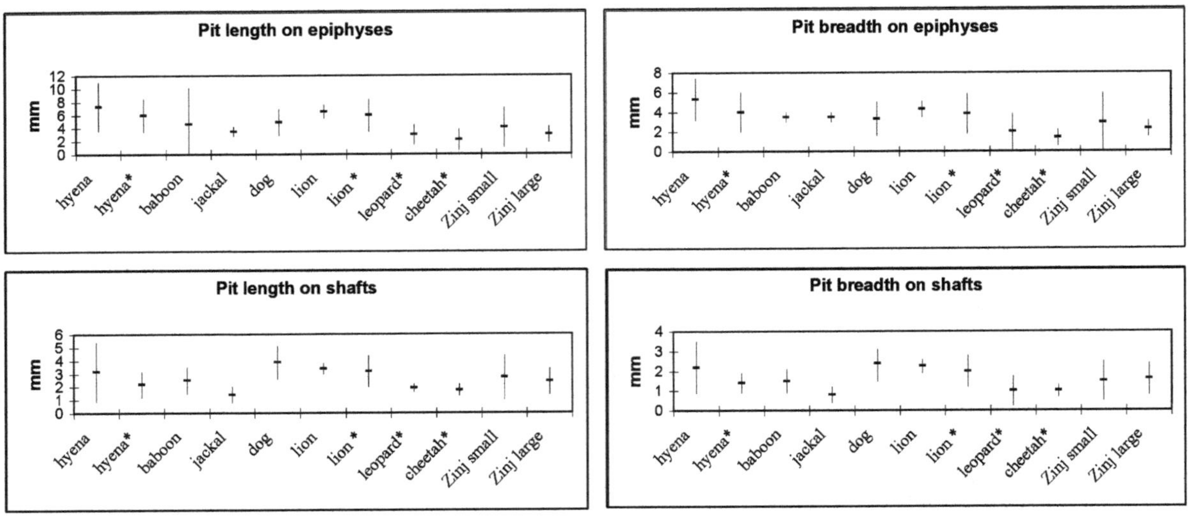

Figure 29. Distribution of means plus one standard deviation of breadth and length measurements of pits from the FLK *Zinj* assemblage. Measurements are given separately for epiphyses and shafts. Means plus a one standard deviation range are also provided for experimental assemblages created by different carnivore types. (See data in Domínguez-Rodrigo and Piqueras [2003].) (*Data from Selvaggio [1994].)

conducted for measuring size of tooth pits created by hyenas were small, with few bones, heavily ravaged by hyenas. Furrowing was frequent and tooth marks conspicuous. As we have noted earlier, most FLK *Zinj* epiphyses are fairly complete with almost no furrowing preserved. This suggests that if hyenas modified this fossil assemblage, they left far more inconspicuous traces than those seen in experiments. Given that epiphyses are marginally modified by carnivores, they are not robust indicators of carnivore type. Rather, the size ranges of tooth pits on shafts (cortical tissue) seem to be more indicative of carnivore type.

Selvaggio and Wilder (2001) reported data on tooth-mark sizes for a wide variety of carnivores and made interpretations about FLK *Zinj*. They observed that the "mean area of tooth pits on cancellous bone in the FLK *Zinj* sample is similar to those inflicted by hyenas and lions," and that "for cortical bone, the *Zinj* sample is most similar in area to pits inflicted by cheetahs, leopards, and spotted hyenas" (Selvaggio and Wilder, 2001: 467). According to these data, carnivore involvement at FLK *Zinj* could be the effect of: (1) hyenas postravaging bones abandoned by hominids or (2) medium-sized felids defleshing carcasses, followed by demarrowing by hominids and finally, hyenas ravaging the discarded assemblage. In her earlier study, Selvaggio (1994) did not distinguish biochemical marks from those inflicted by carnivores, and her measurements in the later publication probably include specimens of both types and are not useful for carnivore-type attribution. The present analysis shows that when securely identified tooth pits on shafts are measured, the size range is too large for intermediate carnivores; rather, the data suggest that hyenas created the pits. In sum, tooth pit size cannot be used to support the carnivore–hominid–carnivore model as argued by Selvaggio and Wilder (2001).

PERCUSSION MARKS BY BONE PORTION

Data in Table 2 show that the frequencies of percussion marks identified by Blumenschine (1995) and by us are very similar. Blumenschine identified percussion marks as those in which microstriations could be observed. However, we observed that his identification of percussion marks at FLK *Zinj* also included percussion notches; some specimens identified by Blumenschine as bearing percussion marks did not, upon closer inspection by us, show a single percussion or biochemical mark, but only a percussion notch. Blumenschine's percussion-mark frequencies therefore lump together these distinct bone modifications, which he identified subjectively. Quantitative methods for identifying notches were later developed by Capaldo and Blumenschine (1994), but they have not been applied to FLK *Zinj* until the present study.

The frequencies of percussion marks (i.e., those generated by hammerstones and mostly showing microstriations) that we identified in the assemblage are shown in Table 7 and Figure 30. In the following analysis, we compare our data with experimentally derived referential frameworks developed by Blumenschine (1988, 1995) and Capaldo (1995) via human (hammerstone)-only and human-to-carnivore experiments. For both carcass sizes, the frequency of percussion marks is within the range for human-to-carnivore experiments. In small carcasses, frequencies of both percussion marks alone and percussion marks plus notches (which we present to enable comparison with Blumenschine's data) are similar to the mean frequency of percussion marks from hammerstone-only samples. In larger carcasses, the frequency of percussion marks is similar to that from hammerstone-only samples, except for near-epiphyseal fragments, which exhibit a low frequency of percussion marks, even when notches are included. This different frequency for this bone portion could be explained if some specimens with very little cancellous tissue on the medullary surface were included by us in the midshaft sample, which might explain differences between Blumenschine's and our near-epiphyseal sample size. That irregularity aside the remaining bone portions from large carcasses with percussion marks and notches closely match the frequencies derived from human–carnivore experiments. This should theoretically support Blumenschine's claim that most long bones at FLK *Zinj* were broken by hominids during marrow exploitation.

ANALYSIS OF PERCUSSION MARKS BY BONE SECTION AND ELEMENT

In addition to analyzing percussion marks by bone portion, we examined them by bone section and element and noted their precise location on the bone. This offers greater resolution than bone portion analysis alone. Our results suggest that Plio-Pleistocene hominids had an understanding of bone breakage dynamics which could only be acquired if they were regularly involved in carcass exploitation; patterns in cut-mark distribution (Chapter 6) suggest the same. Figures 31 and 32 show the anatomical distribution of percussion marks on long limb bones for small and large carcasses. Humeri for large-sized carcasses are preferentially broken on the distal cranial and medial shafts, which would face up when the humerus lies in its most stable position on an anvil, and which have thicknesses that can be broken with a single blow. The latter is supported by the fact that percussion marks appear isolated and there are few specimens containing more than one mark. Similar observations are made for femora of all carcass sizes. Marks preferentially occur on the medial shaft and, less abundantly, on the cranial proximal shaft, which are the most convenient spots for femur breakage if the bone rests on an anvil, since the condyles and femoral head condition how the bone is placed and how the shaft is

Table 7. Mean percentages of percussion-marked specimens in relation to the total number of specimens in experimental assemblages and in the FLK Zinj assemblage, according to bone portion: epiphysis (EPIPH), near-epiphysis (NEP), and midshafts (MSH)

	EPIPH	NEP	MSH	All
Small-sized carcasses				
Hammerstone only*				
Mean %	54.4	43.3	26.6	36.6
95% CI	(35.7–73.1)	(0–91.6)	(13.1–40.1)	(26–47.2)
Hammerstone-to-carnivore*				
Mean %	25	17.7	15.9	31.1
95% CI	(0–100)	(0–35.9)	(2.6–29.2)	(17–45.2)
Hammerstone-to-carnivore**				
Mean %	–	30.2	23.6	29.6
95% CI	–	(18.9–41.5)	(18.4–28.8)	(23.2–36)
FLK *Zinj* (PM only)				
Percent	20	23.5	22.7	22.1
No. marked/total	(10/50)	(4/17)	(28/123)	(42/190)
FLK *Zinj* (PM and estimated percussion notches)				
Percent	26	29.4	37.3	17.3
No. marked/total	(13/50)	(5/17)	(46/123)	(33/190)
Large-sized carcasses				
Hammerstone only*				
Mean %	75	100	33.3	50
95% CI	(–)	(–)	(–)	(–)
Hammerstone-to-carnivore*				
Mean %	35.7	45.5	20.6	26
95% CI	(0–98.8)	(3.2–87.8)	(6.4–34.8)	(8.6–43.4)
Hammerstone-to-carnivore**				
Mean %	–	22	13.1	16.8
95% CI	–	(9.4–34.6)	(4.7–21.5)	(8.7–24.9)
FLK *Zinj* (PM only)				
Percent	20	3.7	14	12.5
No. marked/total	(12/60)	(2/54)	(55/392)	(63/506)
FLK *Zinj* (PM and estimated percussion notches)				
Percent	28.3	7.4	20.4	20
No. marked/total	(17/60)	(4/54)	(80/392)	(101/506)

Analyses of 95% CI for sets of experiments are included. They were calculated using the t distribution, where 0.025 is the critical value of t with n–1 degrees of freedom
* Blumenschine (1988, 1995)
** Capaldo (1997)

broken. Again specimens with multiple percussion marks are rare, indicating that hominids were successful at opening bones with few blows. This interpretation is stronger for bones from large carcasses, given the larger sample size available to us.

Intermediate bones often bear percussion marks on the proximal cranial shaft, especially in large carcasses, where most of the blows concentrate on the cranial and caudal sides, since the bone's thickness makes it difficult to break open from the medial–lateral sides. This is not a

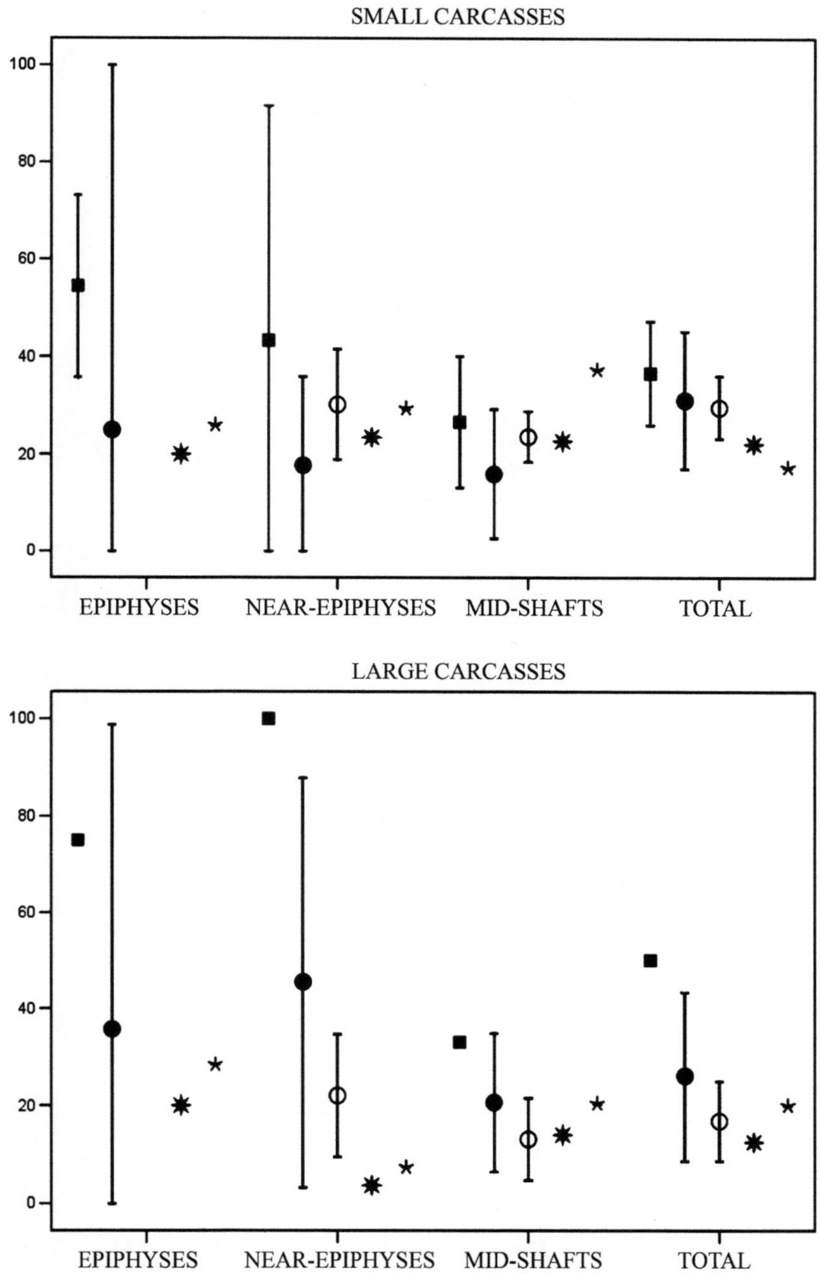

Figure 30. 95% CI (confidence intervals) for the frequency of percussion-marked long bone portions from experimental assemblages and from the FLK *Zinj* assemblage. ■, Hammerstone only (Blumenschine, 1995); ●, human (hammerstone)-to-carnivore (Blumenschine, 1995); ○, human (hammerstone)-to-carnivore (Capaldo, 1995, 1997, 1998a, 1998b); ✳, FLK *Zinj* percussion marks; ★, FLK *Zinj* percussion marks plus notches.

problem with thinner radii from small-sized carcasses, where blows on the medial side are also documented. Tibiae from large carcasses are preferentially broken on the cranial and caudal shafts, which are less dense; the same is true for radii. Tibiae from small carcasses are less dense and may be broken from any angle. Percussion marks on metapodials from both sizes occur on the proximal half of the shaft, where the marrow cavity is broadest. Marks occur on all four sides,

SMALL CARCASSES

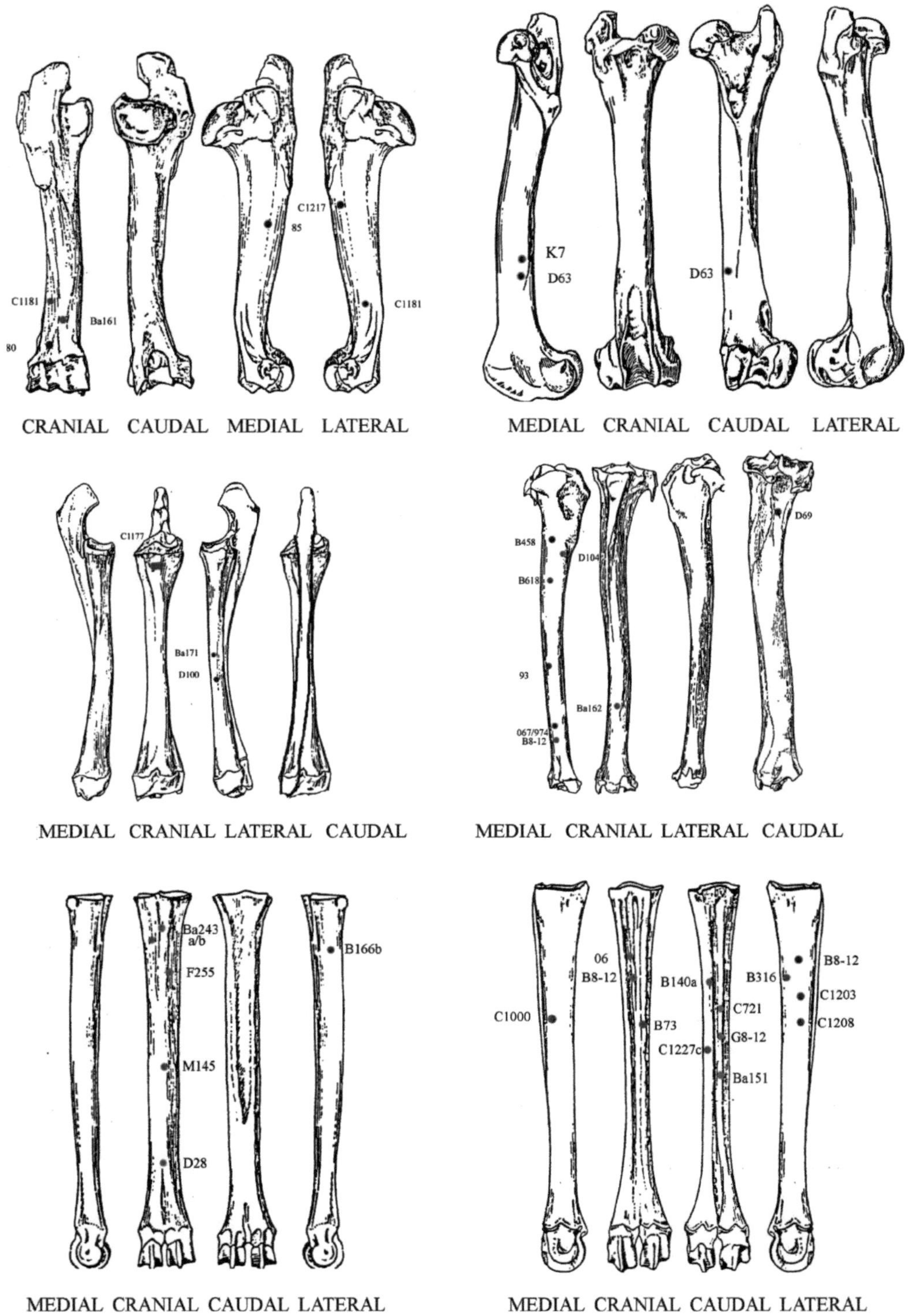

Figure 31. Distribution of percussion marks on long bones from small carcasses. Letters and numbers indicate specimen number from the FLK *Zinj* assemblage. Specimens-bearing marks which could not be anatomically located were not incorporated into this figure. (Bones are redrawn from Pales and Lambert [1971].)

LARGE CARCASSES

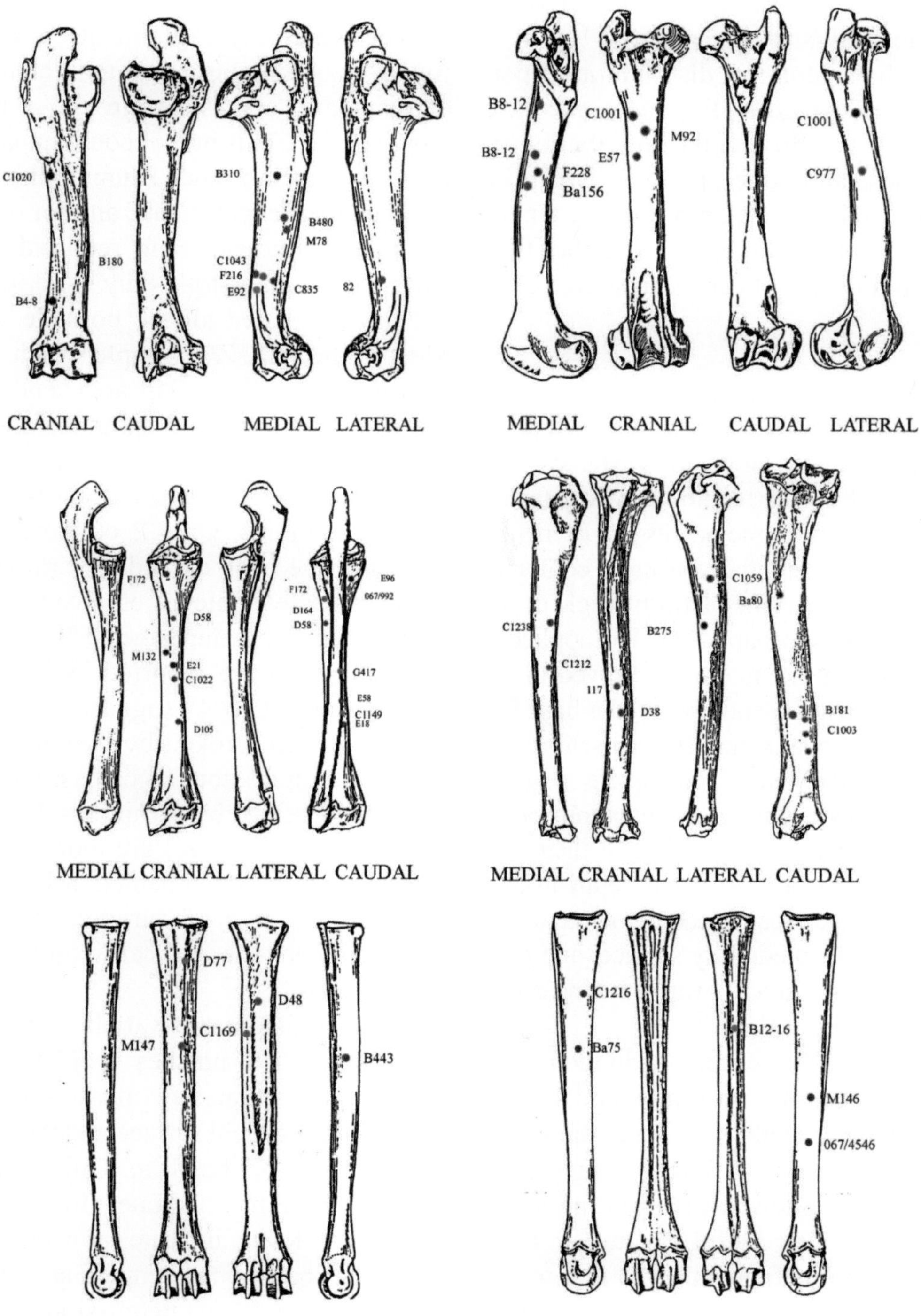

Figure 32. Distribution of percussion marks on long bones from large carcasses. Letters and numbers indicate specimen number from the FLK *Zinj* assemblage. Specimens-bearing marks which could not be anatomically located were not incorporated into this figure. (Bones are redrawn from Pales and Lambert [1971].)

although they abound on the caudal side of metatarsals, the thinnest one of the four sides, especially in small-sized carcasses.

In sum, the anatomical distribution of percussion marks suggests that hominids had knowledge of the differential distribution of bone thickness and bone breakage dynamics, which could only be acquired through repetitive carcass exploitation episodes, rather than through sporadic use of carcass resources.

ANALYSIS OF NOTCHES

Capaldo and Blumenschine (1994) suggest that notches created by carnivore gnawing (static loading) and hammerstone percussion (dynamic loading) can be differentiated using a combination of two ratios, plus the platform angle of the notch (as discussed Chapter 3). We applied a similar analysis to the notches preserved in the FLK *Zinj* archeofauna, and observed a high frequency of what Capaldo and Blumenschine call "incipient notches" and "incomplete notches types A–C." However, since our intention was to test the utility of the combination of their five variables (notch breadth to notch depth from a cortical view, flake scar breadth to maximum notch depth on the medullary surface, and platform angle of the notch), we only measured complete notches.

Capaldo and Blumenschine divided their experimental sample into metapodial and non-metapodial assemblages, given their different thicknesses and differential resistance to breakage; we followed the same protocol. This division is also useful because it separates meaty (non-metapodial) and non-meaty (metapodial) limb bones. In the sample of large-sized meaty bones from FLK *Zinj*, our ratios of notch breadth to notch depth and flake scar breadth to maximum notch depth substantially overlap with the 95% CI ranges provided by Capaldo and Blumenschine (1994), and about half of the sample falls outside their ranges both for hammerstone- and carnivore-only experiments (Figure 33). However, for this carcass size, their sample for carnivore-only experiments is about four times bigger than for hammerstone-only experiments, and therefore the former are more reliable. This may account in part for the overlap Capaldo and Blumenschine found between their experiments, and for the narrow notch breadth:depth ratio reported by them, since the hammerstone-only experiment may not have included all the possible variation. Most of the FLK *Zinj* sample which falls outside the 95% CI overlap area tends towards the size and shape of hammerstone-created notches, suggesting that most notches on meaty bones from large carcasses were broken by hominids. Only 2 out of 30 notches seem to have been caused by carnivores. For small carcasses, notches on meaty bones also seem to cluster around the 95% CI for hammerstone breakage reported by Capaldo and Blumenschine (1994), suggesting that most of the bones were broken by hominids. Only 3 out of 15 notches for this carcass size seem to have been created by carnivores. Metapodials from both carcass sizes are more ambiguous, since the ranges documented at FLK *Zinj* are outside the range of variation reported for both hammerstone and carnivore experimental scenarios.

Another diagnostic variable of the agent that created the notches is platform angle (Figure 34), obtained by measuring the angle between the cortical surface and the medullary negative scar of the flake. Only 4 out of 30 notches from meaty bones from large carcasses at FLK *Zinj* indicate carnivore breakage; more than half of the remaining sample suggests hammerstone percussion, based on comparison with values reported by Capaldo and Blumenschine (1994). Metapodials once again are more ambiguous. For smaller carcasses, only 3 out of 15 meaty bones might have been the result of carnivore bone gnawing. More than 50% of the sample clearly suggests hammerstone percussion.

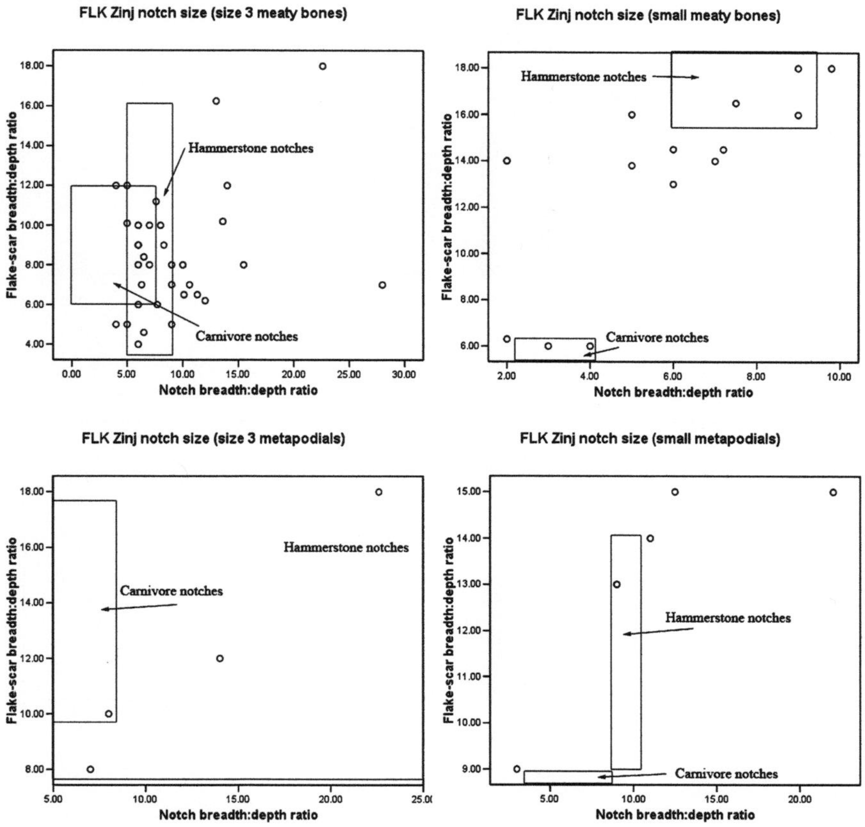

Figure 33. Bivariate analyses (including Pearson's index) of the two linear measurements of notches. Squares and rectangles show the ranges of variation (95% CI) of the notch breadth:depth ratios and the flake-scar breadth:depth ratios for carnivore-made and hammerstone-made notches produced in experimental assemblages (Capaldo and Blumenschine, 1994).

Results from this type of analysis must be interpreted with caution, since the experimental framework is still incomplete. The sample size for large carcasses in Capaldo and Blumenschine's experiment (1994) was small and ranges of variation remain poorly understood. In their analysis, the number of notches created by hyenas on large carcasses is very small. Our observation of some of the notches created by hyenas suggests that they occasionally produce a different morphology from notches created by other carnivores or by simulating bone breakage in static loading through pressure. The reason may be that hyenas sometimes initiate bone breakage not just by applying pressure on the bone surface but also by first biting with an open jaw and hitting the bone forcefully. The end result is a notch with a more open morphology and more acute/obtuse angles on the platform flake and flake scar. This could account for some of the morphological variation reported in the FLK *Zinj* notches versus those more homogeneous notches reported from experimental assemblages.

In sum, the study of notches, when combining the ratios and platform angles, suggests that both carnivores and hominids may have been involved in bone breakage, although hammerstone percussion is predominant, especially for large-sized carcasses. This further supports inferences drawn from the analysis of percussion marks earlier.

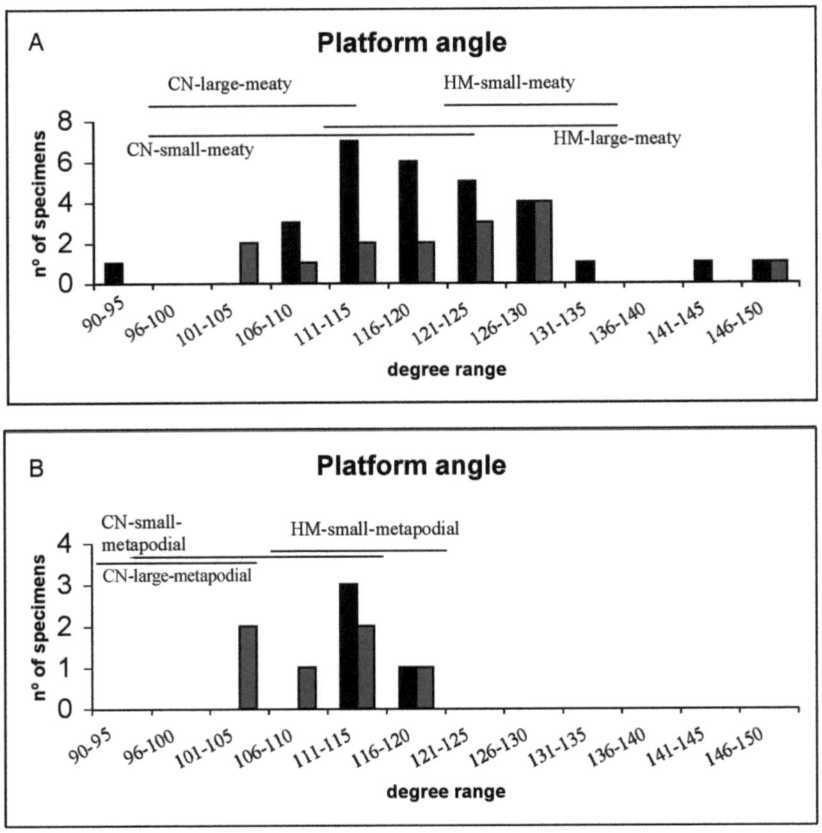

Figure 34. Frequencies of platform angle measurements for the FLK *Zinj* notches. Horizontal lines show the ranges of variation (95% CI) for carnivore-made (CN) notches and hammerstone-made (HM) notches for meaty (A) and non-meaty (B) long limb bones produced in experimental assemblages. (Data for experiments are from Capaldo and Blumenschine [1994].)

BREAKAGE PLANES AND ANGLES

Breakage planes at FLK *Zinj* are predominantly oblique and longitudinal. Alcantara *et al.* (2006) showed that oblique planes were more informative of the type of loading (dynamic versus static) than longitudinal and transversal planes (see Chapter 3); therefore, this study has focused on the former, although it has also included longitudinal planes as a complement. Angle measurements were carried out on a random selection of one out of every five specimens analyzed, since it was thought that such a large sample would be representative. Figure 35 shows the number of angle specimens distributed by angle range. Most specimens from all carcass sizes are clustered within the range for dynamic loading (hammerstone percussion), based on experimental frameworks (Alcantara *et al.*, 2006). This clustering is most clear for specimens from large-sized carcasses and for oblique breakage planes.

In small carcasses, most breakage plane angles fall within the range of dynamic loading rather than static loading, and most of the remaining sample extends beyond the limits of the dynamic loading range. However, given that the outlying specimens have either very acute or very obtuse fracture angles, they are likely the result of hammerstone percussion (dynamic loading), since fracture planes created by carnivores tend more towards right angles (Alcantara *et al.*, 2006). In larger carcasses, most breakage

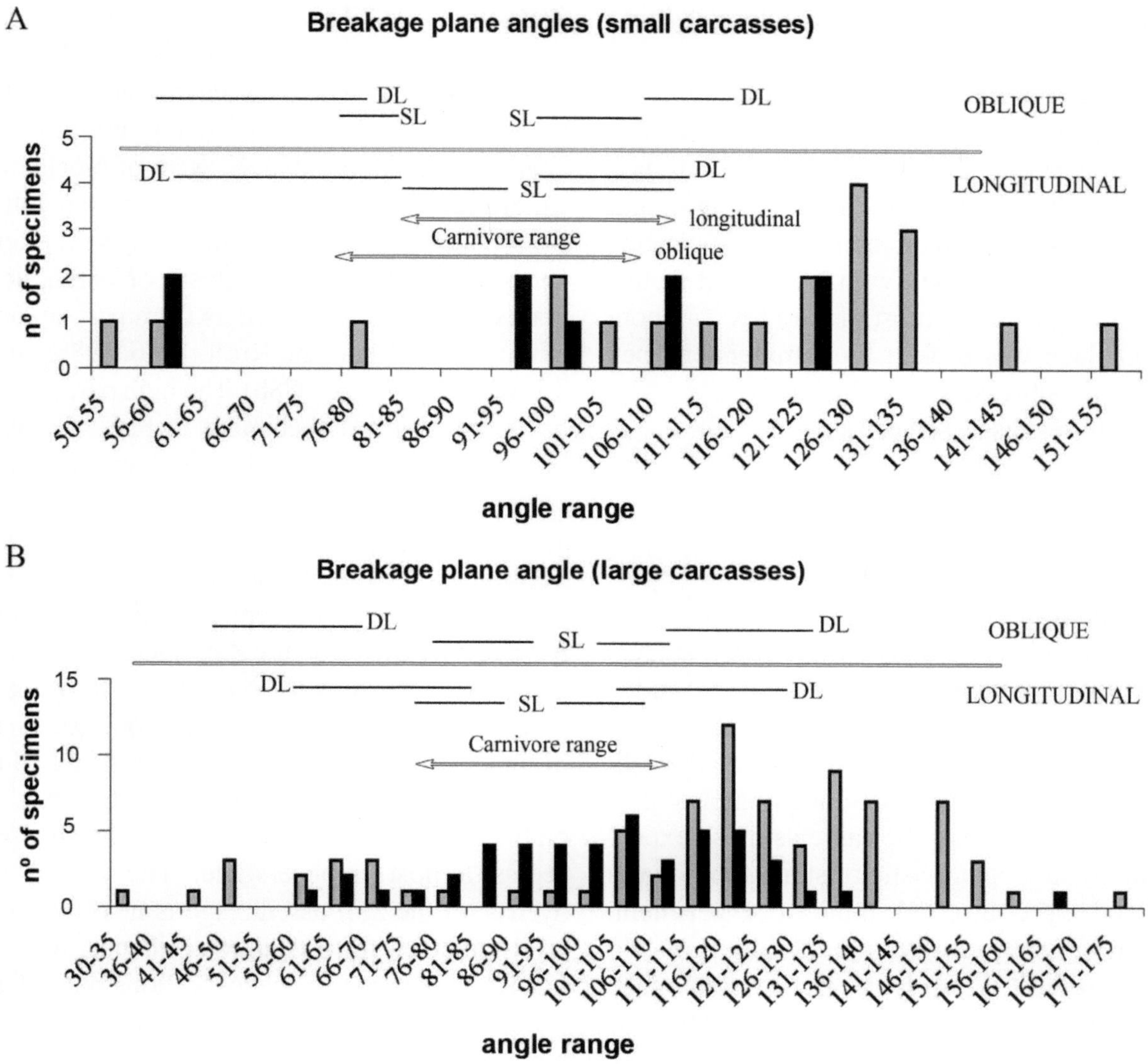

Figure 35. Distribution of green fracture breakage plane angle measurements for both oblique and longitudinal planes. Horizontal lines indicate the ranges of variation (95% CI) for experimental assemblages modeling hammerstone percussion (dynamic loading = DL) and carnivore pressure (static loading = SL) for small carcasses (A) and large carcasses (B).

plane angles cluster outside the static loading range, and of these many fall within the dynamic loading range. Most specimens outlying the dynamic loading range show extreme obtuse or acute angles suggestive of percussion rather than tooth pressure, as noted earlier for small carcasses.

The fact that the experimental ranges of variation are smaller than the ranges of variation for archaeological specimens suggests that the experiments have not reproduced all the possible ranges of variation. A large number of bones in Capaldo and Blumenschine's (1994) sample as well as in Alcantara *et al.*'s (2006) sample were broken by applying artificial pressure. This models most but not all carnivore behavior when breaking bones. As described in the previous section, hyenas sometimes close their jaws on the bone before breaking it, creating notches and breakage planes that may be similar to those generated by hammerstone percussion. This may account for the narrower variation ranges seen in the experimental data. Despite the limitations of the experimental samples, the abundant number of specimens with breakage planes with acute and obtuse angles suggests that most of the FLK *Zinj* bone assemblage was broken by hammerstone percussion, buttressing conclusions drawn from the analyses discussed earlier.

Discussion

Taphonomic referential frameworks should be built from observational–experimental studies. The use of referential frameworks to understand archaeological processes should always consider potential equifinality, especially if the agents that have modified bone in archaeological contexts are not the same as those experimentally modeled. Distinguishing tooth marks created in controlled modern experiments is easy. Every single modification on the bone surface is likely to be a tooth mark. Archaeological bone surfaces are hardly ever as clear as those of modern bones – and this is especially true in the FLK *Zinj* assemblage, despite claims of excellent cortical preservation (Blumenschine, 1995). At the microobservational level, the surfaces of bones in the FLK *Zinj* assemblage – as in many other archaeological assemblages – have been subjected to various destructive processes creating many modifications. Inconspicuous tooth marks and biochemical marks may be easily confused, unless the criteria outlined in this study are considered.

This confusion accounts for erroneous estimates of tooth marks at FLK *Zinj* (Selvaggio, 1994; Blumenschine, 1995; Capaldo, 1995, 1998b). Blumenschine (1995: 30) claimed that a "high confidence in mark attributions was achieved during the analysis of the FLK *Zinjanthropus*: 87.3% of the percussion mark identification and 94.6% of the tooth mark identifications were scored as unambiguous." The present study shows that many purportedly unambiguous "tooth marks" were of biochemical origin. When true tooth marks are separated from biochemical marks, they support inferences drawn from our analysis of percussion marks and notches, as well as from past studies of cut marks (Bunn and Kroll, 1986, 1988; *contra* Binford, 1988a, 1988b). The high percentage of "tooth"-marked specimens at FLK *Zinj* has been used as a classic example of strong carnivore ravaging on hominid-made archaeological assemblages (Blumenschine and Marean, 1993; Selvaggio, 1994; Capaldo, 1995). The revised data show that only moderate carnivore damage has been preserved on the *Zinj* bones. Most of it is too inconspicuous to be attributed to hyenas. It seems evident that much cancellous bone has disappeared from the assemblage, probably due to hyena action. However, the impact of hyenas on the surviving long bone shafts and epiphyses seems almost unobservable. The action of hyenas on the *Zinj* assemblage is nearly exclusively inferred from negative evidence, that is, from the absence of some cancellous bone. The frequencies of bone surface modifications reported in this study are similar to those documented by Oliver (1994) in his analysis of the same assemblage.

From a methodological standpoint, this study raises the following issues:

1. Inconspicuous tooth marks are not a useful analytical tool when biochemical modifying agents have been active on bones. However, natural processes do not mimic the diagnostic features used to identify percussion marks, as described by Blumenshine and Selvaggio (1988). Also, biochemical processes and other destructive processes (e.g., digging tool damage, abrasion, trampling) do not resemble cut marks when cortical surfaces are in good condition – even in the

presence of natural marks created by biochemical processes.

2. The "bone portion" method (Blumenschine, 1988, 1995) used to tally bone surface modifications has been criticized and deemed unhelpful to interpret frequencies of cut marks on bones since it leads to equifinality (Domínguez-Rodrigo, 1999b, 2002). This study shows that the bone portion method is also unhelpful with tooth marks: when their distribution is considered by bone section (including their precise location) and element, we derive information which is unavailable using the bone portion method. Our analysis by bone section and element helps determine if marks concentrate on midshafts as meat-bearing areas (expected if carnivores deflesh carcasses first) or on grease-bearing areas (expected if humans have primary access). This approach shows that at FLK *Zinj* midshaft (meat-bearing) sections were minimally tooth-marked and that most tooth marks concentrate on the grease-bearing ends of long bones. This negates primary access to carcasses by carnivores.

Having relegated carnivore activity at FLK *Zinj* to the consumption of grease-bearing bones and bone sections, which are frequently abandoned intact by modern human foragers to the action of scavengers, our taphonomic study of FLK *Zinj* indicates that hominids acquired carcasses that were fully fleshed, as has already been suggested by the frequency and anatomical distribution of cut marks (Bunn and Kroll, 1986, 1988; Oliver, 1994; and discussion in Domínguez-Rodrigo, 1997a, 2002). Given that tooth marks on midshaft sections are minimal, it can also be argued that no carnivore processed any of the meat from carcasses prior to hominid defleshing. Contrary to Domínguez-Rodrigo's (1999b) original claim, percussion marks and notches indicate that most limb bones were broken by hominids. Carnivores intervened only at the last stage of bone modification and destruction, although there are a few specimens suggesting that carnivores may have marginally broken some limb bones (Figure 36). The results presented here are in agreement with those reported by Oliver (1994) in his study of the *Zinj* fauna, where he documented a high frequency of cut marks on meat-bearing bones, a lower frequency of tooth marks than reported by Blumenschine (1995) and an almost exclusively hominid bone authorship in bone breakage. Oliver (1994) also documented scarce evidence of hyaenid modification of bones.

Skeletal part profiles show that all anatomical units from most bovid carcasses are fully represented by several elements. In some cases, epiphyseal biometry and opposite side elements (work in progress), together with the high presence of axial elements at FLK *Zinj*, indicate that carcasses may occasionally have been transported complete or almost complete, as Capaldo suggested (1995, 1998b). This would suggest that hominids were frequently gaining early

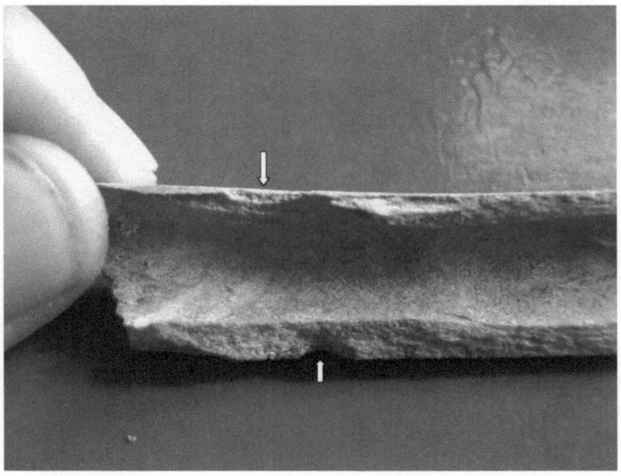

Figure 36. A double opposing notch. Opposite notches along an oblique axis appear frequently in carnivore-broken bones.

access to carcasses. This type of access resurrects the hunting hypothesis.

Several critics have categorized supporters of the hunting hypothesis as outdated (O'Connell *et al.*, 2003). However, the passive scavenging hypothesis is clearly rejected by careful taphonomic analysis of Plio-Pleistocene sites. There is not a single Plio-Pleistocene site in which long limb bone shaft fragments are tooth-marked at the rates expected if the remains were scavenged from carnivore kills (Selvaggio, 1994; Blumenschine, 1995; Capaldo, 1995, 1997). Furthermore, confrontational scavenging at its earliest stage is currently untestable (see Domínguez-Rodrigo, 2003; *contra* Lupo and O'Connell, 2002). Confronting Plio-Pleistocene felids seems to be an extremely risky strategy and there is no current evidence that could be used as a proxy for confrontational scavenging by early hominids. The ability of the Hadza to drive off predators depends on their use of heavy bows (effective 40 m away from the kill). Bushmen are less effective because they have smaller bows. Recent reports from Uganda on death tolls for humans scavenging from predators show that aggressive scavenging is risky even with the use of modern technology (Treves and Naughton-Treves, 1999). If the logic is that the smaller the bow, the less effective this strategy is, then a complete lack of bow would make confrontational scavenging fairly unlikely. Plio-Pleistocene hominids would be left to confront large felids at close encounter and that is not observed among modern foragers.

Given that the carnivore–hominid–carnivore hypothesis, the epitome of the passive scavenging models, has been falsified by the present study, it is appropriate to raise the hypothesis of hunting, which might, after all, have been one of the strategies deployed by hominids to gain early access to fleshed carcasses. This is supported by all sources of taphonomic and zooarchaeological evidence currently available at the FLK *Zinj* level.

One of these sources is the study of cut marks. In the present chapter, we have seen that the high frequency of tooth marks previously identified on midshafts at FLK *Zinj*, interpreted as the result of felids defleshing carcasses prior to hominids, was an artifact of method. Biochemical marks were confused with tooth marks. The actual frequency of tooth marks at FLK *Zinj* is very low. This result supports, for the first time, inferences made on the meaning of cut-mark frequencies and distribution at FLK *Zinj*, which suggest primary access by hominids to fleshed carcasses (Domínguez-Rodrigo, 1997a, 1997b, 2002). The next chapter shows the evidence drawn from the study of cut marks, interpreted within the experimentally created referential frameworks that allow archaeologists to confidently differentiate primary from secondary access to carcasses by hominids.

6. The behavioral meaning of cut marks at the FLK *Zinj* level: the carnivore-hominid-carnivore hypothesis falsified (II)

M. DOMÍNGUEZ-RODRIGO AND R. BARBA

Introduction

Several authors envision Plio-Pleistocene hominids as scavengers of animal carcass resources (Binford, 1981; Shipman, 1983, 1986; Blumenschine, 1991; Blumenschine and Madrigal, 1993; Blumenschine *et al.*, 1994; Selvaggio, 1994; Capaldo, 1995), a view that runs counter to traditional interpretations of hominids as efficient hunters or, at the very least, primary consumers of fully fleshed carcasses. Scavenging arguments are based partly on the observation that felids tend to abandon their prey with large quantities of intact marrow. Passively scavenging hominids would therefore have been forced to focus on within-bone nutrients. Such a scenario would have created tooth marks resulting from felid defleshing and percussion marks from hominid marrow extraction. Regarding cut marks, supporters of scavenging models suggest that these marks reflect the removal of the marginal flesh scraps that commonly survive initial carnivore consumption. However, this interpretation of cut marks remains either unsupported by experimental or observational studies (Blumenschine, 1991) or incorrectly modeled (see, for example, the critique of Selvaggio [1994] in Domínguez-Rodrigo [1997a]).

For example, although Blumenschine (1986: 446) argues that cut marks "might simply reflect removal of these scraps by scavenging hominids before the primary nutritional goal (marrow) was realized," he also notes that "experiments are needed to determine if differences exist in the quantity and location of cut marks inflicted while defleshing whole muscle masses or small scraps of flesh" and "[u]ntil these experiments are conducted, the cut mark data from FLK *Zinj* can be interpreted unequivocally to mean only that processing by hominids was concentrated on limb parts, the goal(s) of this processing beyond marrow extraction being uncertain." The experiments of Selvaggio (1994) and Capaldo (1995) presumably demonstrated that cut marks alone could not reliably distinguish primary from secondary access to carcasses. Capaldo (1995: 321) in particular claims that "given the ambiguous nature of experimentally derived cut mark criteria, we must avoid using them to make inferences on the amount of flesh acquired from carcasses by Plio-Pleistocene hominids, the timing of hominid carcass processing and the mode of carcass acquisition." These studies led to a widespread misconception that such equifinality limited the interpretive utility of cut marks. This was unfortunate because "if animal-food sharing was anywhere as regular and economically central as it is in modern hunter-gatherers, the cut mark data will have to show that flesh was the mainstay of the system" (Blumenschine, 1991: 58).

In 1986, Bunn and Kroll reported cut mark data from the FLK Level 22 "*Zinjanthropus*" level that seemingly could not be explained by hominid butchery of largely defleshed

M. Domínguez-Rodrigo, R. Barba, and C.P. Egeland, Deconstructing Olduvai: A Taphonomic Study of the Bed I Sites, 75–100.

carcasses. The distribution of cut marks appeared to reflect a wide variety of activities including skinning, dismembering, and filleting. In addition, cut mark frequency was argued to co-vary with the amount of flesh initially present on bones. Finally, and despite arguments to the contrary (e.g., that cut-marking was linked to the removal of marginal flesh scraps from largely defleshed carcasses [Binford, 1988a]), the anatomical distribution of cut marks, particularly their presence on limb bone midshafts, was consistent with the butchery of fairly complete carcasses (Domínguez-Rodrigo, 1997a, 1997b).

Such disagreement and the cut mark equifinality established by supporters of scavenging models are largely an outcome of methodology. Importantly, quantifying surface marks using Blumenschine's (1988) bone portion system, as he, Selvaggio, and Capaldo have done, does not inform on the *actual* location of a mark (see also Bunn, 2001). For example, a bone fragment defined as an epiphyseal specimen (with part of an articular end) may (or may not) have various lengths of near-epiphyseal or midshaft sections. Therefore, a cut-marked epiphyseal fragment may give the false impression that the cut mark appears on the epiphysis itself, when it in fact appears on the near-epiphysis or midshaft section (Figure 37). This methodology does not allow cut mark frequencies to be linked to the anatomical distribution of flesh. Such information is critical for evaluating different models of hominid carcass acquisition and processing. For instance, it has been shown that scraps of flesh on midshaft sections of limb bones are very rare after lion consumption (see below). The simple prediction emanating from this observation is that if hominids are butchering carcasses defleshed by felids, cut marks should not appear where flesh is absent. A more precise classification of cut mark distribution that takes into account bone type (axial, upper, intermediate, or lower limb) and the actual bone section on which they appear is required to test such a hypothesis.

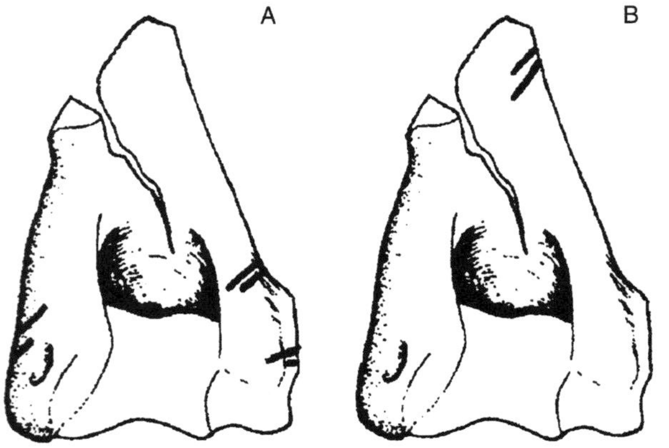

Figure 37. Behavioral meaning of cut marks on various areas of limb bones. (A) disarticulation; (B) defleshing. Such differences are masked when cut marks are quantified by bone portion (epiphyseal, near-epiphyseal, midshaft) rather than by the actual location of the mark.

Domínguez-Rodrigo (1997a, 1997b) carried out experiments on medium-sized carcasses (in this chapter also referred to as "large" given that *Zinj* only has two size ranges: small – Bunn's [1982] sizes 1–2 – and large – Bunn's [1982] sizes 3–4) using this method of cut mark analysis. These studies compared cut mark frequency and distribution resulting from the butchery of defleshed felid kills and fully fleshed carcasses. It became evident that primary access could be distinguished based on three factors (see original papers for complete statistical datasets):

1. Cut mark frequencies (in terms of number of identified specimens [NISP]) greater than 15%
2. High frequencies of cut-marked meaty upper limb bones (humerus and femur) followed by intermediate (radius and tibia) and lower (metapodial) limb bones
3. High frequencies of cut marks occurring on midshaft sections of limb bones (between 43% and 50% of the total cut-marked NISP).

In contrast, human butchery of felids kills exhibits a much different pattern of cut mark representation:

1. Cut mark frequencies of less than 10%
2. Low frequencies of cut-marked upper limb bones, while intermediate and especially lower limb bones show higher cut mark frequencies
3. With the exception of metapodials, no limb bone displays any significant number of cut marks on midshaft section

Experiments carried out on fully fleshed small-sized carcasses obtained similar results (Domínguez-Rodrigo and Barba, 2005). Again, upper limb bones were highly cut-marked (>50%), followed by intermediate and lower limb bones. Midshaft sections also preserved the highest frequency of cut marks. However, relative to medium-sized carcasses, small carcasses showed an overall lower frequency of cut-marked specimens by both bone type and section. This chapter re-evaluates cut mark frequencies from FLK Level 22 (*"Zinjanthropus"*) in light of this experimental referential framework.

Cut Mark Frequencies at FLK *Zinj*

Bunn and Kroll (1986) report that 62% of the cut-marked limb bone specimens from smaller carcasses and 40% from larger carcasses are from meaty bones. Moreover, as many as 73% of cut-marked meaty limb bone specimens from smaller carcasses and 59% from larger carcasses appear on midshaft sections. For all carcass sizes, upper and intermediate limb bone specimens comprise a majority (88%) of the cut-marked limb bone sample. Allowing that simple marrow extraction could account for cut-marked metapodials – skinning is usually necessary (Blumenschine, 1986) – the high frequencies of cut-marked upper and intermediate specimens, particularly on midshafts, are consistent with the butchery of fully fleshed carcasses. Oliver's (1994) revision of the cut mark data from FLK *Zinj* also supports this contention. Again, upper limbs show the highest cut mark frequencies, followed by intermediates and, lastly, metapodials. All these data mirror closely the results of Domínguez-Rodrigo's (1997a, 1997b, 1999a) primary access experiments. If these comparisons are valid, then Blumenschine's (1986, 1991) hypothesis that hominids were focusing on extracting marrow from limb bones already defleshed by felids and that cut marks reflect scrap removal cannot be correct (Chapter 2, Figure 2).

Another important issue is the difference in cut mark frequencies between small and large carcasses, a pattern that has been documented both experimentally and archaeologically (Figure 38). Although carcass size is probably an important factor in determining cut mark frequency, differential levels of fragmentation must first be taken into consideration. For example, in their comparisons of Hadza assemblages and a number of experimental samples, O'Connell and Lupo (2003) ignore the possible influence of bone fragmentation on their results. This is potentially important given that the Hadza sample is composed of boiled bones that would have been less attractive to ravaging carnivores, which will in turn lead to lower fragmentation levels. Arguments have been made to suggest that increased levels of fragmentation are accompanied by decreased surface mark frequencies (Bartram, 1993; Domínguez-Rodrigo, 2003). Domínguez-Rodrigo and Barba (2005) have shown that in experimental assemblages of butchered bone, a significant negative correlation exists between the midshaft NISP:MNE ratio –that is, the ratio derived by comparing the proportion of midshaft specimens and the number of complete elements to which they belong – and cut mark frequencies (Figure 39). Because smaller-sized carcasses are almost twice as fragmented as their larger-sized

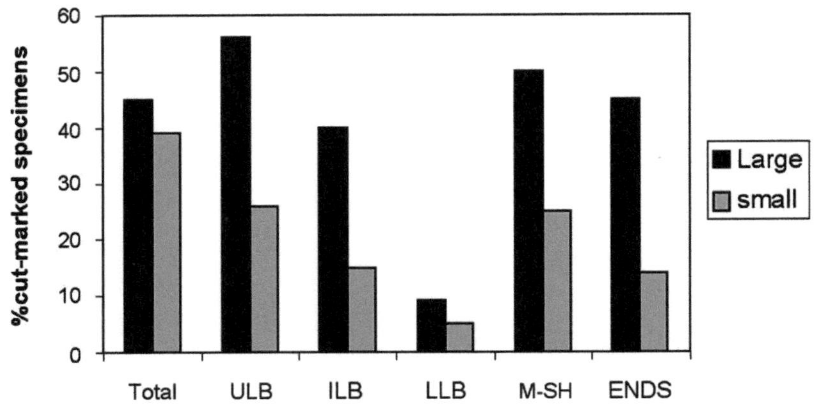

Figure 38. Mean percentage of cut marks on each appendicular anatomical area for large and small carcasses in experimental assemblages. Abbreviations: ULB, upper limb bone; ILB, intermediate limb bone; LLB, lower limb bone; M-SH, midshafts; ENDS, epiphyses. (Data from Domínguez-Rodrigo and Barba, 2005.)

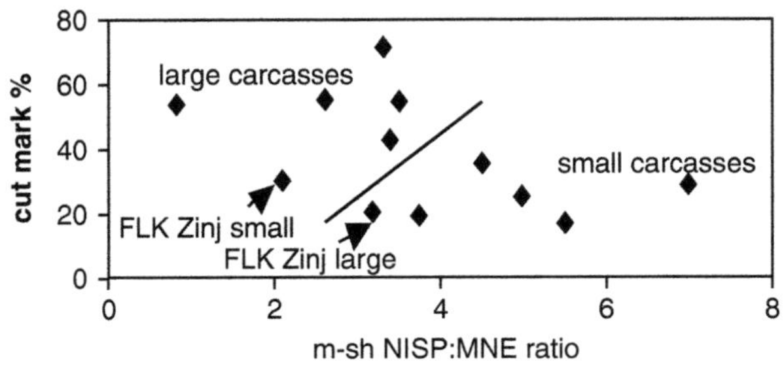

Figure 39. Relationship between cut mark frequencies and fragmentation (midshaft NISP:MNE), including the cut-marked sample from FLK *Zinj* for small and medium carcasses. (Data from Domínguez-Rodrigo and Barba, 2005.)

counterparts, they show cut mark frequencies almost 50% lower. Fragmentation also seems to explain cut mark frequencies at FLK Level 22, where larger carcasses (midshaft NISP:MNE = 3.2) show lower cut mark frequencies (20.3%) than smaller carcasses (midshaft NISP:MNE = 2.1; cut mark frequency = 30.4%). Whether differential fragmentation between carcass sizes is due to bone size (i.e., ease of breakage) or to other taphonomic processes is still not clear. Nevertheless, it is clear that both carcass size and differential fragmentation (either through biotic agents or diagenesis) affect cut mark frequencies (*contra* O'Connell and Lupo, 2003).

On larger carcasses, this analysis found cut mark frequencies very similar to those reported by Bunn and Kroll (1986) in their macro-sample (Table 8). Frequencies are slightly lower for metapodials, where a higher number of abrasion marks were identified. Cut mark frequencies on smaller carcasses are lower than Bunn and Kroll's (1986) macro-sample but similar to their micro-sample. Again, many smaller-carcass specimens preserved either ambiguous cut marks or abrasion marks. Nevertheless, this conservative approach probably underestimates the actual number of cut-marked specimens and we believe that cut mark frequencies reported by Bunn and Kroll (1986), although less conservative, are probably closer to the original frequencies. Cut marks are more abundant on midshaft sections for all four meaty limb bones (Table 8 and Figure 40). In more than 88% of all cut-marked limb bone specimens from smaller carcasses, cut marks occur on midshafts. A total of 62.5% of cut-marked limb bone specimens from larger carcasses also show the marks on midshaft sections (Table 8). This supports primary access by hominids to fleshed carcasses (Domínguez-Rodrigo, 1997a) and further refutes Binford's (1988a)

Table 8. Cut mark frequencies per long limb bone section at FLK Zinj. Numbers in numerator are total numbers of cut-marked specimens. Numbers in denominator are total numbers of specimens. Numbers in parentheses are percentages

	FLK Zinj	
CM	Size 1/2	Size 3/4
HUM PSH	0/5(0)	1/2(50)
HUM MSH	5/16(31.3)	9/44(20.5)
HUM DSH	2/9(22.2)	8/11(72.7)
RAD PSH	3/6(50)	8/15(53.3)
RAD MSH	2/21(9.5)	14/51(27.5)
RAD DSH	0/0(0)	0/3(0)
MC PSH	1/6(16.7)	1/7(14.3)
MC MSH	1/20(5)	1/15(6.7)
MC DSH	0/5(0)	0/2(0)
FEM PSH	0/1(0)	2/2(100)
FEM MSH	1/14(7.1)	6/41(14.6)
FEM DSH	2/5(40)	0/2(0)
TIB PSH	2/4(50)	0/1(0)
TIB MSH	5/32(15.6)	13/86(15.1)
TIB DSH	1/3(33.3)	2/4(50)
MT PSH	1/9(11.1)	1/6(16.7)
MT MSH	1/23(4.3)	2/24(8.3)
MT DSH	0/2(0)	2/3(66.7)
Total	**27/181(14.9)**	**72/319(22.6)**
ULB*	0/2(0)	2/13(15.4)
ILB*	2/2(100)	0/3(0)
LLB*	0/0(0)	2/3(66.7)
MSH	15/126(11.9)	45/261(17.2)
ENDS	12/55(21.8)	27/58(46.6)
Meaty MSH	13/83(15.7)	42/222(18.9)
NCMSP:NISP	27/181(14.9)	72/319(22.6)
NCMMSSP:NCMSP	15/27(88.2)	45/72(62.5)

HUM, humerus; RAD, radius; MC, metacarpal; FEM, femur; TIB, tibia; MT, metatarsal; PSH, proximal end or shaft; DSH, distal end or shaft; MSH, midshaft; ULB, upper limb bone; ILB, intermediate limb bone; LLB, lower limb bone; NCMSP, number of cut-marked specimens; NCMMSSP, number of cut-marked mid-shaft specimens.
* Identified to limb anatomical section but not to element type.

claim that Bunn and Kroll (1986) had misidentified many cut marks.

In order to further support the contention that hominids were butchering fully fleshed carcasses, this study presents detailed data on the anatomical distribution of flesh scraps on a large sample of lion kills. As argued above, cut marks simply should not appear on areas regularly devoid of flesh if hominids are passively scavenging from felid kills.

Anatomical Distribution of Flesh on Felid Kills

PHASE 1: MAPPING THE ANATOMICAL LOCATION OF FLESH SCRAPS

Although research has been conducted on the feeding behavior of hyenas (see review in Blumenschine and Marean, 1993) and leopards (Brain, 1981; Cavallo and Blumenschine, 1989; Cavallo, 1998), no studies have systematically addressed bone modification and destruction among lions. These data are important given the presence of both lions and similarly sized false-sabertooth cats in Plio-Pleistocene African savannas, all of which could have provided marrow bones from abandoned carcasses (Blumenschine, 1986, 1991, 1995; Blumenschine and Madrigal, 1993; Marean and Ehrhardt, 1995). Documenting general patterns of bone modification and the

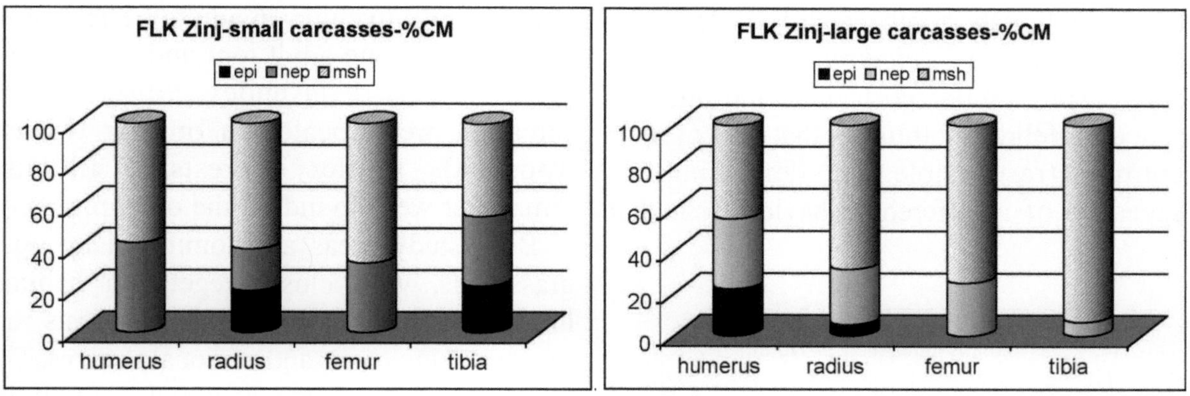

Figure 40. Distribution of all cut-marked specimens according to carcass size and element type in the three main limb bone sections. Abbreviations: EPI, epiphyses; NEP, near-epiphyses, MSH, midshafts.

location of flesh scraps[1] subsequent to lion consumption will therefore provide critical data for assessing the timing of hominid access to carcasses.

It could be argued that modeling large felid carcass consumption behavior from the pattern documented in modern lions is not necessarily accurate, since sabertooth felids might have abandoned larger amounts of flesh on the bones of the carcasses they preyed upon (Marean, 1989). This hypothesis currently lacks scientific support. It is known that the cranio-dental and postcranial anatomy of sabertooth felids was different from the more generalized carnivores present today; however, it is unknown how this may have affected their behavior and, more importantly, how this might have affected the amount of flesh they may have abandoned on carcasses. Marean and Ehrhardt (1995) provided a convincing case – against previous assumptions in Marean (1989) – that carcass defleshing by *Homotherium* must have been fairly thorough. The behavioral differences argued for sabertooth felids versus modern large felids have been well illustrated by Lewis (1997), who reconstructed the sabertooth adaptive patterns in a similar way to the diversity of adaptive patterns exhibited by modern felids in African savannas: some were ambush predators like leopards (*Dinofelis*) and other were more cursorial (*Homotherium*).

Most of the carcasses accumulated in the Olduvai Bed I sites are small to middle-sized and the only sabertooth candidates for their accumulation (*Megantereon* and *Dinofelis*) show overall similar postcanine morphology to modern felids. Nothing in that part of their dentition (responsible for defleshing) is suggestive of a different behavior from that documented in similarly sized extant felids. The most parsimonious assumption, therefore, is that unless a different defleshing behavior by sabertooth felids could be shown in the future, their defleshing behavior ought to have been similar to that documented in modern felids. If not, then the scraps of flesh abandoned by them must have followed the same structural properties of the scraps abandoned by modern felids, because it is more strongly related to the inherent structure of muscle attachment on the bones than to how felids might have pulled the flesh off of them. For this reason, a study of the anatomical locations of flesh scraps is of interest, since placement of cut marks, which is conditioned by the anatomical location of flesh scraps remaining after felid kills, should reflect either primary or secondary access to carcasses by hominids.

Methods and Sample

Observations of lion kills were made over the course of 5 months during three consecutive dry seasons in two areas of the Maasai Mara National Reserve, Kenya. Area 1 is situated in the central part of the reserve between the Mara and the Olare Orok Rivers, and Area 2 is located in the north of the reserve, between the Mara and Olchorro le Musiara Rivers, which is outside the official park boundaries (for details see Domínguez-Rodrigo, 1999a). The sample consists of 28 medium-sized carcasses, mostly wildebeest (3 juveniles, 14 adults) and zebra (2 juveniles, 6 adults). The sample also includes one adult topi and two buffalo (1 infant and 1 juvenile). Fifteen of the carcasses were located in or near riparian woodlands (gallery forests) while the remainder were found on the open plains.

Both study areas are dominated by open grasslands, with closed-vegetation habitats limited largely to narrow gallery forests that run parallel to rivers and seasonal streams. The structure of the large mammal community reflects the open vegetation of Maasai Mara.

[1] This study was initially published in Domínguez-Rodrigo (1999a). Flesh availability and bone modification in carcasses consumed by lions. *Paleogeography, Palaeocology and Palaeoclimatology*, 149: 373–388.

Both herbivores and carnivores (especially lions) live in large groups. Carnivore diversity and density is high, and intense interspecific competition accounts for the short duration of carcasses (Domínguez-Rodrigo, 1994a). However, levels of competition were found to vary by habitat type. Closed habitats exhibited much lower levels of competition, and there was never more than one species of carnivore feeding on carcasses. In contrast, carcasses occurring on the open plains were often visited by multiple species at the same time (Domínguez-Rodrigo, 1999a). This study focuses on documenting general behavior patterns of lions adapted to an open, semiarid grassland savanna. Therefore, to prevent distortion of the overall patterns of flesh availability and bone modification, the following unusual behaviors were omitted:

1. Brief episodes of "surplus killing" (Kruuk, 1972; Schaller, 1972) or the hunting of more than one prey during the same day. These most often occurred in areas where literally thousands of herbivores were migrating from the Serengeti (Domínguez-Rodrigo, 1999a).
2. Episodic kills made by one or two lions in which the rest of the pride did not participate in carcass consumption. Such kills were usually made by solitary subadult males or when the other members of the pride were in other locations.
3. Episodes of carcass consumption that were interrupted by the intervention of other predators such as hyenas.

Surplus killings, because the lions cannot defend multiple kills from other competitors, promote the availability of large flesh packets and even semi-complete carcasses. After multiple kills, lions may also "abandon a kill after having gorged, thereby leaving a considerable amount of meat for scavengers" (Schaller, 1972: 271). Single kills were always defleshed (Schaller, 1972; see also Figure 41). However, surplus killing is highly dependent on three factors: (i) geographical setting: it does not occur in the same place but changes according to the advancement of migratory fauna; (ii) time of year: it only happens during the long dry season; and (iii) ecological conditions: multiple kills were only documented in the Serengeti and Maasai Mara where long-distance migrations occur. It is therefore not surprising that this behavior was observed only once in this study and other ethological studies describe this behavior as highly unusual (Kruuk, 1972; Schaller, 1972). For example, Schaller (1972) reports only 1.5% of

Figure 41. Carcasses from Maasai Mara (Kenya) subsequent to lion consumption. Note the almost complete lack of flesh on both the zebra (left) and wildebeest (right) (Domínguez-Rodrigo, 1999a).

zebra kills and 4% of wildebeest kills were multiple kills made by different individuals from the same pride. Overall, the infrequent occurrence of surplus killing makes it unsuitable for building general models of lion carcass consumption.

The second behavior was not included in this study because the consumption of medium-sized carcasses by single individuals (or pairs) is also very rare in areas like the Maasai Mara and Serengeti. In these areas, lions live in prides or nomad groups and prey acquisition and consumption are usually collective activities (Schaller, 1972). Although Schaller (1972) observed that prides are not necessarily permanently cohesive social units, foraging groups in the Seronera study area were made up of four to six individuals. In the Maasai Mara, foraging group size averaged seven individuals. Over 5 years of observation in the Maasai Mara (Domínguez-Rodrigo, 1994a), this study found that less than 10% of all kills were made and consumed by a single (or sometimes two) individual(s). Under these rare circumstances flesh scraps were much more abundant. It also important to note, however, that small-group hunting occurs on a seasonal basis only; in addition, solitary hunters often subsist on smaller size class 1 and 2 prey. This behavior has been observed in the Serengeti:

> For instance, when the seven lionesses in the Seronera pride fed mainly on such large prey as wildebeest between January and May, 1969, an average of 6.4 associated wherever I encountered them. But from June to August, when Thompson's gazelle constituted their principal food, an average of only 3.6 were together. Then in September, after zebra moved into the area, all seven tended to form a group again (Schaller, 1972: 37).

It is also likely that carcass processing behavior differs between such small game and the more commonly acquired medium-sized prey. The observations from this study carried out in Maasai Mara were made during the summer (July through September) when the plains supported plentiful fauna. This accounts for the fact that lions foraged in larger groups most of the time and that the consumption of middle-sized carcasses often involved more than four individuals. Restricted consumption, like surplus killing, is therefore a sporadic occurrence in the Maasai Mara and was also excluded from this study.

A total of 86 complete limbs and 28 axial skeletons were analyzed. Scraps of meat smaller than 2–3 cm are not included in this study, although they were considered in a previous study that analyzed 11 of the carcasses from this sample (Domínguez-Rodrigo, 1997c). The anatomical distribution of flesh scraps was analyzed in relation to three basic units: head, trunk, and limbs. The trunk was further divided into three sections: Section 1 comprises the neck (atlas, axis, and other cervical vertebrae); Section 2 includes the rib cage and thoracic vertebrae; and Section 3 combines the pelvis and the lumbar and caudal vertebrae. Limbs were also divided into upper (humerus, femur), intermediate (radio–ulna, tibia), and lower (metapodials) bones (Figure 42). Both ribs and limb bones were further subdivided into proximal, distal, and midshaft sections. On limb bones, proximal and distal equate with their respective near-epiphyseal sections. In this study, "fleshed bone" is defined as those elements or element sections that retain more than 10% of their original flesh mass. If less than 10% remained, the element was considered "defleshed" and evaluated in terms of scraps (if any remained).

Rather than simple presence/absence, scraps were analyzed in terms of three categories: total absence (lack of flesh or the occasional marginal scrap), practical or near absence (flesh weight is less than 150 g and all joints and most of the bone surface or periosteum can be seen on a particular bone section), and presence (flesh weight is more than 150 g). Meat weights were approximated visually.

Results

Lions are primarily social hunters and, like other social predators, once a kill is made,

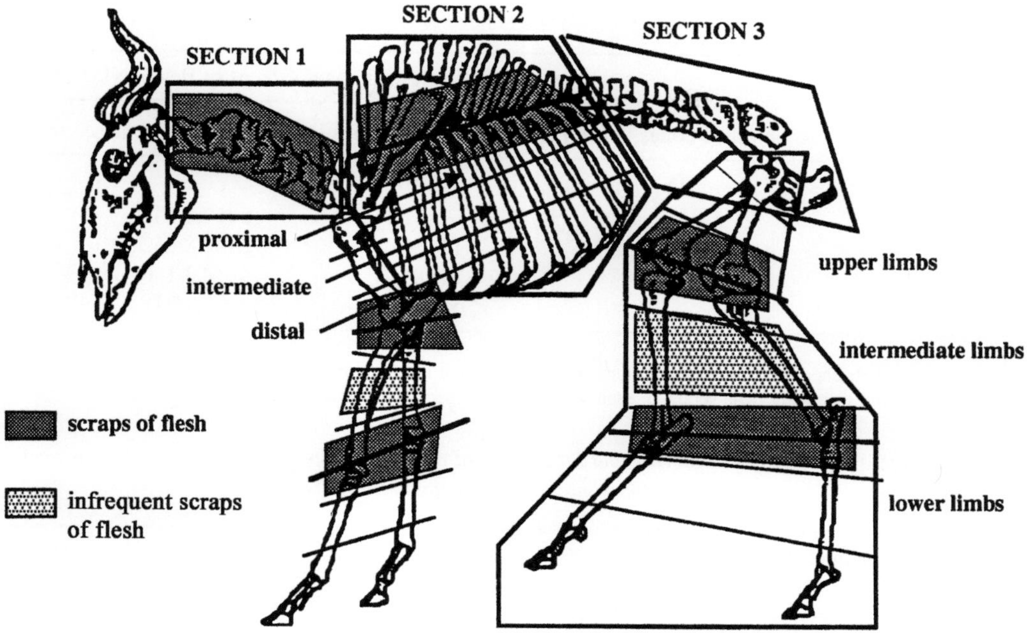

Figure 42. Location of each anatomical unit (skull, axial, limbs) and sections thereof (neck, thorax, lumbar for axials; upper, intermediate, and lower for limbs). After lion consumption, scraps of flesh are mainly restricted to the neck, dorsal vertebrae, proximal section of ribs, distal ends of upper limbs, and proximal and distal ends of intermediate limb bones. Scraps infrequently appear on the midshafts of intermediate limb bones. Metapodials are left unmodified. (Data from Domínguez-Rodrigo, 1999a.)

they compete for access to carcass resources (Domínguez-Rodrigo, 1994a). High levels of intraspecific competition result in frequent disarticulation, although limb bones usually remain in close proximity to the rest of the carcass. During the wet season when prey abundance is relatively high, lower competition means it is possible to find carcasses with little conspicuous evidence for lion involvement (Domínguez-Rodrigo, 1994b, 1999a). Nevertheless, apart from the aberrant behaviors cited above, lions show a consistent pattern of carcass processing, flesh consumption, and bone damage.

As shown in Tables 9 through 11 and Figure 42, lions process the flesh of carcasses very thoroughly, leaving few edible flesh remains. Often, only the cranium preserves flesh after lions abandon their kills. Because metapodials lack meat to begin with, they are also left intact in most instances. Overall, therefore, the occurrence of flesh scraps is very rare and a majority of observed carcasses preserved no scraps at all. On the trunk, flesh scraps were most abundant on the neck (Section 1), while in 50% and 86% of the carcasses, respectively, rib cages (Section 2) and the lumbar vertebrae and pelves (Section 3) were completely defleshed. Almost 89% of all bone sections preserving any scrap were placed in the "practical absence" category. The neck unit had relatively high frequencies of scraps, and in three cases scraps were abundant in large chunks of meat at the base of the cranium and upper section of the cervical vertebrae. On Sections 2 (rib cage and thoracics) and 3 (lumbars and pelves) flesh scraps occurred on thoracics or the proximal sections of ribs (although tiny scraps less than 2 cm were occasionally observed on the intermediate and distal sections of ribs).

Eighty-two percent (32/39) of the limb bones that preserved scraps were intermediates, while only 8% were uppers. Only 10% of

Table 9. Distribution of scraps of flesh according to anatomical section

	Samples (n)	Fleshed bones, NISP (%)	Defleshed bones (scraps of flesh), NISP (%)		
			Total absence	Practical absence	Presence
Cranium	28	28 (100)	–	–	–
Section 1	28	–	9 (32.1)	16 (57.1/84.2)	3 (10.7/15.8)
Section 2	28	–	14 (50)	12 (42.8/85.7)	2 (7.2/14.3)
Section 3	28	–	24 (85.7)	4 (14.3/100)	–
Limbs	86	–	51 (59.3)	32 (37.2/91)	3 (3.5/9)

Percentages in the "practical absence" and "presence" categories are given first for the number of bones with respect to the total sample, and second for the number of bones showing flesh scraps in each section with respect to the total number of bones with flesh scraps.

Table 10. Distribution of scraps of flesh on ribs and limbs, according to bone type (upper, intermediate, lower) and bone section (proximal/distal ends and midshafts)

	Samples (n)	P	MS	D
Ribs	14	14 (100)	–	–
Limbs				
Upper	35	1 (1.1/2.8/14.2)	–	6 (6.9/17/85.8)
Intermediate	35	12 (13.9/34.2/37.5)	4 (4.6/11.4/12.5)	16 (12.6/45.7/50)
Lower	35	n.m.	n.m.	n.m.

Percentages (in parentheses) are first for each section/bone type with respect to the total sample, second for each section/bone type with respect to the sample of bones showing flesh scraps and, finally, for each section/bone type with respect to the total number of bones with scraps of flesh of the same anatomical section; n.m. = not modified; P = proximal end; D = distal end; MS = midshaft section.

Table 11. Distribution of scraps of flesh according to anatomical section and according to habitat type (riparian woodland vs. open plain)

	Samples (n)	Fleshed bones, NISP (%)	Defleshed bones (scraps of flesh), NISP (%)		
			Total absence	Practical absence	Presence
Riparian woodland					
Cranium	15	15 (100)	–	–	–
Section 1	15	–	9 (60)	6 (40/100)	–
Section 2	15	–	12 (80)	4 (26.6/100)	–
Section 3	15	–	13 (86.6)	2 (13.4/100)	–
Limbs	49	–	44 (89.9)	10 (20/100)	–
Open plain					
Cranium	13	13 (100)	–	–	–
Section 1	13	–	–	10 (76.9/76.9)	3 (13/13)
Section 2	13	–	2 (15.3)	8 (61.5/80)	2 (15.2/20)
Section 3	13	–	11 (84.6)	2 (15.4/100)	–
Limbs	37	–	7 (19)	22 (59.4/88)	3 (8.2/2)

Percentages in the "practical absence" and "presence" categories are first for the number of bones with respect to the total sample and second, for the number of bones showing flesh scraps in each section with respect to the total number of bones with flesh scraps.

the limb bones bearing any flesh preserved scraps on the midshaft section. No midshaft sections from any upper limb bone retained any scraps of meat. On intermediate bones, distal sections preserve the highest frequency of scraps (50%) followed by the proximal (37.5%) and finally midshaft sections (12.5%). Overall, *90% of all scraps occurred on proximal or distal sections*. Moreover, 80% of all scraps on upper limb bones occurred on the distal section.

There was an interesting contrast in the pattern of flesh scrap availability between human- and lion-processed carcasses. Because lions pull flesh from their prey, they often remove entire muscle masses. Thus, bones are more completely defleshed and scraps cluster near muscle origins and insertions (i.e., near-epiphyses) (Domínguez-Rodrigo, 1999a). On the other hand, humans butcher carcasses by cutting through flesh masses and are therefore less likely to remove muscle masses as complete units (Domínguez-Rodrigo, 1997a). This may account for the common presence of scraps on vertebrae and the midshaft sections of limb bones.

An unexpected observation from this study was the fact that most carcasses in riparian woodlands exhibited almost no flesh scraps (Table 11). Nearly all scraps were found on carcasses in the open plains. Because of low visibility in riparian habitats, lions are able to spend up to several days unmolested by competitors and thus prey consumption is very thorough. In the open plains, intense competition prevents lions from remaining with carcasses for more than about 2 h, sometimes resulting in incomplete carcass processing (Domínguez-Rodrigo, 2001).

Discussion and Conclusions

This study demonstrates that the scraps of flesh that commonly survive large felid consumption are less common than previously thought (Blumenschine, 1986, 1991). When they do occur, they are preferentially distributed on the neck and the proximal sections of rib cages. Lumbar vertebrae and pelves are usually defleshed completely. Most of the scraps on limb bones occur on intermediate bones, with upper limb bones almost always completely defleshed. The few scraps that are retained on upper limb bones occur mainly on distal sections. *No upper limb midshaft sections preserved any flesh scraps*. Only 4.5% of intermediate limb bones preserved scraps on midshaft sections.

This study has also shown that a consistent pattern characterizes lion carcass processing, where several individuals participate in prey consumption and carcasses are almost always completely defleshed. The former is true irrespective of habitat type (open versus closed), while the level to which carcasses are consumed depends on ecological setting. A previous study on variation in competition (Domínguez-Rodrigo, 2001) found that all lion kills on the open plains were also visited by other carnivores (especially vultures, hyenas, and jackals). Only in rare instances when the kill was witnessed did the researcher precede the arrival of competitors who, nonetheless, took little time to discover the kill and wait until the lions completed their initial bout of feeding. In the open plains, a significant number of carcasses still being fed upon by lions (42.8%) were located with more than one competitor species present at the kill. The most common associations were vultures and hyenas and vultures and jackals. During the first 3 h of carcass consumption, hyenas were present on all occasions, while vultures and jackals occurred at 85.6% and 65.6% of the kills, respectively. Overall, consumable resources lasted less than 3 h from the time carcasses were located, and hyenas were often responsible for their complete deletion (Domínguez-Rodrigo, 2001).

The situation in riparian woodlands was very different. Competitor species showed no time overlap at carcasses. This likely stems from the lack of vultures, which signal the location of carcasses to other carnivores. Jackals were seen at 59.9% of carcasses, and almost half the time they discovered kills in less than 12 h. Hyenas, on the other hand, usually discovered carcasses only after 24 h and only 36.7% of the time did they find them within 48 h. The lack of intense hyena intervention allowed carcasses to last many days in riparian habitats (Domínguez-Rodrigo, 2001). All of these results are comparable to those reported in the Serengeti by Blumenschine (1986).

The distribution of flesh scraps on lion kills can thus serve as a useful starting point for modeling early hominid behavior from archaeological faunas. Cut mark distribution according to body section (cranium, trunk, limb), bone type, and bone section reflects the resources available to hominids and thus the order of hominid access. For example, a high frequency of cut-marked upper limb bones and/or midshaft sections would indicate access to fleshed carcasses whereas a high frequency of cut-marked fragments from intermediate bones and metapodials and/or epiphyseal sections would suggest access to defleshed carcasses. Furthermore, this study and others (e.g., Blumenschine, 1986) have demonstrated the contrast in competition and flesh availability in open versus closed habitats. Therefore, bone assemblages in which percussion marks are the only indicator of hominid behavior likely reflect hominid scavenging in closed environments.

PHASE 2: MAPPING THE EXACT ANATOMICAL LOCATION OF FLESH SCRAPS ON LIMB BONES

Despite the data presented above, there remains some potential ambiguity in the interpretation of cut mark distribution, especially in small samples. Therefore, the distribution of scraps should be mapped at an even more precise level. For instance, are surviving scraps preferentially located near muscle, ligament, or tendon insertions? Are they more common on particular aspects of limb bones? Modern felids leave few scraps of flesh after they finish consuming their prey, but did sabertooth felids do the same, or could they have left more flesh on the bones? Since muscles are attached to bones in specific ways according to bone type (e.g., there are more muscle attachments on the caudal side of femora and tibiae than on the cranial side), scraps of flesh (irrespective of their size) are more likely to occur on some parts of bones than on others. Therefore, this section documents in detail the exact anatomical location of the scraps of flesh that survived observed instances of lion consumption in Maasai Mara. This is significant because, as mentioned earlier, cut marks in archaeological assemblages appearing on areas that do not preserve scraps cannot be interpreted as secondary butchery of defleshed carcasses.

Humerus. All the flesh scraps on humeri appeared exclusively on the ends, especially the distal epiphysis (Figure 43). The marginal scraps observed on the proximal epiphyses appeared on the caudal aspect under the humeral head and extended to the lateral rim of the neck. They are thus located at the attachments of the M. biceps brachialis and the M. triceps brachii. In the Maasai Mara sample, the M. supraspinosus, M. pectoralis, and M. subscapularis were never represented as scraps. Scraps on the distal epiphysis were

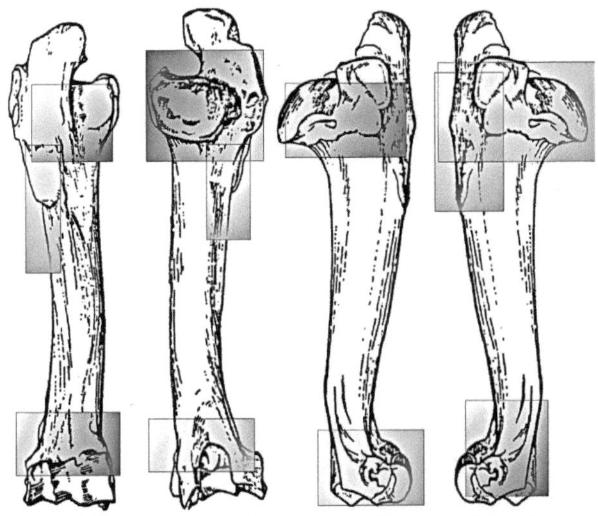

Figure 43. Anatomical locations of flesh scraps (shaded square areas) on humeri from lion kills documented in Maasai Mara (Domínguez-Rodrigo, 1999a).

located mainly on the lateral aspect and belong to the M. extensoris carpi ulnaris (partially covered by the Ligamentum radioulnare proximale laterale). Small scraps of the M. biceps brachii were also documented on a few occasions. The cranial aspect preserved small pieces of the M. extensoris digitorum communis, although the underlying Ligamentum collaterale laterale could prevent cut marks from appearing. On the caudal aspect, some small scraps of the M. anconeus appeared around the olecranon fossa. Cut marks generated through the removal of these scraps could mimic those labeled Hd-3 and Hd-2 by Binford (1981: 123).

Radio–Ulna. Like the humerus, flesh scraps were rare on radio–ulnae (Figure 44). On the proximal end, scraps from the M. biceps brachialis appeared on the cranial and lateral aspects. Small scraps belonging to M. flexor carpi radialis, M. flexor carpi ulnaris, and M. extensoris carpi ulnaris were occasionally observed on the posterolateral aspect of the radius. The proximal ulna preserved a majority of the flesh scraps, where scraps of M. triceps brachii and M. anconeus were common. The cranial aspect and much of the caudal aspect of the radius shaft were clear of scraps. Scraps clustering near the distal epiphysis belonged to the M. extensor carpi radialis, M. abductor digiti I longus, M. flexor carpi radialis, and M. flexoris digitorum superficialis.

Femur. Lions thoroughly consumed the flesh on the proximal ends of the femora (Figure 45). Shafts were also completely defleshed. The few scraps that were observed appeared on the caudal, medial, and lateral aspects of distal ends and belonged mainly to the Caput mediale m. gastrocnemii and Caput laterale m. grastrocnemii.

Tibia. The few scraps that were observed on tibiae were located on the cranial and caudal aspects of both ends (Figure 46). These belonged to the M. flexor digitalis and M. extensor digitalis. Some scraps of M. flexor

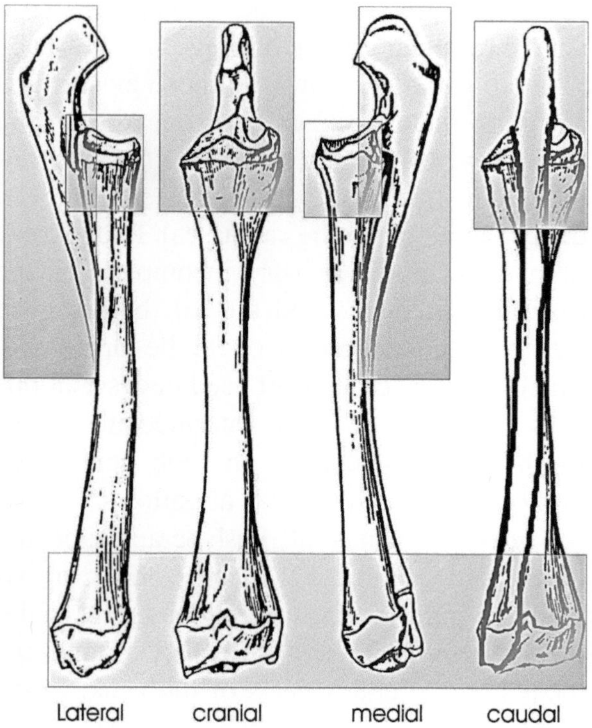

Figure 44. Anatomical locations of flesh scraps (shaded square areas) on radii from lion kills documented in Maasai Mara (Domínguez-Rodrigo, 1999a).

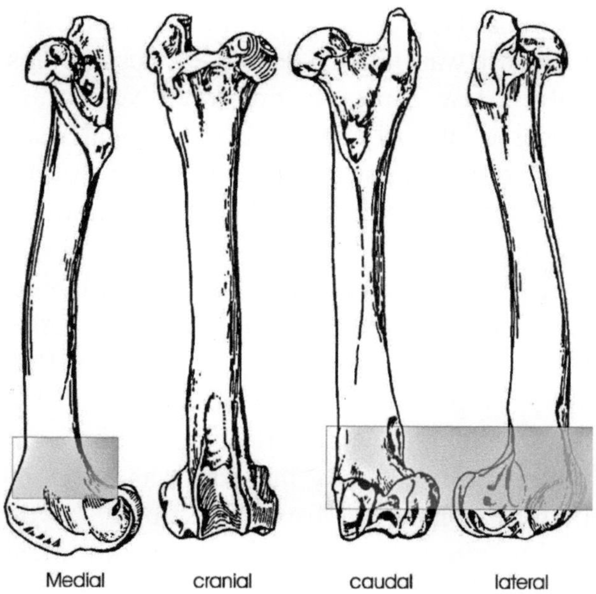

Figure 45. Anatomical locations of flesh scraps (shaded square areas) on femora from lion kills documented in Maasai Mara (Domínguez-Rodrigo, 1999a).

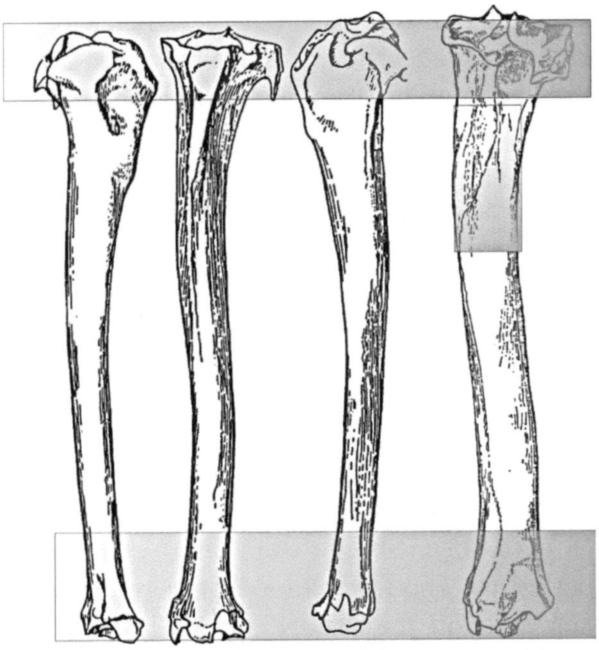

Figure 46. Anatomical locations of flesh scraps (shaded square areas) on tibiae from lion kills documented in Maasai Mara (Domínguez-Rodrigo, 1999a).

were also occasionally observed on the proximal part of the tibial crest. The medial side of the tibial crest was always free of scraps. The midshaft was largely devoid of flesh scraps; only a few carcasses preserved very small pieces of the M. flexor on the caudal shaft. The only consistently appearing scraps occurred on the distal end emerging from the tarsal area.

Similar observations of flesh availability on lion kills have been made by other researchers. For example, Selvaggio (1994: 54–55) reports that "limbs were abandoned by lions with little or no flesh on humeri and femora. Occasionally the skin was not completely consumed from the tibiae or the radius–ulnae and small scraps of flesh remained near the *distal* epiphyses" (emphasis added). She also notes that "long bones abandoned by large groups of carnivores were usually disarticulated from the carcass and while the marrow cavity remained intact, the bones were usually encountered completely defleshed" (Selvaggio, 1994: 124). Therefore, we are confident that the sample presented in this study, coupled with the observations of others, reflects an adequate degree of variability in lion carcass consumption behavior.

Nevertheless, we also present data on partially consumed carcasses to address inferences of flesh availability on carcasses potentially abandoned by sabertooth felids. This sample consists of three Maasai cows that were killed by lions in the Galana and Kulalu Game Reserves in Kenya. All three kills were made in the middle of the night by small prides (two to three individuals) and subsequently dragged away from the Maasai camps. Maasai men would rediscover the carcass 2–3 h later, often driving the lions off. The data from these particular carcasses have not been presented before, given the incomplete level of carcass consumption. Therefore, this sample provides data on scrap distribution on carcasses that were hastily and incompletely consumed by small groups of lions.

In all cases, the cows had been eviscerated and limb bones from at least one side of the carcass were partially defleshed (limbs from the side resting on the ground had not yet been defleshed in two of the carcasses). In one case, limb defleshing was very incomplete as the Maasai men discovered the kill before even the intermediate bones could be processed. Therefore, this case is not used in descriptions of scrap distribution on intermediate bones. Flesh scrap distribution on limb bones partially consumed by lions is as follows.

Humerus. Almost all flesh scraps appeared on the ends (Figure 47). Scraps on the proximal end appeared on the caudal aspect under the humeral head and extended around the medial and lateral aspects of the neck. Small pieces of the M. triceps brachialis, M. biceps brachii, M. supraspinosus, M. pectoralis, and M. subescapularis appeared on the proximal end of one or more humerus. Scraps of M. pronator teres, and M. teres major appeared

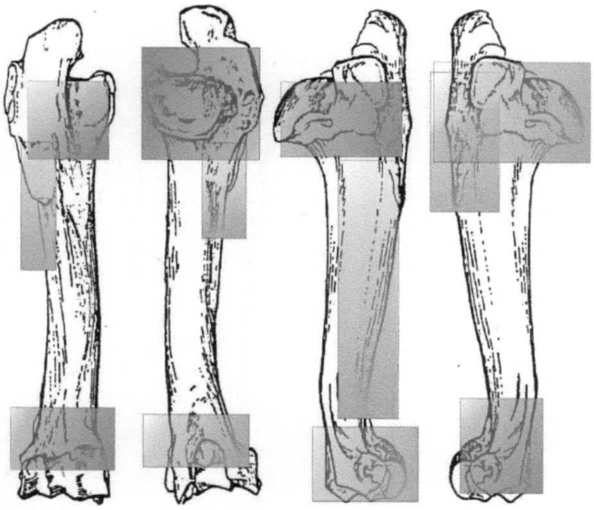

Figure 47. Anatomical locations of flesh scraps, that is, Cold Zone 1 (shaded areas), on humeri from lion kills documented in Galana and Kulalu. Hot Zone 1 is the unshaded area where flesh scraps were not observed. Cut marks found in this zone on bones from archaeological assemblages should be indicative of primary access to carcasses.

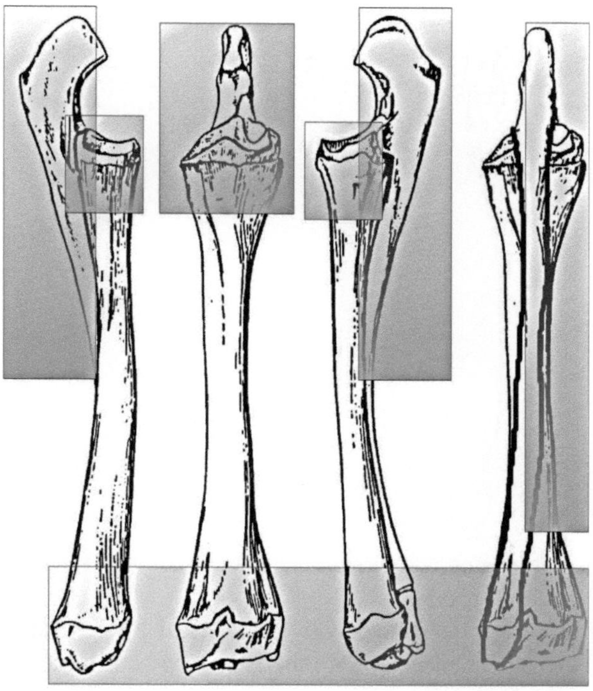

Figure 48. Anatomical locations of flesh scraps, that is, Cold Zones 2 and 3 (shaded areas), on radii from lion kills documented in Galana and Kulalu. Hot Zones 2 and 3 are the unshaded areas where flesh scraps were not observed. Cut marks found in these zones on bones from archaeological assemblages should be indicative of primary access to carcasses.

on the near-epiphyseal area. Compared to the sample of completely processed carcasses described above, scraps appeared further down the deltoid crest, although the shaft area around the deltoid crest was largely free of flesh scraps. Flesh scraps on the distal epiphyses were documented on the caudal (M. anconeus), medial and lateral (M. extensoris carpi ulnaris, M. pronator teres, M. flexor carpi radialis), and cranial aspects (M. biceps brachii).

Radio-Ulna. Scraps of flesh were more abundant on radio-ulnae (Figure 48). On the proximal end, they appeared on the cranial (M. biceps brachii, M. extensor carpi radialis, M. extensor digitorum) and lateral aspects (M. biceps brachii). Small scraps of the M. flexor carpi radialis, M. flexor carpi ulnaris, and M extensoris carpi ulnaris were also observed on the caudal aspect of the radius. In some cases, they would extend all the way down the lateral aspect of the ulna almost to the distal end. Abundant flesh occurred on the proximal ulna where scraps of M. triceps brachii and M. anconeus were common. The cranial aspect of the shaft was completely clean of scraps. The caudal shaft, on the other hand, preserved frequent scraps on the lateral rim. Flesh scraps belonging to the M. extensor carpi radialis, M. abductor digiti I longus, M. flexor carpi radialis and M. flexoris digitorum superficialis also appeared clustered at the distal end.

Femur. Femora preserved more abundant flesh scraps relative to the sample of completely consumed carcasses (Figure 49). These were distributed on the caudal aspect of the distal shaft (M. gastrocnemius) and proximal shaft (M. vastus lateralis). Scraps belonging to M. gluteus accesoris were observed in the intertrochanteric fossa. Scraps from the

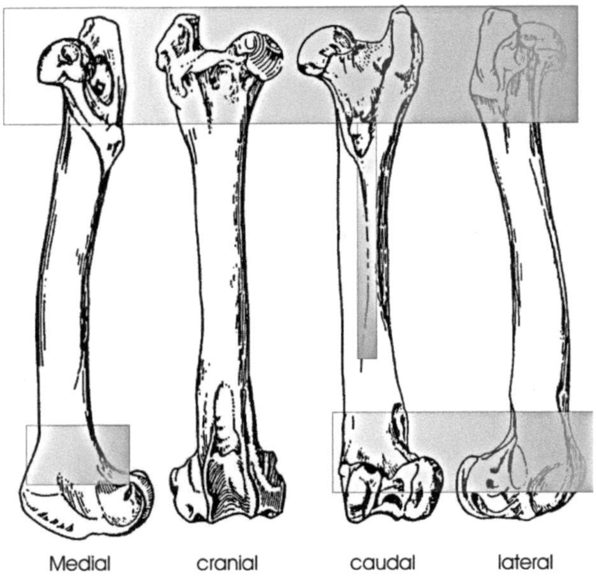

Figure 49. Anatomical locations of flesh scraps, that is, Cold Zone 4 (shaded areas), on femora from lion kills documented in Galana and Kulalu. Hot Zone 4 is the unshaded area where flesh scraps were not observed. Cut marks found in this zone on archaeological assemblages should be indicative of primary access to carcasses.

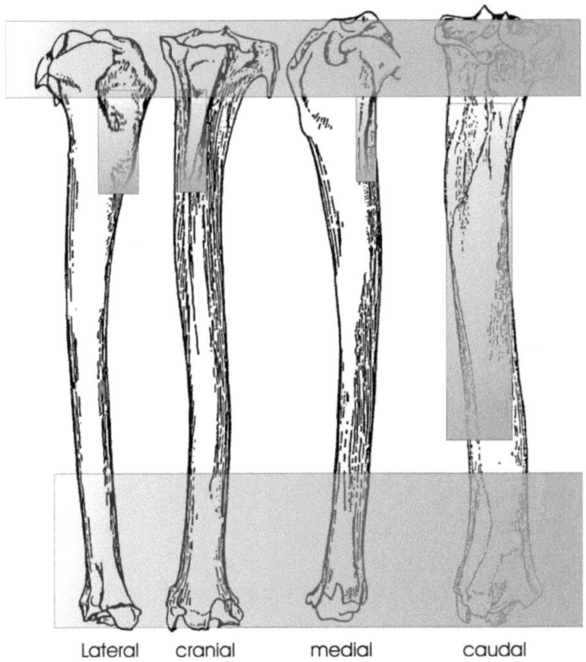

Figure 50. Anatomical locations of flesh scraps, that is, Cold Zones 5 and 6 (shaded areas), on tibiae from lion kills documented in Galana and Kulalu. Hot Zones 5 and 6 are the unshaded areas where flesh scraps were not observed. Cut marks found in these zones on bones from archaeological assemblages should be indicative of primary access to carcasses.

M. iliacus were also documented on the lesser trochanter. No flesh scraps occurred on the cranial aspect of the shaft. The caudal aspect of the shaft along the linea aspera showed pieces from the M. vastii intermedii.

Tibia. Most of the flesh scraps on tibiae were located on the caudal aspect (Figure 50), where pieces of the M. extensor digitalis ran from the proximal to distal ends. Scraps of M. flexor digitalis were observed on the cranial aspect of the proximal end, extending along the near-epiphyseal shaft on the tibial crest. The cranial aspect of the midshaft was largely devoid of flesh scraps. Scraps were regularly observed on the distal end.

Conclusions

Overall, the combined Maasai Mara and Kulalu and Galana samples provide the range of variability required to test secondary access scenarios. Specifically, the definition of anatomical zones where scraps of flesh are never present meshed with the distribution of cut marks will provide a sensitive indicator of the order of hominid access to flesh. This section concludes by describing "Hot Zones," which are areas on limb bones where flesh scraps never survived lion consumption in our study. The distributional analysis of cut marks in relation to these Hot Zones combined with the analysis of cut mark distribution by bone and bone section (Domínguez-Rodrigo, 1997a) will determine the order of hominid access to carcasses. The "Hot Zones" are defined below.

Hot Zone 1 (Figure 47). Hot Zone 1 is found on the cranial, caudal, and lateral (excepting the deltoid crest) aspects of the humerus midshaft. The medial aspect of the midshaft is not included because very small (2–4 mm) scraps (much smaller than what has been defined as scrap – see above) of the M. pronator teres adhere to the shaft surface. The zone begins at the lower third of the deltoid crest and finishes when the shaft begins to twist at the lateral epicondyle, halfway between the nutrient foramen and the olecranon fossa on the caudal aspect.

Hot Zone 2 (Figure 48). This zone spans the cranial aspect of the radius shaft from the proximal near-epiphysis at the insertion of the M. biceps brachii to the distal near-epiphysis. Although there are no muscle insertions on the cranial aspect of the distal near-epiphysis, the scraps emerging from the carpal area may survive lion consumption.

Hot Zone 3 (Figure 48). This zone encompasses the medial aspect of the radial midshaft in addition to part of the cranial curvature and the flat caudal surface opposite the ulna.

Hot Zone 4 (Figure 49). The cranial, medial, and lateral aspects of the femur midshaft define this zone. It runs from the smaller trochanter to the medial insertion of the M. gastrocnemi and the midpoint of the supracondylar fossa, but does not include its perimeter and the area occupied by the linea aspera.

Hot Zone 5 (Figure 50). This zone includes the cranial aspect of the tibial midshaft and begins proximally at the lower third of the tibial crest, just above the posterolateral nutrient foramen, and runs to the distal near-epiphysis. Like the radius, although no muscle insertions appear on the cranial aspect of the distal shaft, scraps adhere to the tarsal area.

Given these defined Hot Zones, hominid-carnivore access can be established in fossil faunas using the following procedure:

1. Quantify cut marks by bone type and section following Domínguez-Rodrigo (1997a).
2. Analyze cut mark distribution in relation to the Hot Zones defined above.
3. Identify those cut marks created through dismemberment. None of the cut marks that would be inflicted during the removal of flesh scraps could be mistaken for disarticulation. This is extremely important, as disarticulation is an unlikely activity for hominids focusing on marrow extraction rather than flesh removal. Disarticulation is mainly related, at least among modern hunter-gatherers, to carcass transport and distribution.

Cut Mark Frequencies in Hot Zones: Modern Butchery Experiments

In order to test the efficacy of the Hot Zone approach, cut mark distribution was analyzed from four middle-sized carcasses butchered by local Maasai (Domínguez-Rodrigo, 1997a) and four small-sized carcasses butchered by archaeology students from Complutense University (Barba and Domínguez-Rodrigo, in press). Carcasses were processed with stone tools made either in basalt (for the African sample of middle-sized carcasses) or quartzite (for the small carcasses processed at Complutense University). After disarticulation and filleting, bones were fragmented by placing them on anvils and hitting them with hammerstones. The resulting fragments were boiled in a solution of water and neutral detergent and cleaned. All fragments were then classified according to bone type and section (epiphyseal, near-epiphyseal, and midshaft). Each specimen was also classified as possessing either a Hot Zone or Cold Zones (i.e., those zones commonly preserving flesh scraps after lion consumption). In those cases in which a single specimen preserved both Hot Zones and Cold Zones, both were quantified provided that the zone with the smallest area represented no less than a quarter of the total length of the fragment. Therefore, NISP counts are not equivalent to zone counts, since single specimens frequently had both zones represented.

Analysis of the precise location of cut marks on each limb bone reveals that Hot

Zones are cut-marked in broadly similar frequencies to Cold Zones (Table 12 and Figure 51). For humeri, both zones are cut-marked similarly in small carcasses, while in medium carcasses Cold Zones show higher cut mark frequencies. This is partially accounted for by the M. pronator teres and M. biceps brachialis attachments. For small carcass femora, both zones again show similar cut mark frequencies while Hot Zones are cut-marked more frequently in medium carcasses. The Hot Zones of both carcass sizes are also more frequently cut-marked in radii. Most of the cut marks observed on the radius cluster in Hot Zone 2 (cranial aspect). Cut marks on the tibia are similarly represented in both zones.

Table 12. Distribution of cut mark frequencies per hot and cold zone at FLK Zinj

	Small carcasses		Large carcasses	
Experiments	Hot zone	Cold zone	Hot zone	Cold zone
Zone 1	11/40 (30)	29/40 (70)	16/58 (29)	42/58 (72)
Zone 2/3	14/20 (70)	6/20 (30)	25/33 (75.7)	6/33 (18)
Zone 4	12/24 (50)	12/24 (50)	13/21 (61)	8/21 (38)
Zone 5/6	16/40 (65)	14/40 (35)	12/26 (46)	14/26 (53.8)
FLK *Zinj*				
Zone 1	4/7 (57.1)	3/7 (42.9)	6/18 (33.3)	12/18 (66.7)
Zone 2/3	4/5 (80)	1/5 (20)	14/17 (82.4)	3/17 (17.6)
Zone 4	1/3 (33.3)	2/3 (66.7)	5/8 (62.5)	3/8 (37.5)
Zone 5/6	2/9 (22.2)	7/9 (77.8)	10/15 (66.7)	5/15 (33.3)

Numerator is the number of specimens bearing cut marks on any given zone. Denominator is the total number of cut-marked specimens including hot and cold zones. Numbers in parentheses are percentages.

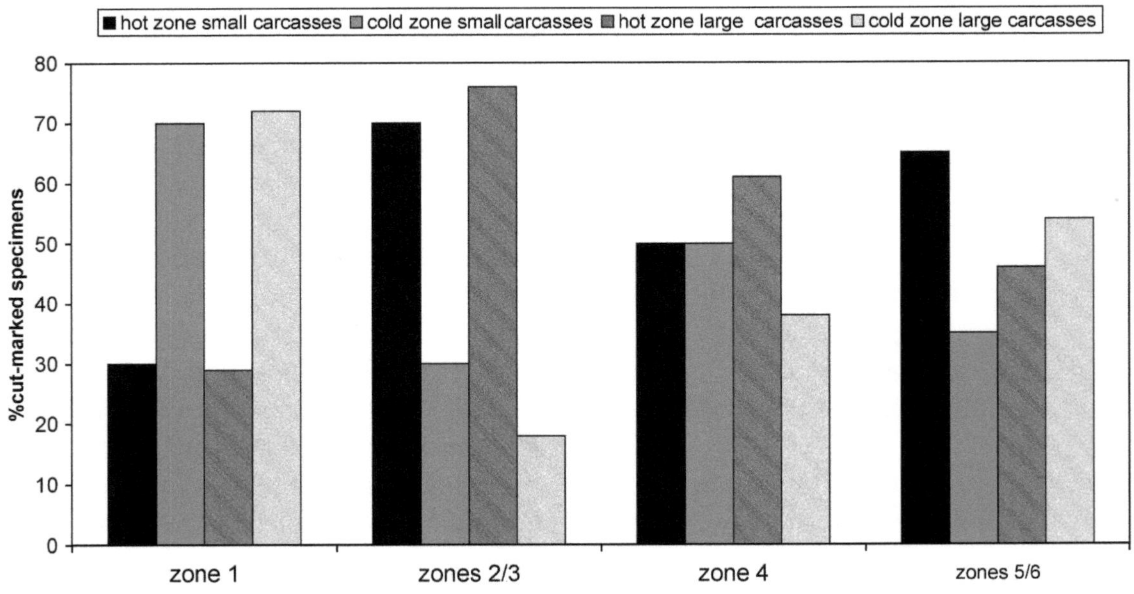

Figure 51. Distribution of mean frequencies of cut marks by carcass size on each limb bone in Hot and Cold Zones in experimental assemblages.

These data demonstrate that cut marks in experiments modeling primary access are found in Hot Zones at similar frequencies than in Cold Zones. Cut marks are never found in Hot Zones after the butchery of lion kills. Therefore, the Hot Zone approach provides a useful framework for interpreting cut marks in fossil assemblages. It also suggests that cut-marking is not necessarily more abundant on areas of muscle, ligament, or tendon attachment.

Cut Mark Frequencies in Hot Zones: The FLK *Zinj* Assemblage

In applying the Hot Zone approach to the FLK *Zinj* sample, it should be pointed out that the overall frequency of cut marks in Hot Zones relative to the modern butchery experiments is less important than simple presence/absence. Cut mark frequencies in both Hot Zones and Cold Zones are contingent on several variables, including fragmentation and the extent to which humans and carnivores are responsible for the accumulation of the fossil assemblage. Unlike the bone type and section approach, the Hot Zone approach is not necessarily dependent on cut mark percentages. Given that Hot Zones are scrap-free on felid prey carcasses, it is the presence/absence of cut marks (i.e., whether or not a high frequency of specimens are cut-marked) on zones that is the critical observation for determining hominid access to carcasses.

Table 12 and Figures 52 and 53 summarize the distribution of cut marks by zone. For small carcasses, cut marks on humeri at FLK *Zinj* occur more frequently in the Hot Zones relative to the experimental samples. Cut marks also cluster in the Hot Zones for radii, a very clear indicator of filleting. Cut marks on femora and tibiae appear more often on Cold Zones, although 40% of femora cut marks and 20% of tibiae cut marks appear on Hot Zones. Although the reason for differences in cut mark frequency by zone is unknown, the systematic presence of cut marks on Hot Zones at FLK *Zinj* strongly suggests that hominids were accessing carcasses before carnivore defleshing. Cut mark location for larger carcasses is even more conclusive. As Figure 53 demonstrates, the distribution of cut-marked specimens on Hot Zones strongly recalls the experimental sample of fully fleshed carcasses. Cut marks on humeri are abundant, and more

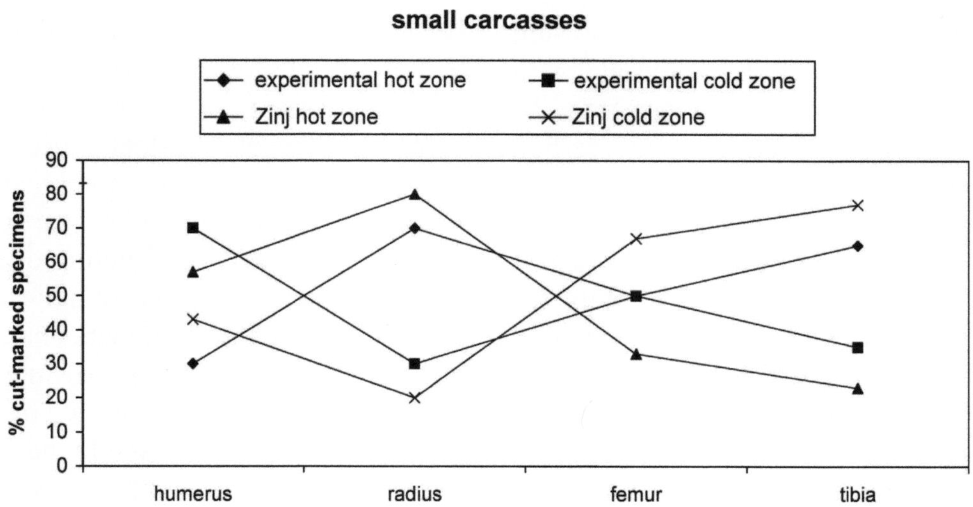

Figure 52. Distribution of cut mark frequencies by element in Hot and Cold Zones from small carcasses at FLK *Zinj*, compared with experimental data.

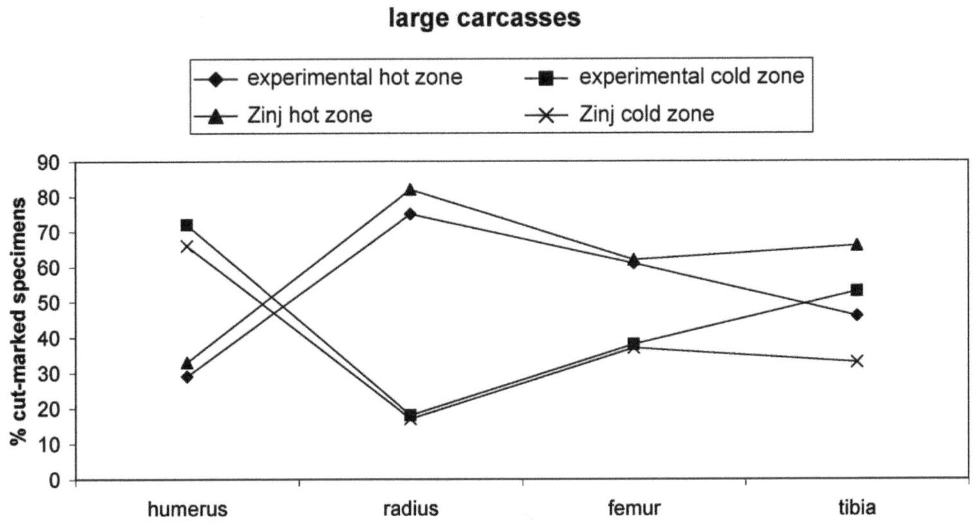

Figure 53. Distribution of cut mark frequencies by element in Hot and Cold Zones from medium carcasses at FLK *Zinj* compared with experimental data.

than 60% of the medium-sized cut-marked specimens from radii, femora, and tibiae have cut marks in Hot Zones.

The high frequencies of cut-marked Hot Zones on femora and tibiae are particularly revealing in that both lions and leopards have been observed to deflesh these elements promptly after prey capture. After evisceration, lions deflesh hindlimbs before any other limb elements (Hill, 1975; Blumenschine, 1986; Domínguez-Rodrigo, 1999a). Cavallo's (1998) observations of 16 size class 1 and 2 carcasses indicate that leopards first consume pelvic and hindlimb flesh, followed by the lower viscera, nasal pulp, facial and mandibular flesh, and upper viscera. If time permits, the remainder of the carcass is then consumed. Therefore, it is unlikely that hominid acquisition of carcasses occurred even subsequent to hurried felid consumption. In terms of the specific scenario of scavenging from leopard kills, the delay by leopards in the consumption of the forelimb flesh and marrow relative to lions is particularly noteworthy. If hominids were scavenging from leopard kills, they would be expected to impart cut marks on the front limb bones. However, the presence of cut-marked femora and tibiae indicate that these elements retained resources when hominids butchered them. In addition, the placement of cut marks on Hot Zones indicates that it was not flesh scraps but rather complete muscle masses that were being removed.

Figures 54 and 55 provide a clear picture of hominid butchery activities at FLK *Zinj*. Cut marks on humeri from larger carcasses appear mainly on the shaft, particularly the cranial, caudal, and medial aspects. Moreover, most cut marks cluster on the midshaft, an indicator of filleting. Cut marks also occur at the ends and reflect both disarticulation (especially on the distal end) and filleting. For femora, cut marks on small-sized carcasses largely reflect filleting while medium carcasses show a combination of filleting and dismemberment (particularly on the proximal end). The filleting of radii from all carcasses sizes is clearly indicated by the presence of abundant cut marks on the cranial midshafts. Disarticulation (using Nilssen's [2000] criteria) is also suggested by cut marks on proximal radii and distal humeri; no disarticulation cut marks were observed on distal radii. Most cut marks on tibiae from all carcass sizes can be attributed to filleting, and, like the radii, no disarticulation cut marks were found on the distal ends.

SMALL CARCASSES

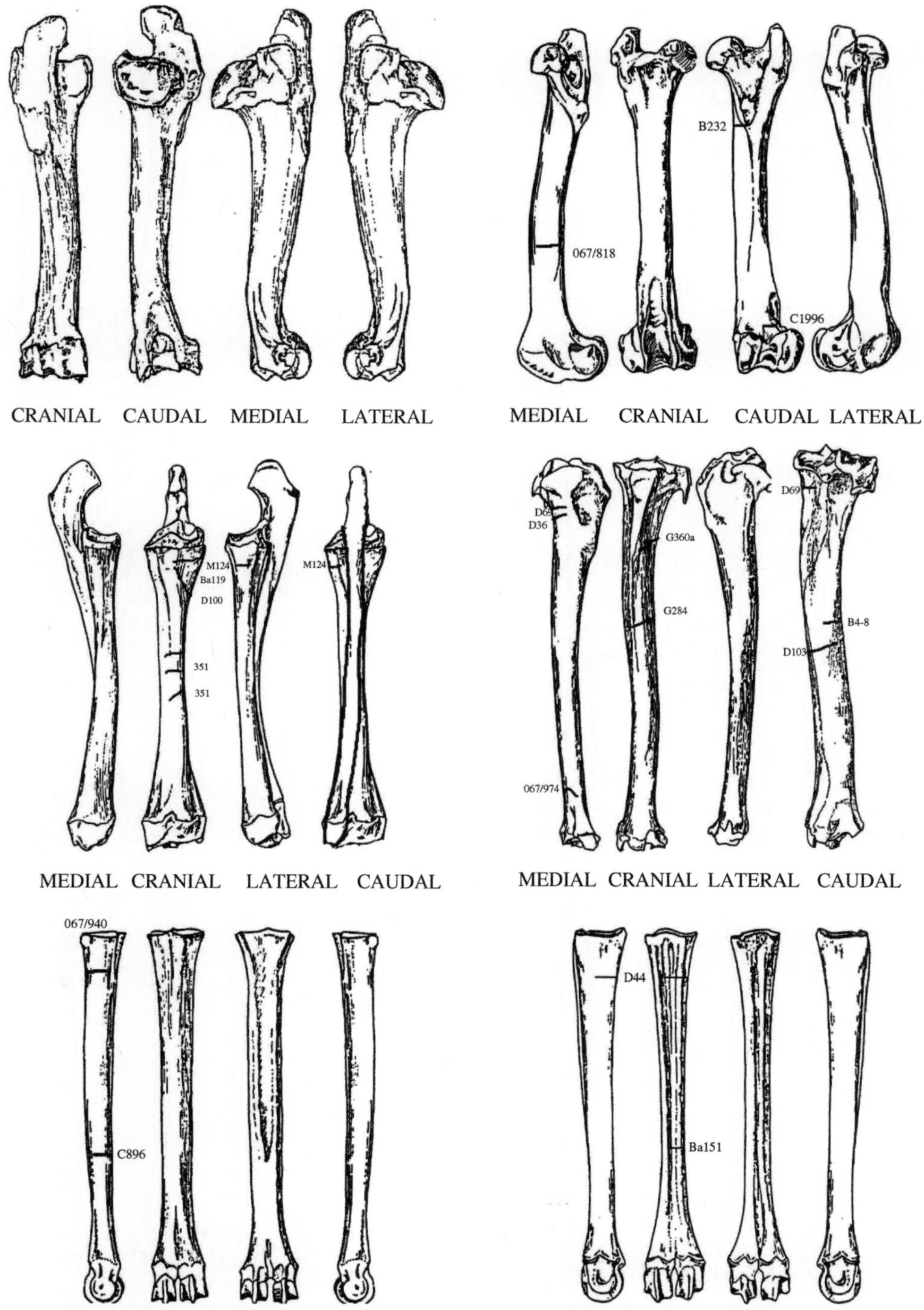

Figure 54. Anatomical distribution of cut marks on limb bones from small carcasses. Letters and numbers indicate specimen number from the FLK *Zinj* assemblage. Each mark represents isolated cut-marked specimens. Even if multiple cut marks were observed on several of these specimens, only one mark was drawn for the purpose of clearer identification of butchering activities. Only in cases of large specimens showing cut marks in locations distant from each other was more than one mark drawn. Specimens bearing cut marks which could not be anatomically placed were left out of this figure. Bones are redrawn from Pales and Lambert (1971).

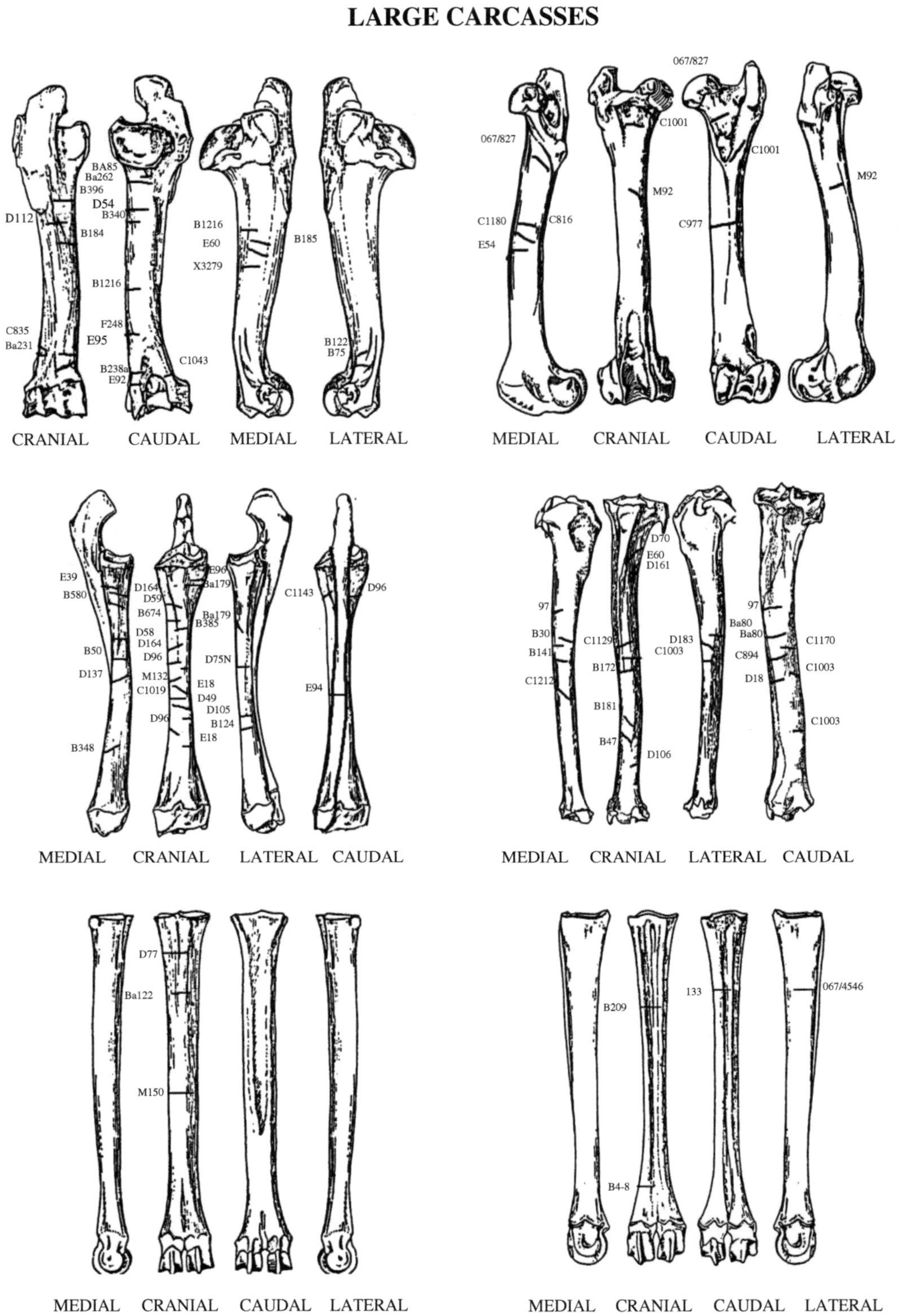

Figure 55. Anatomical distribution of cut marks on limb bones from large carcasses. Letters and numbers indicate specimen number from the FLK *Zinj* assemblage. Each mark represents isolated cut-marked specimens. Even if multiple cut marks were observed on several of these specimens, only one mark was drawn for the purpose of clearer identification of butchering activities. Only in cases of large specimens showing cut marks in locations distant from each other was more than one mark drawn. Specimens bearing cut marks which could not be anatomically placed were left out of this figure. Bones are redrawn from Pales and Lambert (1971).

Cut marks on metapodials are also not consistent with disarticulation, but rather with periosteum removal or skinning. One metatarsal in particular is a clear example of skinning, displaying concentrated slicing marks rather than extensive scraping marks (as would be expected in periosteum removal). Bunn and Kroll (1986) suggested that all cut marks appearing on metapodials likely reflect skinning activities. It is often assumed that skinning cut marks will be left only on distal sections of metapodial shafts. Nearly all scapula cut marks appear next to the glenoid cavity, a location that suggests disarticulation. One specimen from a small carcass preserves filleting cut marks on the caudal aspect of the blade. Filleting cut marks are also present on a pelvic specimen from a medium carcass, although most pelvic cut marks appear around the acetabulum, which is indicative of disarticulation. In sum, disarticulation and the filleting of large muscle masses are convincingly expressed in the location of cut marks on both small and medium carcasses from FLK *Zinj*. It therefore appears that carcass parts were brought on-site in an articulated state, which suggests further that carcasses were transported in large packages or possibly as complete carcass units (see below). Table 13 summarizes cut mark data for the axial skeleton. A total of 10 rib specimens (11% of the rib NISP counting only specimens >3 cm) displayed cut marks, most of which (n = 9) derive from medium-sized carcasses. Three vertebral specimens also displayed cut marks, either on the zygopophyses or the neural arch. Cut marks on vertebrae are not indicative of hominid access (Domínguez-Rodrigo, 1999a); however, the presence of cut-marked ribs can be very informative in this context. As described above, cut marks on mid- and distal sections of ribs likely result from primary access, as evisceration by felids often deletes distal ribs and leaves no edible resources within the rib cage itself (i.e., organs). Cut marks on the medial (ventral) aspect must therefore be the result of evisceration by hominids. Two out of the nine cut-marked ribs from medium carcasses appear on the ventral aspect near the proximal end. The remaining cut marks appear on the shaft, which indicate the removal of flesh that is always absent after felids consume their prey. The relative paucity of cut-marked specimens from the axial skeleton is related to the scarce presence of axial elements compared to cranial or limb elements. Cut marks on skull fragments are most abundant on mandibular pieces (Table 13). Most of these cut marks cluster on the lingual aspect, indicating the removal of the tongue (see also Bunn and Kroll [1986]). Coupled with the limb bone data, it is likely that hominids had primary and – if the preserved portion of the axial data is indicative of what was initially abandoned by hominids before carnivore ravaging – almost sole access to flesh at FLK *Zinj*.

Table 13. Cut mark frequencies for axial elements and for the skull, mandible, scapula, and pelvis for each carcass size at FLK Zinj

	Small	Large
Ribs	1/38 (2.6)	9/47 (19)
Vertebrae		
Cervical	0/9 (0)	2/17 (11.7)
Thoracic	1/9 (11.1)	0/15 (0)
Lumbar	0/6 (0)	1/2 (50)
Sacral	0/0	0/3 (0)
Caudal	0/2 (0)	0/2 (0)
Indet.	0/2 (0)	0/2 (0)
Skull	0/37 (0)	0/46 (0)
Mandible	1/17 (5.8)	6/89 (6.7)
Scapula	1/3 (33.3)	3/10 (30)
Pelvis	2/13 (15.3)	4/13 (30.8)

Numerator is total number of cut-marked specimens. Denominator is total number of specimens. Numbers in parentheses are percentages.

Conclusions

HOW MANY CARCASSES WERE CONSUMED BY HOMINIDS AT FLK *ZINJ*?

Bunn and Kroll (1986) report an MNI of over 29 bovids, 5 suids, and 5 equids at the site. These figures were obtained using dental pieces, many of which were derived from isolated teeth. There is thus a significant mismatch between the dental MNI and the MNIs represented by the postcranial skeleton (which are less than half). Klein (1986) suggests that isolated teeth appear in such abundance either because postdepositional sediment compaction destroyed once present maxillae and mandibles, or because they represent a background scatter unrelated to the rest of the assemblage. Given the possibility that isolated teeth may not correspond to the same taphonomic history as the rest of the FLK *Zinj* fauna, this study includes only those teeth still in maxillae and mandibles to calculate MNIs. Although this clearly provides an absolute minimum number of carcasses, it probably provides more realistic comparisons between cranial and postcranial representation. Complete carcass transport should be reflected by similar cranial (including mandibles) and postcranial MNIs.

When the bovid MNI is estimated using only fragments of mandibles with teeth, 21 individuals are represented (Tables 14 and 15). Seven small and thirteen medium bovid individuals could be identified to species and belong to *Connochaetes* sp., *Parmularius altidens,* and *Kobus sigmoidalis*. This yields MNI estimates much closer to the postcranial MNI inferred from Bunn and Kroll's (1986) MNEs and MAUs (not taking into account element side). Considering limb bones only, a minimum of six carcasses are necessary to account for the small-sized fragments (Table 16). This is very similar to the MNI derived from mandibular fragments (n = 7). When taking into account element side, the MNI for small carcasses is 7 based on metatarsals. Bunn and Kroll's (1986) data on medium-sized carcasses require an MNI of 10 to account for all limb bones, which also mirrors closely the MNI estimate based on mandibles. When element side is taken into account, a total of nine individuals are represented as indicated by the epiphyses of radii and humeri. These counts are very similar to those obtained using shaft fragments, and it therefore seems that epiphyses and mandibles provide similar MNI estimates. Overall, it can be argued that a minimum of 16 carcasses are represented by several anatomical regions by the FLK *Zinj* remains. If taxonomic criteria are taken into account, two more individuals bearing hominid surface modifications (one *Oryx* and one *Syncerus*) can be added, for a total of 18.

Only two undisputed hominid marks (a cut mark on a carpal and a percussion mark on a radius) were found among the suid remains, most of which are teeth. The few postcranial elements that were identified show heavily leached cortical surfaces not amenable to surface mark analysis. Perhaps the suid assemblage represents an accumulation independent of the bovid remains. The equid bones were not available for study in 2004–2005.

Table 14. Minimum number of elements (MNE) from small and large carcasses at FLK Zinj

	Carcasses	
	Small	Large
Skull	3	7
Mandible	7	14
Vertebra	27	43
Rib	38	47
Scapula	3	10
Pelvis	8	7
Humerus	6	14
Radius	6	16
Metacarpal	6	10
Femur	8	14
Tibia	12	19
Metatarsal	10	6

Table 15. Minimum number of individuals represented by cranial/mandibular elements

Antidorcas/antilopini	7
Connochaetes	2
Parmularius	4
Kobus	7
Syncerus	1

Table 16. Minimum number of elements (MNE) for each long limb bone at FLK Zinj based on ends, according to carcass size and to side

	Small carcasses				Large carcasses			
	Proximal		Distal		Proximal		Distal	
	Right	Left	Right	Left	Right	Left	Right	Left
Humerus	1	2	3	1	3	0	4	9
Radius	1	2	0	0	9	6	1	0
Metacarpal	2	4	2	5	4	4	1	3
Femur	1	0	2	1	2	1	2	1
Tibia	1	5	4	0	1	2	2	5
Metatarsal	5	7	4	2	5	2	2	2

Despite claims to the contrary (Capaldo, personal communication), most of the bovid individuals represented by teeth are adults, as argued by Bunn and Kroll (1986). Among small-sized carcasses, there is only one juvenile and one very old individual, together with three aged adults and one prime adult. A similar pattern is found in medium-sized animals (10 out 13 are prime adults). It is true that some of the isolated teeth belong to subadults (19 out of 98). However, it is impossible to ascertain whether their presence at the site is due to bone destruction or to different depositional events. The number of unfused epiphyses in the assemblage is very small (5 out of 59 complete epiphyses or 124 specimens). Therefore, prime adults seem to be most widely represented in the carcasses that make up the assemblage. In Bunn's (1982) estimates, prime adults comprised 14 individuals and subadults amounted to only six individuals when counting only the four best represented bovid taxa.

If it is inferred that carcasses were transported complete, axial remains at FLK *Zinj* are significantly underrepresented. No more than 4 individuals are represented by ribs and vertebrae. Despite this, these elements make up more than 46% of all skeletal elements represented at the site. Given that ribs are often highly fragmented and the fact that they are difficult to quantify through MNE estimates, their representation at FLK *Zinj* is quite remarkable. This study supports previous claims by Bunn and Kroll (1986) that axial elements make up one third of the assemblage and thereby rejects Capaldo's (1995, 1997) assertion that only one in every six elements was axial.

Carnivore ravaging at the FLK *Zinj* has been clearly documented by tooth mark frequencies and distribution on limb bones (see Chapter 5). The paucity of axial elements relative to the number of carcasses represented can be explained either by hominids selectively transporting heads (specifically, mandibles) and limb bones, or by the preferential deletion of these elements by scavenging hyenas. Capaldo (1995, 1998b) reports that between 90% and 100% of axial bones discarded by humans are completely deleted by hyenas. It is therefore likely that the original number of axial elements (prior to hyena ravaging) deposited on-site would be sufficient to account for all the carcasses represented at the site.

Cut mark frequencies show that evisceration, disarticulation, and defleshing were all systematically carried out on both small and medium carcasses at FLK *Zinj*. The number of axial elements surviving hyena ravaging also suggests a much greater original representation of these elements. These data, together with the tooth mark data presented in Chapter 5, clearly indicate that hominids regularly enjoyed primary and almost exclusive access to fully fleshed carcasses. Further, it seems that FLK *Zinj* served as a focal point on the landscape to which hominids regularly transported and exploited carcasses.

This behavioral reconstruction, based on data from the FLK *Zinj*, is not universally applicable to other sites at Olduvai, where stone tools and bones show a similar spatial association. The interpretation that hominids butchered carcasses at FLK *Zinj* and abandoned stone tools in those spots can only be supported because we were able to establish a link between tools and bones through the hominid-imparted modifications on the faunal remains accumulated. The following chapter is the first in a series of individual site analyses that show that this is not the case in most of the other Olduvai sites. FLK North 6 was long defended as one of the oldest large fauna butchery sites. One could not expect a more perfect spatial association of stone tools and bones than is found at this site; however, when applying the physical attribute taphonomic approach to the FLK North 6 faunal assemblage, the results are very different from those documented in the present and previous chapters.

7. A cautionary tale about early archaeological sites: a reanalysis of FLK North 6

M. DOMÍNGUEZ-RODRIGO, R. BARBA, I. DE LA TORRE, AND R. MORA

Introduction

As the previous chapters have shown, FLK Level *Zinj* represents the clearest example of a "central place," in which it is taphonomically possible to ascertain that most of the carcass remains were transported, processed, and exploited by hominids. For the past 25 years, this site has undergone taphonomic studies which have increased confidence that the other "Type C" sites in Olduvai Gorge could also represent similar "central places" or "Stone caches." As discussed in Chapter 1, "Type C" sites have faunal remains from several animals in association with stone tools; the fauna have been commonly assumed to be the result of hominid transport, given that the materials are more densely clustered compared with background concentrations, and the diversity of niches represented by the fauna are not observed in natural scatters in modern savannas (Potts, 1988). Most zooarchaeological research on the Olduvai sites tends to support the idea that "Type C" sites are hominid-made (Bunn, 1982; Potts, 1988; Rose and Marshall, 1996). More than 20 years ago, Binford (1981) suggested that these sites could represent palimpsests in which various agents intervened independently, hominids being the most marginal of them. While this hypothesis has been proven wrong for sites such as FLK *Zinj* (Chapters 5 and 6), it still needs to be tested on other archaeological sites. We believe that untested assumptions still make up an important part of widespread interpretations of early sites. We also believe that behavioral reconstructions obtained for specific sites cannot be uncritically applied to all sites in the same category.

Taphonomy has developed powerful analytical tools to understand ways that humans and other biotic agents accumulate and modify bones. Assuming hominids butchered carcasses as modern foragers do, we would expect that butchery may have left traces in the form of cut marks during dismembering and filleting, distributed accordingly (Binford, 1981; Domínguez-Rodrigo, 1997a; Nilssen, 2000). After bones are defleshed, marrow extraction is expected. As discussed in Chapter 3, breaking open bones for marrow leaves highly diagnostic traces in the form of percussion marks (Blumenschine and Selvaggio, 1988), percussion notches (Capaldo and Blumenschine, 1994), bone fragments with acute/obtuse oblique breakage planes (Pickering *et al.*, 2005b; Alcántara *et al.*, 2006), and a differential representation of shaft fragments from limb bones, in which those that show less than 50% of the shaft circumference are the most abundant (Bunn, 1983a; Marean *et al.*, 2004). From an archaeological point of view, a clear functional link between carcass exploitation and the types of lithic tools represented in the

M. Dominguez-Rodrigo, R. Barba, and C.P. Egeland, Deconstructing Olduvai: A Taphonomic Study of the Bed I Sites, 101–125.

assemblage would be expected. That is, if butchery is taphonomically inferred, a lack of stone flakes would be discordant with this interpretation, unless hominids were transporting their tools instead of discarding them.

FLK North 6 (i.e., level 6 of the FLK North site) typifies how traditional behavioral interpretations of "Type C" sites have been made. FLK North was found on an erosive front spanning approximately 8 m. It contained five fossil and artifact-bearing levels below the uppermost Bed I tuff, and some other levels in the lower part of Bed II. The upper sequence belonging to Bed II consisted of (from top to bottom): a stratum of brown silty clay (2 m), a sandy conglomerate (30 cm), a horizon of green silty clay (30 cm), a chocolate-brown clay (30–60 cm), a second layer of green silty clay (20–60 cm) which contained the partial skeleton of a *Deinotherium*, a layer of nodular limestone, another level of green silty clay (120 cm), and a coarse-grained tuff. Underlying these strata, Tuff IF (dated to 1.7 Ma) marked the top of Bed I. Below it, levels 1, 2, and 3 consisted of a brown silty clay horizon (90 cm), following level 4, which consisted of a dark-brown silty clay (40–120 cm). Level 5 was found under it in the form of greenish-yellow clay (20–50 cm). Level 6, the focus of this chapter, was underlying the whole sequence (for further information see Leakey, 1971: 62–63). All levels below Tuff IF contained fossils and stone tools. All of them are situated in a very low-energy depositional environment, suggesting a lack of physical processes of bone and stone tool transport (Hay, 1963, 1971, 1976).

The FLK North 6 deposit is approximately 60 cm thick, throughout which materials appear vertically distributed. Presented by Leakey (1971) as a butchery site – "Type B" following Isaac (1978) – in which hominids exploited a proboscidean carcass (Figure 56), this level includes a fairly complete elephant (*Elephas recki*) missing only the tusks and some bones from the feet. It was found in a small area (37 m^2) forming a cluster together with 130 stone artifacts (Figure 56). However, the complete elephant and the associated stone tools lie in the middle of a scatter of other bones belonging to smaller fauna. A functional relationship between the elephant and the stone tools was emphasized (Clark, 1970; Leakey, 1971; Isaac and Crader, 1981), relegating the other few bones found in the site to a background assemblage of bones scattered over the landscape.

Curiously, an alternative view was presented by Binford (1981), who claimed that the artifacts at the site were inversely correlated with the elephant remains and that they were spatially associated with the background scatter of non-elephant bones. Contrary to Binford, these non-elephant bones were argued by Potts (1988: 147) to exhibit none of the criteria that can be attributed to landscape assemblages, death sites, or palimpsests. For Potts, this level represents a repeated transport of bones and artifacts by hominids to the same spot in the landscape. This assemblage includes at least 35 individuals (Bunn, 1982) and may represent a sort of "central place" created after the elephant was found, using the spot as a focus of activity to which carcasses were repeatedly transported or, alternatively, the elephant butchery episode may be separated from the background scatter, which might represent "a temporarily distinct Type C occurrence" (Bunn, 1982: 129). Cut marks found both on the elephant bones and on the background fauna would functionally link the fauna to the stone tools and could be used as an argument to support any of these interpretations.

It is clear that interpretations of FLK North 6 are as varied and abundant as the researchers who have studied the site. The present work intends to augment earlier taphonomic approaches (e.g., Potts, 1988) with the new methodological approach that has been described in Chapter 3. The quantitative inclusion of surface damage as well as a more thorough analysis of skeletal part representation and bone completeness has led us to conclude that hominids were not in fact

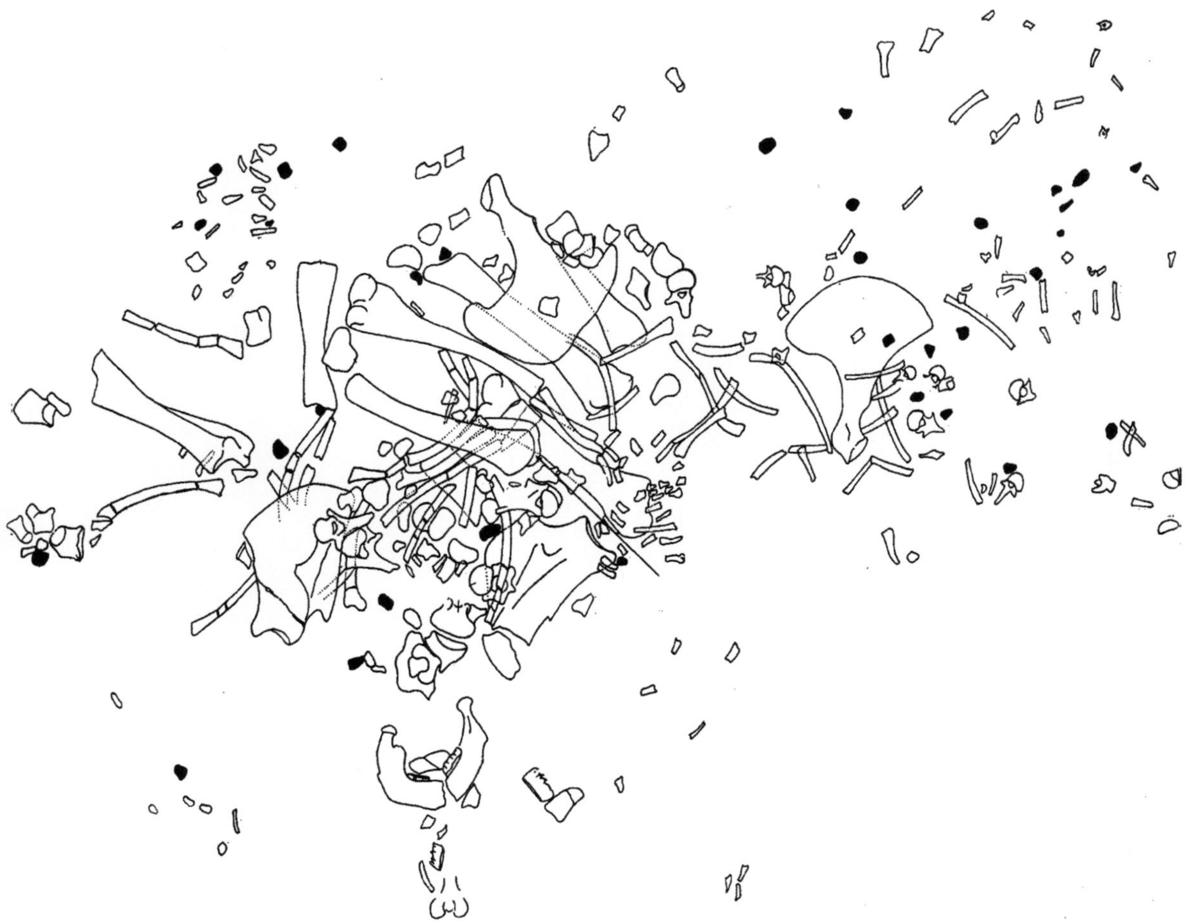

Figure 56. Drawing showing the spatial distribution of the bones and stone tools of the FLK North 6. (Redrawn from Leakey [1971: 65].) Black objects are stone tools.

involved in the formation of the FLK North 6 fauna. It will also be shown that Type C sites can substantially differ in taphonomic histories and behavioral meaning. Finally, we conclude that only a complete archaeological analysis, which combines the study of fauna and lithics, can shed light on the functional association of bone remains and stone tools, and thereby on the behavioral meaning of any given site.

The FLK North 6 Faunal Assemblage

Previous estimates of the faunal composition of FLK North 6 vary by researcher. Bunn (1982) provided an estimate of 211 elephant bones and 411 bones belonging to other taxa. This was slightly larger than the figure provided by Leakey (1971). Potts (1988) reported a total of 357 proboscidean bones (146 more than Bunn) and 383 bones belonging to other taxa. In our study, we analyzed a total of 616 bones, 329 of which were proboscidean (Table 17). Many recent fractures were observed, especially on elephant rib fragments, which may account for the difference in numbers reported among researchers.

Although most proboscidean bones belong to a fairly complete individual, Leakey (1971: 64) reported the presence of "a few bones belonging to a second elephant." Bunn (1982: 130) realized that the high number of proboscidean bones was inflated because "many of the bones, especially ribs, were broken into smaller pieces in getting them out of the

Table 17. NISP of the faunal assemblage at FLK North 6

	Bovid	Suid	Proboscidean	Rhinoceros	Hippo	Giraffid	Carnivore
Horn	2						
Skull	9	1	6				1
Teeth	45	35	9				1
Mandible	8	2	2				
Vertebra	27	8	63	2			1
Rib	13		174				
Pelvis	8	1	4				
Scapula	2	1	2				1
Humerus	8	1	2				
Radius-ulna	18		8			1	1
Carpals	9		14		1		
Metacarpal	7		0	1			
Femur	3		2				1
Tibia-fibula	13	1	4				1
Tarsals	10		11				
Metatarsal	7		4	1	1		
Patella	6		1				
Phalanges	12		14				1
Other	6		0			2	
Indeterminate	7		9				
Total	220	50	329	4	2	3	8

ground or subsequent to that time. The unfused condition of most limbs, ribs and vertebrae represents another inflation factor." Most of the "background" fauna are bovid and suid remains. A look at Table 17 reveals that most suid remains, belonging to at least eight individuals, are teeth. Only bovid skeletons are represented more completely. In this case, the number of individuals accounted for by dentition is 20 (Bunn identified 21), much more abundant than the minimum number of individuals (MNI) estimated from postcranial bones, which is less than half that amount (eight individuals based on radius counts, which is the most abundant long bone element). One individual each of a rhinoceros, hippo, giraffe, and carnivore are represented by very few bones, probably also resulting from the underlying landscape background.

The differential representation of bone remains from bovids and suids suggests different taphonomic histories for each faunal group. The fact that most suids are represented by teeth and by immature individuals indicates a different reason for accumulation from bovids, which are represented more evenly in skeletal terms (although axials are very scarce), displaying a mixed set of adults and juveniles. Alternatively, it could be thought that the differential preservation of elements in both groups can be attributed to different age profiles, if it were not that subadult bovids are better preserved than subadult suids. However, the few numbers of bovid individuals represented by postcranial elements would suggest that at least half of the bovid bones share a similar taphonomic history to the suid bones. There are some bovid taxa (e.g., hippotragini) only represented by teeth; following the logic of Leakey's argument, these would also represent a background scatter. The number of bones (≤4 NISP) for the rhino, the hippo, and the giraffe suggests that they may also belong to a background scatter. More information about landscape bone scatters at Olduvai is needed (see later) to better understand the assemblage.

The bovid taxa present at the site are *Antidorcas recki* (MNI = 4), *Tragelaphus strepsiceros* (MNI = 5), *Connochaetes* sp. (MNI = 3), *Synceros acoelotus* (MNI = 1), *Parmularius altidens* (MNI = 4), Hippotragini (MNI = 1), Tragelaphini size 3 (MNI = 1), and Alcelaphini size 3A (MNI = 1) (as also identified by Bunn, 1982).

Most of the non-elephant postcranial bones preserved belong to the appendicular skeleton. Estimates for the number of identified specimens (NISP) and minimum number of elements (MNE) for bovids (the most abundant group) are provided in Tables 18 and 19, respectively.

Table 18. NISP of each bovid long limb bone portion according to element type at FLK North 6

	NISP
Humerus complete	3
Humerus prox.	0
Humerus shaft	2
Humerus dist.	1
Humerus dist. + shaft	2
Radius complete	4
Radius prox.	3
Radius shaft	1
Radius dist.	1
Radius dist. + shaft	4
Metacarpal complete	2
Metacarpal prox.	1
Metacarpal shaft	0
Metacarpal dist.	0
Metacarpal prox. + shaft	2
Metacarpal dist. + shaft	2
Femur complete	2
Femur prox.	1
Femur shaft	0
Femur dist.	0
Femur prox. + shaft	0
Tibia complete	1
Tibia prox.	0
Tibia shaft	3
Tibia dist.	4
Tbia prox. + shaft	1
Tibia dist. + shaft	2
Metatarsal complete	2
Metatarsal prox.	0
Metatarsal shaft	1
Metatarsal dist.	1
Metatarsal prox. + shaft	1
Metatarsal dist. + shaft	2

Table 19. MNE of bovid limb bones represented at FLK North 6

	Small carcasses		Large carcasses	
	Right	Left	Right	Left
Humerus	1	1	3	3
Radius	1	1	2	5
Metacarpal	1	1	1	3
Femur	1	0	1	1
Tibia	1	0	1	3
Metatarsal	1	0	1	1

As can be observed, half of the bones have been preserved in complete state. Nobody, hominids or carnivores, has broken them: one of the most surprising taphonomic features in the assemblage is the virtual lack of shaft fragments, despite complete recovery of bone fragments by the excavators. The NISP:MNE ratio is 1.0 for humeri, 1.4 for radii, 1.1 for metacarpal, 1.0 for femora, 2.6 for tibiae, and 2.3 for metatarsal. This means that the degree of fragmentation is very low and there is parity in NISP/MNE estimates for several elements. If MNI values had to be established on postcranial elements alone, bovids would be represented by six middle-sized carcasses and two small ones. Carcass size classes follow Bunn (1982): small carcasses refer to Bunn's sizes 1 and 2 (e.g., gazelle and impala, respectively). Middle-sized carcases are Bunn's sizes 3A and 3B (e.g., lesser kudu and zebra, respectively). Large carcasses are Bunn's sizes 4–6 (buffalo to elephant).

Taphonomic Analysis of FLK North 6

If the fauna and stone tools from FLK North 6 are functionally associated, then from a taphonomic point of view, the preservation of traces of butchery and demarrowing processes would be expected. Also, if carnivores were involved in the accumulation and modification of the faunal assemblage, one would expect signatures of their intervention. Next follows an analysis of bone breakage and modification processes at the site, using the analytical criteria outlined in Chapter 3.

BONE BREAKAGE AT FLK NORTH 6

When the taphonomic analytical frameworks of the "physical attribute" approach were applied to the study of the FLK North 6 fauna, it was discovered that not a single percussion mark was present, despite the very good preservation of all the limb bone cortical surfaces. Only three specimens had poor cortical preservation, five of them showed moderate cortical preservation and the remainder showed excellent bone surfaces, on which any cortical modification could have been preserved. One third of the long bone specimens showed inconspicuous biochemical marks as were described for FLK *Zinj*, although the frequency of those marks in the latter site is close to 75% (Domínguez-Rodrigo and Barba, 2006). Therefore, comparatively, bone surfaces at the FLK North 6 are even better than those of FLK *Zinj*. If no percussion marks were documented in the former, it is because hominids did not seem to break any bones and not because of preservation or taphonomic biases. This is further supported by the fact that not a single percussion notch was reported from the site; in fact, there were no notches in any of the appendicular specimens at the site, not even notches that could be attributed to the action of carnivores.

When analyzing the distribution of Bunn's (1982) circumference types on long bones, another contrasting picture emerges regarding the comparative experimental frameworks. Figure 57 shows how most assemblages created by humans, carnivores, or the interaction of both have a predominant frequency of Type 1 circumference specimens. FLK North 6 is mainly composed of Type 3 specimens, that is, specimens which preserved 100% of the shaft section. This explains why most of the appendicular bones seem to be fairly complete (Figure 58). Only six specimens offered breakage planes which could be measured and the angle range was 82°–105°, which falls completely within the range for breakage caused by static loading and therefore can confidently be attributed to the action of carnivores. Most limb bones thus seem to have been preserved complete, and in those which were missing sections, this was because carnivores

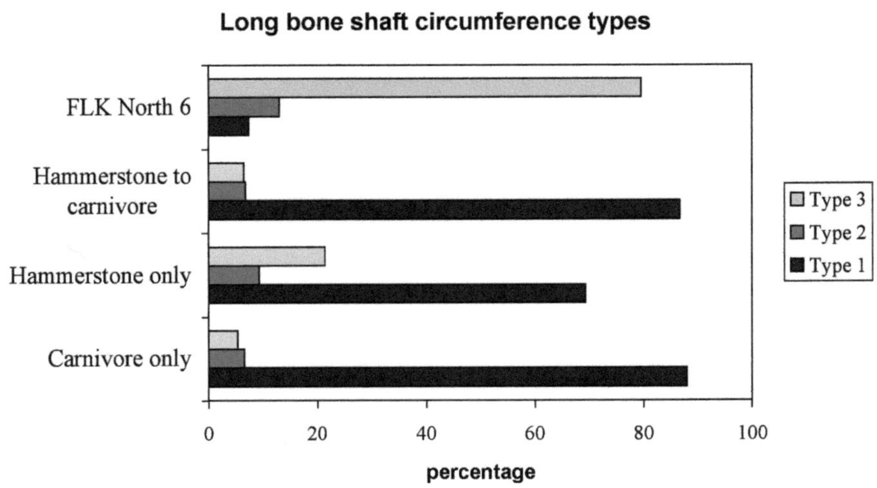

Figure 57. Distribution (in percentages) of the different types of bovid long bone circumference types (Bunn, 1982) in experimental assemblages and at FLK North 6. (Data for experimental assemblages are from Marean and Spencer [1991] and Marean *et al.* [2004].)

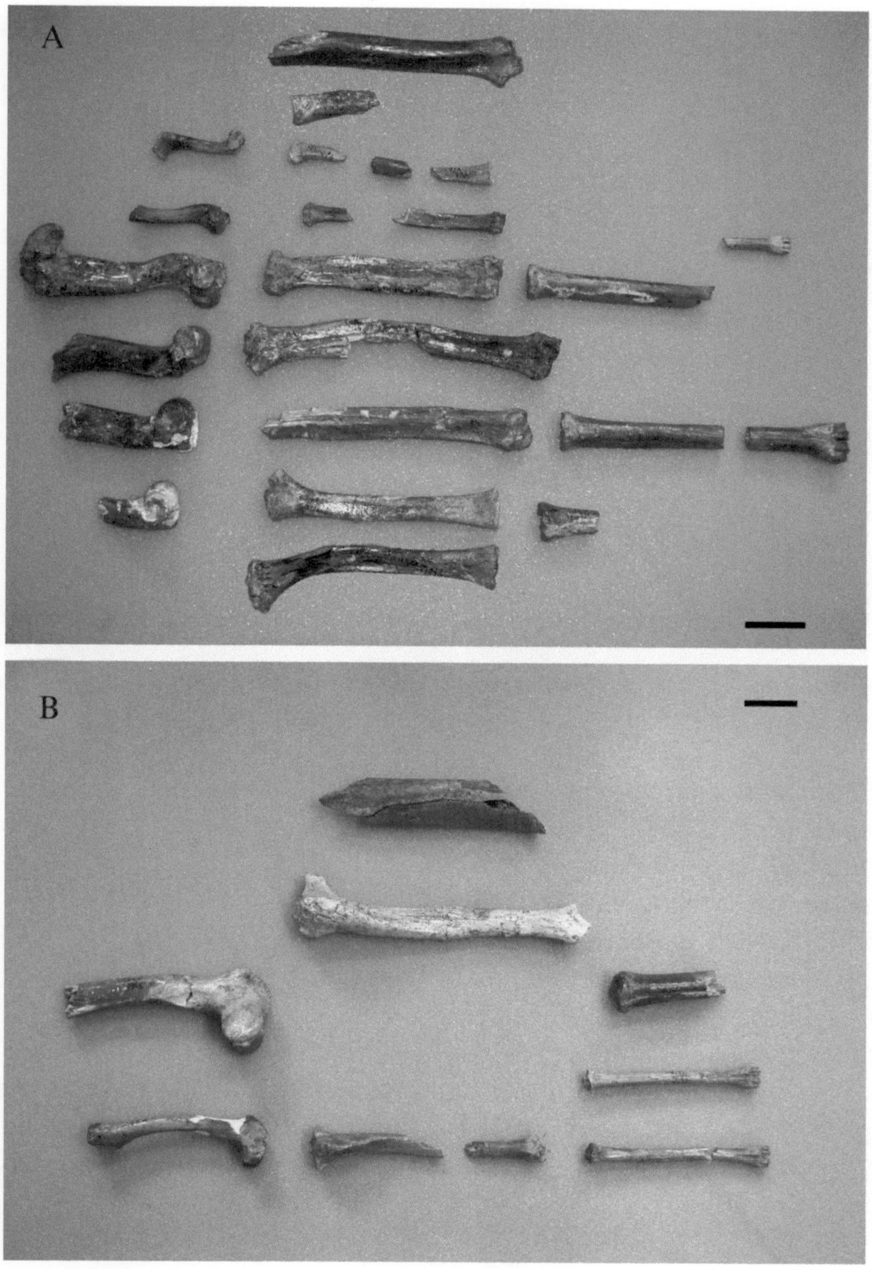

Figure 58. Aspects of most of the postcranial non-elephant elements at the FLK North 6. (A) Front limbs (radii are the most abundant element type) (scale = 5 cm) and (B) hind limbs. Some elements present a twisted aspect due to sediment compaction but most of them are fairly complete (scale = 3 cm). Most shaft fractures are diagenetic. Fractures on proximal ends of humeri and tibiae are fresh.

had ravaged them (Figure 59). Bone sections have been preserved evenly for metapodials, femur, and radius. The proximal epiphyses of humerus and tibia are underrepresented. Most of the bone shafts are complete, which suggests that carnivore ravaging was not very intense, although it is conspicuously shown in some elements such as on humeri (see Figure 59).

Tooth marks are scarce. Only eight specimens of the long bone sample and one phalanx are tooth-marked in the whole assemblage

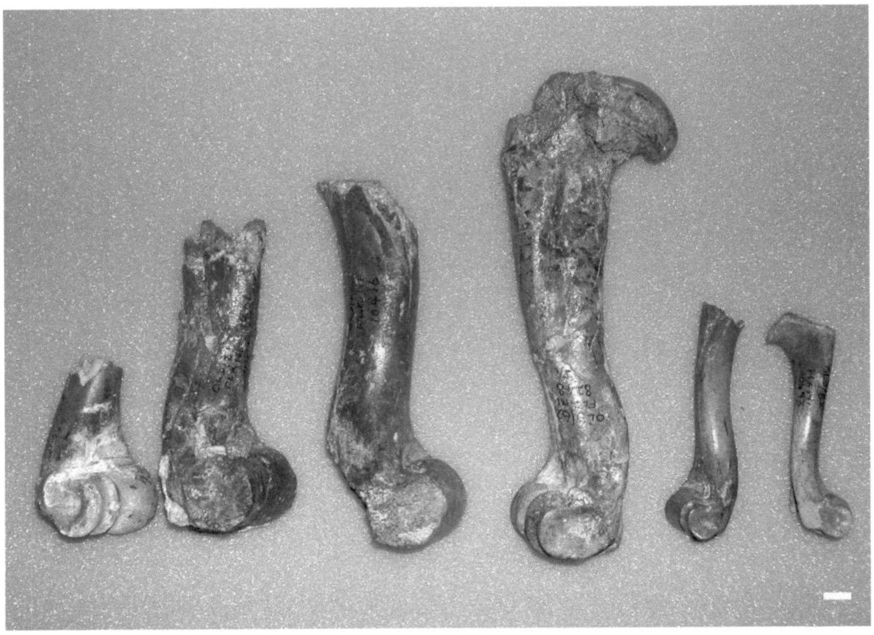

Figure 59. The entire humerus collection, showing intense furrowing on proximal epiphyses, which could be caused by hyaenids or felids (scale = 1 cm).

(Table 20). Tooth marks are preferentially distributed on upper limb bones. Humeri and femora show a high rate of tooth-marking, whereas tibiae are less tooth-marked. Radii show no marks at all, whereas 40% of ulnae do. Tooth marks on ulnae are restricted to the olecranon. The high rate of tooth marks on humeri and femora cluster around proximal ends can be explained either by hyaenid ravaging or by felid consumption of the carcass. No metapodial bears any tooth mark. Almost 75% of all the tooth marks occur on ends, either on epiphyseal or near-epiphyseal sections.

CARCASS DEFLESHING AT FLK NORTH 6

As mentioned earlier, the spatial association of stone tools and the faunal remains at FLK North 6, especially those belonging to the elephant, prompted the assumption of a functional

Table 20. Frequencies of tooth-marked specimens by bone portion on long limb bones

		Epiphysis	Near-epiphysis	Midshaft
Humerus	50 (4/8)	1	2	1
Radius	0 (0/13)	0	0	0
Ulna	40 (2/5)	2	0	0
Metacarpal	0 (0/7)	0	0	0
Femur	33.3 (1/3)	1	0	0
Tibia	9 (1/11)	0	0	1
Metatarsal	0 (0/7)	0	0	0
Indeterminate	0 (0/5)	0	0	0
Total	13.5 (8/59)	4	2	2

Percentage of tooth-marked specimens for each element is given first, followed by parentheses containing first the number of tooth-marked specimens, then the total number of specimens link between them. To reinforce this link, it has been claimed that cut marks have been found on the elephant bones (Bunn, 1982; Potts, 1988) and on several bovid bones (Bunn, 1982).

A number of modern cut-marked assemblages are available, derived from both ethnoarchaeological and experimental studies for comparative purposes with archaeological assemblages. Importantly, cut-mark frequencies on limb bone specimens are remarkably consistent across these varied data sets, ranging between 15% and 30% of all specimens recovered after hammerstone bone breakage (e.g., Bunn, 1982, in identifiable bones only; Lupo and O'Connell, 2002). Although cut-mark frequencies in Domínguez-Rodrigo's (1997a) experimental butchery data set are beyond the range cited earlier, we note that those experiments were aimed explicitly at "complete" flesh removal. Substantial scraps of flesh still adhere to defleshed bones after a more "typical" episode of human butchery. Most commonly (e.g., in ethnoarchaeological situations with hunter-gatherers), butchers do not bother to remove these adhering scraps. Indeed, in the one experiment by Domínguez-Rodrigo (1997a) in which "typical" butchery was carried out (H1/S1/2d), cut-mark frequencies, at 29%, agree with the "typical" expected values.

It was claimed that cut marks at FLK North 6 occurred on several rib fragments and on the distal end of the right humerus. Bunn (1982: 131) also mentioned that "many of the elephant bones have a very thick layer of glue which, while preserving the bones, poses an obvious obstruction in the search for marks." Bunn (1982) also argued that cut marks on bovid bones could be found on the following specimens: a left distal radius plus shaft of *P. altidens*, a left proximal radius plus shaft of a size 3 alcelaphini, a metatarsal shaft of a size 3A bovid, a proximal rib plus shaft of a size 3A bovid, and right third metacarpal of *Hippopotamus gorgops*. All these specimens were found, since their tags described the presence of cut marks on them, except the hippo metacarpal which could not be located. Following procedures described in Chapter 3, each specimen was analyzed first by naked eye and then by hand lens, and all the purported cut marks were molded in high-resolution silicone and inspected under electronic microscope at Complutense University in Madrid. Only the elephant humeri were not molded since the marks were covered in glue.

A contextual approach to bone surface analysis revealed in most cases the presence of microabrasion on the surface of bones (Figures 60 and 61). This microabrasion is

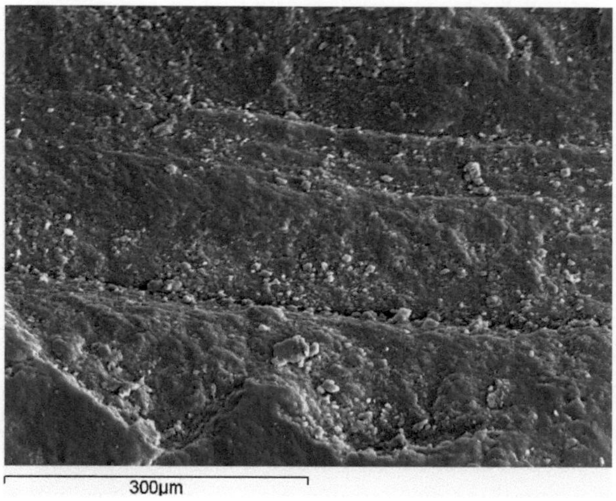

Figure 60. SEM photograph (200X) of the purported cut mark found on the proboscidean rib. The photograph shows the middle of the striation and exhibits a broad groove. The sides of the groove are shaped by a step at a right angle, produced by the disappearance of the upper cortical layers, rather than by an inclined wall as would be expected in the V-shaped groove of a cut mark. There is a virtual lack of internal microstriations in part of the mark. On the upper section only, a few microstriations can be observed (some of them discontinuous), which have a winding outline. This is documented in experimentally trampled bones, since sediment particles undergo a friction process that makes them roll briefly in the groove they create, which is reflected in a microscopically winding outline. This is opposite of microstriations created in cut marks by the use of stone tools, which are more straight.

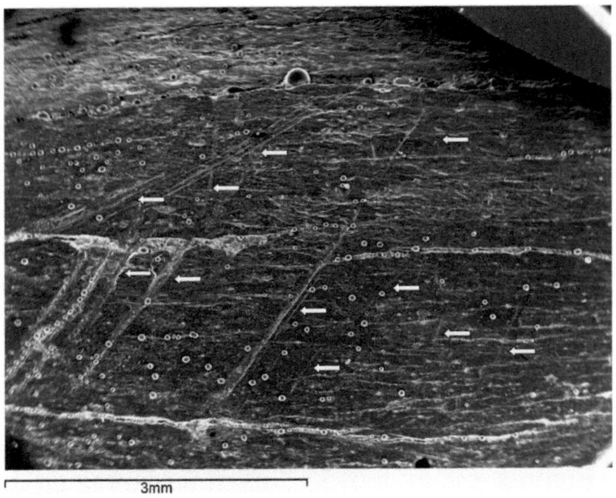

Figure 61. SEM photograph (20X) of various marks on a bovid bone (indicated by arrows), previously reported as cut marks. All of them show very broad grooves that are shallow and contain multiple microstriations. The only exception in the long V-shaped mark in the middle, which if observed at higher magnification, shows a winding shape rather than a straight one which suggests that it was not created with a stone tool. Parallel to this mark on the right side and on a large area, multiple, shallower marks appear suggesting that the whole set was created by trampling rather than by a single stone tool stroke. On the left side, several broad marks with multiple orientations and lengths indicate a similar cause.

caused by the friction of the bone surface against the ground by trampling. The presence of this feature on each of the specimens bearing a purported cut mark is cautionary. A couple of the specimens also showed multiple striations in patches with diverse orientation. One of the most surprising features is that most of the purported cut marks were extremely shallow. These abrasive marks on proboscidean elements, easily confounded with cut marks, were clearly described by Haynes (1991). They also have been reported in Olduvai Bed II assemblages (Monahan, 1996, see Appendix).

The elephant bones were carefully inspected and it was found that a substantial amount of abrasion occurred on some of them. Both humeri were scrutinized for cut marks and only sporadic patches of very shallow striations indistinguishable from abrasion were located. The fact that they appeared on parts of the bone that, by its form, may have rested directly on the ground, further reinforced such an inference. Several rib specimens also showed inconspicuous abrasion. We found not a single elephant or bovid bone specimen with any mark with a deep "V"-shaped profile, nor with the presence of secondary classical cutmark features (Hertzian cones, shoulder effects). When observed through the SEM, some of the purported marks did not even have microstriations. The proboscidean rib specimen on which all previous researchers coincided in identifying a cut mark (Bunn, 1982; Potts, 1988; Shipman, n.d.) was also carefully analyzed (Figure 60). The mark is a very broad groove, shallow and containing multiple microstriations. Its morphology is reminiscent of cut marks but it has all the diagnostic features that occur in marks caused by trampling (Andrews and Cook, 1985). Most trampling marks show a wide shallow profile and the grooves usually have no shoulder effects. There is not a single criterion that could be used to isolate this mark and the remainder of purported cut marks from those caused by trampling and sediment abrasion.

Given that most marks are extremely shallow, surrounded by other inconspicuous striations caused by abrasion, and that 38% of the bone specimens show traces of these inconspicuous microstriations (Figure 61), it is very likely that, despite previous claims to the contrary, the "indisputable" cut marks can be best explained as abrasion marks.

Review of the Arguments Supporting an Anthropogenic Origin of FLK North 6

Potts (1988) has convincingly argued that the dense bone concentrations at Olduvai sites could never be mistaken with background landscape scatters because they are denser in various ways than natural bone clusters occurring in modern savannas. Bunn (1982: 132) summarized this by stating that "if the non-elephant remains represent the local natural background, then the background is denser by several orders of magnitude than in the closest modern analogue environment." Both Potts and Bunn base their arguments on modern actualistic work on bone dispersal and concentration in modern savannas. Analysis of bone distribution in the Amboseli National Park in six different habitats, including bush, open woodland, dense woodland, plains, swamp edge, and dry lake bed, led Behrensmeyer and Boaz (1980), Behrensmeyer (1983), and Hill and Behrensmeyer (1984) to conclude that the densities of bones in these environments were extremely low when compared to archaeological sites. The number of bone remains, the MNI represented, the skeletal part profiles and the ecological niche range exhibited by the taxa represented were the arguments used by Potts (1988) to claim the anthropogenic origin of the Olduvai sites. Bed I sites at Olduvai exhibited a higher number of bone remains, with the number of individuals represented ranging between 20 and 40, in contrast with the maximum number of individuals represented in modern landscape bone accumulations per unit sampled (MNI = 3). In these modern landscape accumulations, sampling units spanned 1,500 m^2 in contrast with the < 300 m^2 in most Olduvai sites, which makes the difference even bigger. In addition, most clusters of bone in the savanna landscape are composed of cranial and axial elements, whereas at Olduvai sites, appendicular bones are predominant, as in purported transported assemblages. Also, the mix of taxa documented in Olduvai sites cannot be observed in modern bone clusters in savanna ecosystems.

This analytical approach requires complete dependence on the adequacy of ecological sampling and the use of skeletal part profiles, taxonomic range, and density of individuals and bones accumulated at specific *loci*. Inferences about hominid intervention at any given site are secondarily derived from these comparisons instead of directly observed on bones. These frameworks were obtained through observation. Although Potts also used bone surface modifications in his study, he did so as a secondary argument for one reason: at that time no referential backgrounds existed to understand the meaning of frequencies and anatomical distributions of cut marks, tooth marks, and percussion marks. Potts' referential frameworks were created from ecologically dependent studies in which no standarized quantification of damage to bones in the form of fragmentation and bone surface modifications existed. For this reason, it would be appropriate to term this historic taphonomic approach as an "ecological" or "ecotaphonomic" variant of the traditional "paleontological approach" described in Chapter 3. The next stage of research supplied much-needed analogs, which were mainly obtained though experimentation rather than observation; researchers were concerned with studying the physical modifications of bones imparted by each agent, since these would leave traces that could be directly observed on bones. This is what we have called the "physical attribute" approach in Chapter 3.

The ecological arguments used to support the anthropogenic origin of Olduvai sites made perfect sense from an actualistic point of view. However, there are frequent mismatches between actualistic observations and the application of the resulting referential frameworks to the past. The first thing that should be emphasized is that most of the individuals documented in some Olduvai sites are represented by teeth; sometimes only one tooth. It could be argued that modern bone counts in savannas could not possibly include all the teeth lying on the ground, since the vegetation cover would have made them hard to see. Only the major bones would have been spotted. It could also be argued that the referential framework created by Behrensmeyer (1983),

Behrensmeyer and Boaz (1980), and Hill and Behrensmeyer (1984) was insufficient to incorporate all the variability in long-term ecosystem dynamics. The low diversity of ecological niches represented by the taxa documented in each bone scatter/cluster could be due to the studies having been carried out in a phase of ecological stability. Recently, Behrensmeyer (2005) and Cutler *et al.* (1999) have documented the existence of an ecological change in the Amboseli ecosystem, in which fluctuating conditions prompted changes in ecological niches that could be reflected in bone scatters in the same environments as in those studied in the 1980s, if they were studied today.

Furthermore, the studies published by Behrensmeyer and Boaz (1980) and Behrensmeyer (1983) may be incomplete in documenting the exceptional ecological *loci* that propitiate bone concentration. For instance, they reported a maximum number of three individuals in the densest transect that they discovered. Applying a similar size transect, Domínguez-Rodrigo (1993) documented a higher number of individuals (MNI = 5) around a pond in the Serengeti. He also recorded more than 12 buffalo carcass remains in less than 1,500 m^2 around another pond in Kulalu (unpublished), dead as a result of a drought. Capaldo and Peters (1995) also document even bigger clusters of individuals created by natural processes in absence of transport. This indicates that there are specific environments in which bones accumulate in higher densities than in the rest of the background landscape. These places are usually located in certain areas of alluvial environments and are highly likely to be missed if a random sampling strategy to measure background bone density is applied, given that they are so spatially restricted.

Further support to this assertion could be found in the study of pre-Pliocene paleontological sites, where the intervention of hominids can be, naturally, excluded. A selection of some Miocene sites in Spain could be used as an example. Intensive taphonomic studies were undertaken at four Miocene sites by Alcalá (1994): La Gloria 4 (Lower Ruscinian), Milagros (Upper Turolian), Puente Minero (Lower Turolian), and La Roma 2 (Upper Vallesian). These sites were excavated in areas of 30 m^2 (maximum) and bones were retrieved from single individual layers spanning 15 cm (Puente Minero) to 40 cm (Milagros) deep. The number of individuals represented at these sites is similarly high to Olduvai sites (Figure 62), despite the smaller excavation areas and the less vertically dispersed materials when compared with some Olduvai sites such as FLK

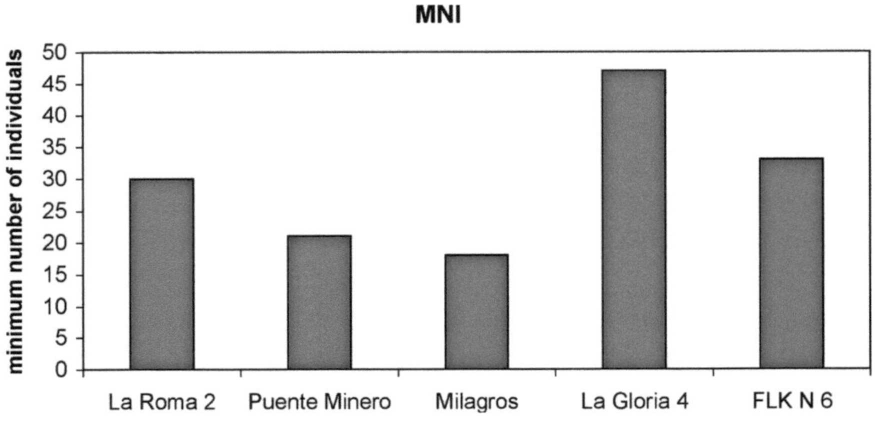

Figure 62. Minimum number of individuals represented at the four Miocene sites discussed in the text. (Data from Alcalá [1994].)

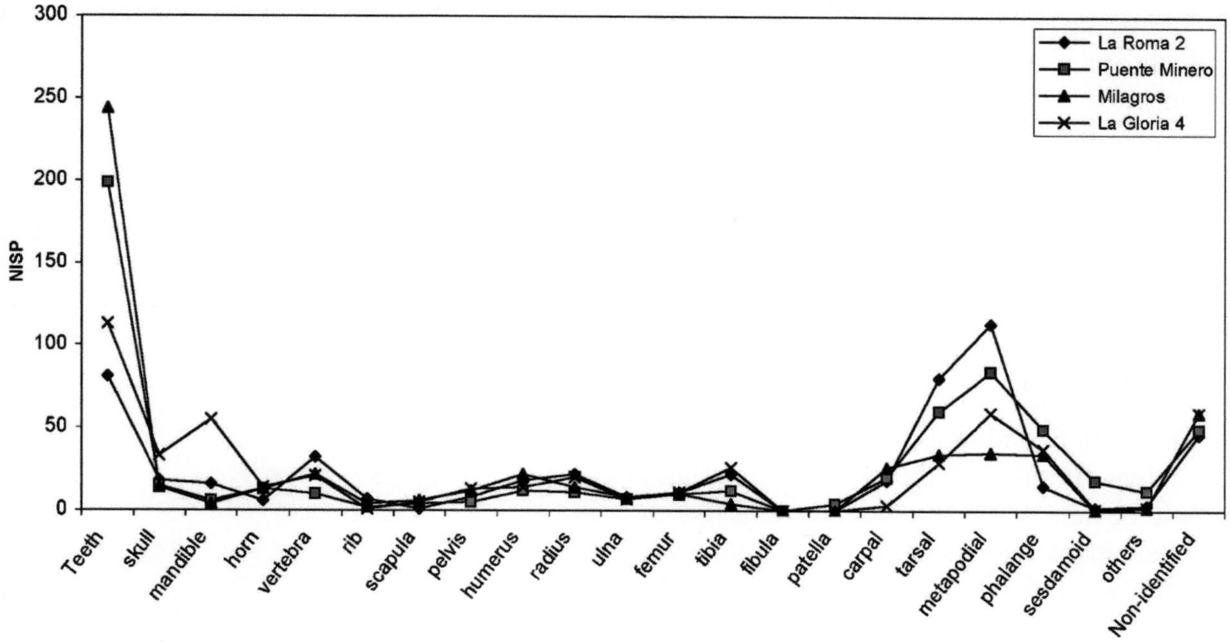

Figure 63. Skeletal part distribution and frequencies for the four Miocene sites discussed in the text. (Data from Alcalá [1994].)

North 6. Not surprisingly, as can be seen in Figure 63, most of the individuals are represented by teeth, which in some sites may be as high as close to half of the element sample. Skeletal part profiles at these sites are dominated by teeth and limb bones. Axials are underrepresented. All sites show a similar pattern of element representation, in which teeth and lower limb bones are the most abundant bones. What is more surprising is that the comparability with Olduvai lies not only in the similarity in number of individuals represented and in skeletal part profiles dominated by skull/teeth elements and limb bones, but also in the number of specimens composing each assemblage. In this case, the similarity with FLK North 6 is even closer in bone composition: Miocene assemblages comprise between 500 and 600 bone fragments. The ecological diversity of the taxa represented is also important. Open landscape species (such as rhino, *Protoryx*, and gazelle) are mixed with closed vegetation taxa (such as *Tragoportax*, deer, bear, and elephant) and species adapted to mixed habitats: *Hipparion* and giraffes (Alcalá, 1994). Therefore, the presence of potentially nonoverlapping ecological niches by the fauna represented at these Miocene sites makes comparisons with Olduvai even more appealing.

Most of the individuals represented at the Miocene sites mentioned are prime adults. This should also serve as a caution against claims made for anthropogenic origin of prime-adult accumulations. Alcalá (1994) interprets these sites as the result of diverse processes. Only Milagros (the least dense of the sites) is associated with a high-energy sedimentological environment and could be the result of fluvial transport. The other three sites, however, are found in low-energy depositional environments (lacustrine alluvial/floodplains) and are interpreted as the result of natural faunal concentrations due to resources and the formation of taphocenoses due to natural deaths and repeated predation. This assertion was also based on the fact that little carnivore damage was found on the bones. These sites correspond to the category of sites that has been classified as “deposits of concentration” or

"Fossil-Lagerstätten" (Seilacher *et al.*, 1985) which are very common in alluvial environments.

Therefore, this shows that there are natural processes that can generate the density of bone concentrations (including total number of specimens, number of MNI, skeletal parts represented, and ecological variation of the taxa represented) observed at most Olduvai sites. Given the absence of taphonomic evidence for hominid manipulation of faunal remains at FLK North 6, one should consider the alternative hypothesis that the bone accumulation could be natural. Bone concentrations like the non-elephant one reported at the site could naturally occur as background landscape scatters, especially in circumstances in which carnivores may have been active in participating in the accumulation.

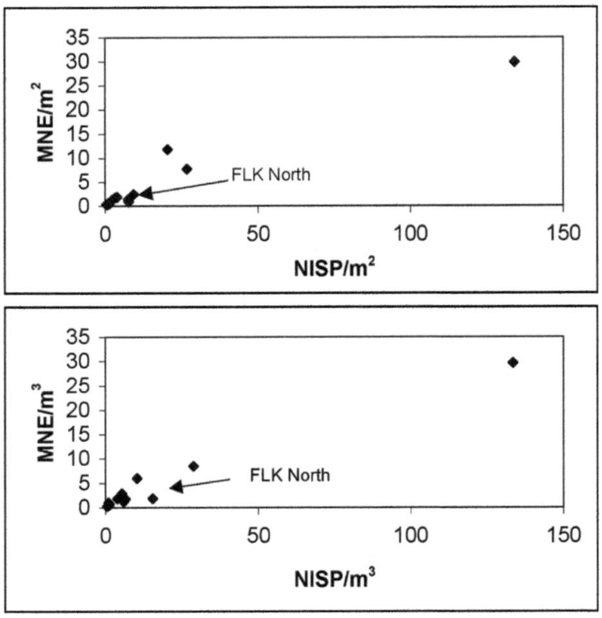

Figure 64. Density of non-elephant bone remains at FLK North 6, compared to the densities found in each of the landscape facets targeted at the Landscape Archaeology project at Olduvai (Cushing, 2002). Densities are shown as ratios per square meter and per cubic meter.

To start addressing this scenario, it is necessary to know the density of bone scatters at Olduvai during the Early Pleistocene. The ongoing landscape archaeology project at Olduvai (Blumenschine and Peters, 1998) has obtained a wealth of information about bone scatters on the landscape during that time. Cushing's (2002) study of these scatters reveals that bone density on the landscape during Lowermost Bed II times varied according to facet (including habitat type) and that natural bone clusters on the landscape were sometimes fairly dense. Cushing's analysis of bone densities (from sampling units no larger than 4 m^2) on geological facets representing various parts of the alluvial and lacustrine plains, including deltaic, channel, and interfluve systems, shows that bone density – measured either by the $NISP/m^2:MNE/m^2$ ratio or the $NISP/m^3:MNE/m^3$ ratio (Figure 64) – is similar to that reported for the FLK North 6 non-elephant bone assemblage (Table 21). Furthermore, several natural landscape bone scatters are denser than the FLK North 6 assemblage. Cushing's taphonomic analysis of these bone assemblages stressed the virtual lack of hominid intervention therein, given the overwhelming majority of specimens bearing no traces of hominid involvement. Of all the trenches analyzed by Cushing, those belonging to Olduvai Bed I Middle Bed – the same stratigraphic interval in which archaeological sites such as FLK *Zinj* are situated – show the highest densities of bone scatters. The channel and interfluve facets show densities that are similar to FLK North 6 (the latter) or of a magnitude of several times higher (the former).

This suggests that there were *loci* on the landscape in which natural bone accumulations were substantially high and in which no hominid participation was required. This is better observed when the data are split by single-test trenches rather than by the mean figure, which was derived from lumping all the trenches in any given facet together, obscuring variation and the occurrence of specific preferential depositional places. For example, there is an extremely high density of bones reported for Trenches 57 and 65 in Bed I.

Table 21. Degrees of bone density in several environmental facets in the Early Pleistocene paleolandscapes at Olduvai Cushing (2002), and density of non-elephant bones at FLK North 6

	NISP/m²	MNE/m²	NISP/m³	MNE/m³	MNI/m²	MNI/m³
Alluvial plain	0.58	0.27	0.57	0.29	0.23	0.23
Upper lacustrine plain	3.88	1.88	4	1.87	0.72	0.72
Middle lacustrine plain	2.33	1.33	1.37	0.78	0.58	0.34
Long K	9.31	2.43	6.63	1.74	1.81	1.29
HWK Complex	26.81	7.68	28.88	8.44	2.21	2.37
FLK Complex	3.37	1.8	5.44	2.9	0.62	1.01
Lower lacustrine plain	0.37	0.37	1	1	0.25	0.77
Northeastern lacustrine plain	1.31	0.5	1.1	0.42	0.31	0.31
Southwestern lacustrine plain	7.66	1.41	6	1.24	0.66	0.65
Delta	20.37	11.75	10.36	5.98	3.37	1.72
Channel	134.1	29.8	133.5	29.68	4.5	4.48
Interfluve	1.22	0.63	0.95	0.54	0.38	0.34
FLK North 6 (non-elephant)	7.7	0.91	15.5	1.83	0.94	1.7

Over 100 specimens per cubic meter and close to 30 elements from larger mammals per cubic meter were documented. This is far denser than the non-elephant bone assemblage at FLK North 6. Cushing noticed the disparity of bone density among trenches and stated that assemblages like those from Trench 57 are "somewhat of an anomaly on the landscape" (Cushing, 2002: 92). It could be thought that such densities might have been the result of fluvial systems favoring transport and accumulation, given that Trench 57 is situated in a channel facet. However, the representation of bones from most of the skeleton, including those easily transported (such as ribs and vertebrae) together with those that are heavier (long bones) argues against that, despite water-worn specimens probably introduced into the assemblage by water flow. The low frequencies of polishing and abrasion characteristic of bone transport and the presence of bones from various animal sizes (including micromammals) support a local origin for the bone accumulation.

Therefore, Cushing's study documents that the Early Pleistocene landscape at Olduvai contained bone scatters that varied according to the ecological dynamics of each habitat but which, in some fluvial and lacustrine environments, were similar and even far higher than the bone density reported for the non-elephant assemblage at FLK North 6. Bone density expressed in number of specimens or elements cannot, therefore, be used as a discriminating agent of hominid authorship. Skeletal part profiles alone are also controversial, since the abundance of appendicular elements in several of the trenches (particularly in Trench 57, for instance) argues against the idea that all the assemblages with high representation of limb elements are transported assemblages due to attrition. Low competition areas, in which bone transport and destruction is reduced, would enable the preservation of limb elements (Blumenschine, 1989; Domínguez-Rodrigo, 1996). The number of individuals per landscape facet reported in the Olduvai landscape project range from 0.23 to 3.37 per square meter (Table 21). The FLK North 6 individual density per square meter is 0.94. The density in the landscape project, expressed in MNI/m³, ranges from 0.23 to 4.48. The non-elephant density at FLK North 6 is 1.7 (Table 21). Therefore, the number of individuals represented cannot be used as a criterion to support hominid authorship of the transport of the bone assemblage at FLK North 6. Taxonomic diversity is also a criterion that

Table 22. Categories and raw material types represented in the FLK North 6 lithic assemblage (de la Torre, 2006)

	Quartz		Lava		Total	
	n	%	n	%	n	%
Test cores	–	–	–	–	–	–
Cores	1	0.9	3	14.3	4	3.1
Retouched pieces	–	–	–	–	1	0.8
Hammerstones	–	–	11	52.4	11	8.6
Anvils	9	8.3	1	4.8	10	7.8
Flakes	6	5.6	3	14.3	9	7
Chips	35	32.4	–	–	35	27.1
Possible flake fragments	17	15.7	1	4.8	18	14
Angular fragments	9	8.3	–	–	9	7
Battered fragments	30	27.8	–	–	30	23.3
Unmodified stones	–	–	2	9.5	2	1.6
Total	107	100	21	100	130*	100

* Including all raw material types.

must be subjected to critique. Some of the landscape facets show >25 MNI from large mammals including most bovid tribes. More specifically, individual trenches (e.g., Trenches 57–65) also show taxonomic diversity (Cushing, 2002).

Bone density expressed in number of remains, number of elements, differential skeletal parts, and taxonomic diversity have previously been used as arguments to claim hominid authorship of several Olduvai Bed I sites (Potts, 1988). We have shown that either using data from Miocene sites (Alcalá, 1994) or from bone scatter densities at Olduvai Bed I and Bed II (Cushing, 2002), similar bone density and representation to FLK North 6 (despite substantially smaller excavation areas) actually suggests that the latter may be a natural site in which hominids – given the lack of taphonomic evidence – did not participate.

The Analysis of the Lithic Assemblage

The FLK North 6 lithic assemblage has been studied by Leakey (1971), Potts (1988), and Ludwig (1999). It consists of a small assemblage of 130 artifacts, which Leakey (1971: 64) interpreted as the result of debitage activities. Potts' (1988: 388) reanalysis reduced the original number of identified choppers to only one and doubled the number of manuports identified by Leakey (1971: 64). De la Torre's (2006) study has yielded fairly different results (Table 22). One of the most remarkable features of the assemblage is that, despite being a small assemblage, no refitting sets have been found. A total of 52 pieces (40.3%) out of the 130 artifacts show battering damage. Taking into account that most of the assemblage is composed of chips smaller than 20 mm, in which it is not possible to identify such traces, it is logical to think that the representation of battered pieces is even higher. If we express this feature per amount of lithic mass represented, 14.4 kg out of the 16.5 kg of raw material represented bear traces of battering. This should be indicative of the functional nature of the assemblage.

In FLK North 6, knapping activities are represented by the presence of four cores suggesting debitage processes. It is significant that the only quartz core also shows traces of battering (Figure 65). Two of the cores are bifacial choppers and the other two were created by reduction through abrupt flaking; one of them is

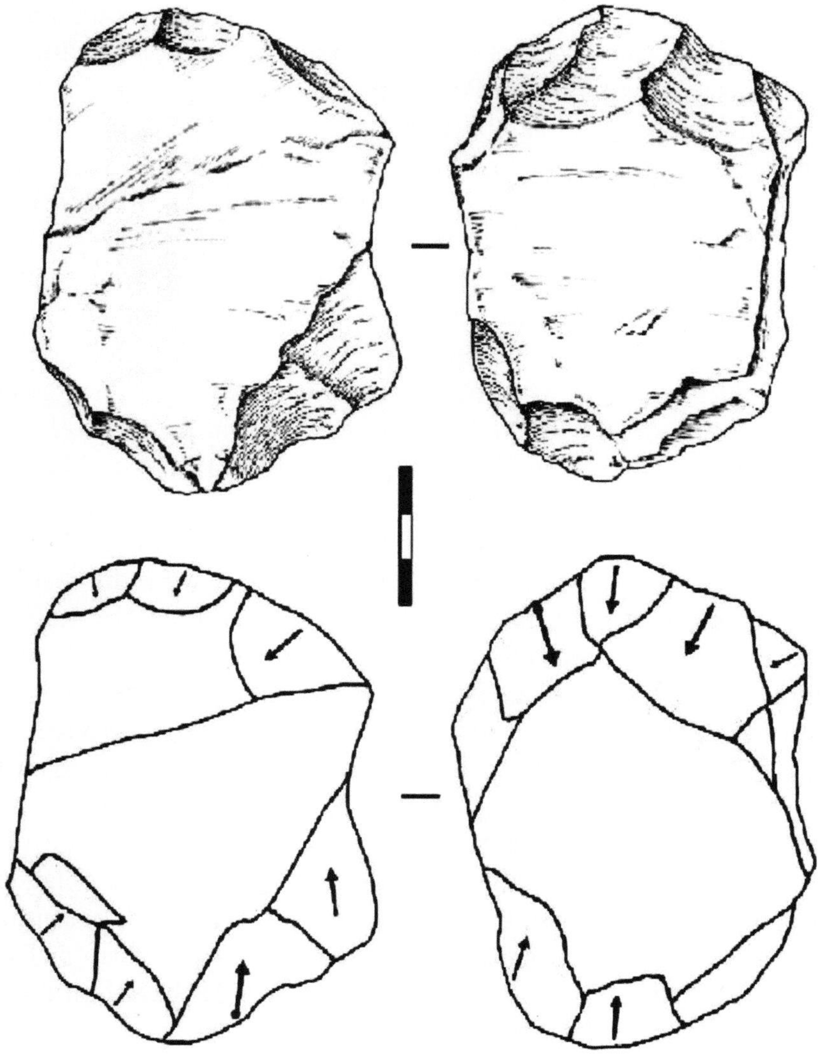

Figure 65. Bifacial partial quartz core (chopper) from FLK North Level 6. (Drawn by Noemi Morán.)

unifacial and the other one is bifacial. None of them show long sequences of reduction; rather, they indicate moderate exploitation. The identification of the flaking products is, however, more problematic. The technomorphological features that identify intentional (i.e., anthropogenic) flakes are hard to observe in the FLK North 6 flakes previously identified as such. Very often, it is difficult to differentiate these features from those produced by other activities. As a matter of fact, most of the previously identified six quartz flakes cannot be clearly defended to be intentional. If it is hard to link possible flakes with any debitage system, it is even more difficult to link most of the purported knapping waste (flake fragments, angular fragments, and chips) to any intentional flaking. A large number of the purported flake fragments, irregular, and fairly small (< 2 mm), even if lacking any traces of battering, could also be the result of processes other than debitage, making the percentages of products of intentional flaking even smaller.

In de la Torre's (2006) study, a total of ten hammerstones have been identified, more than double the number reported by Leakey (1971).

They are high-quality rounded lava cobbles. In several of them, the battering traces are not very conspicuous. This may explain why Potts (1988) identified some of them as manuports. Anvils are the most relevant category of the artifacts represented at FLK North 6. As can be seen in Table 22, these artifacts make up only 7.8% of the assemblage, but they account for almost half the weight of the lithic assemblage (7.6 kg). Together with hammerstones, they represent most of the raw material mass. The surfaces of the anvils are modified as a result of use. They show numerous scars with stepped morphologies around their entire periphery. This clearly shows that several small fragments were detached from anvils, due to percussion during battering activities. This is where the purported "debitage" fits the functional interpretation resulting from our study. Many of the purported flake fragments lack the typical features that would identify them as intentional flakes. They lack bulbs, dorsal faces with previous scars, butts, and so on. Alternatively, the battering traces in some of them, together with the repetition of specific morphologies and sizes observed in the scars on the anvils, would indicate that most of these fragments could be the result of battering activities generated through percussion on the anvils (see Mora and de la Torre, 2006).

With the exception of a few pieces, the entire FLK North 6 lithic assemblage could be interpreted as the result of percussion rather than flaking. This would suggest that if hominids had intentionally butchered the carcasses in the faunal assemblage, they could not have done so with the lithic artifacts represented at the site. Taking into account that a minimum average of 15 or 20 quartz or basalt flakes, respectively, is necessary to butcher a middle-sized carcass (Domínguez-Rodrigo *et al.*, n.d.), a high number of intentional flakes would be expected, given the number of carcasses represented at the site. The intensive battering reported at the site could have been the result of bone breakage, but not a single trace of this activity was found on the bones (see earlier). The lithic assemblage could be the result of hominids using that spot for any activity either before or after the bones were deposited; the former is more likely, given that some pieces were found under the proboscidean elements.

Hominids seem to have repeatedly used certain locations, as suggested by the existence of multiple levels in most Bed I archaeological sites. Mary Leakey (1971) suggested that several of those levels, including most of the 22 levels that make up the FLK site, were the result of natural scatters on the landscape and the presence of occasional lithic artifacts may have been the result of hominids repeatedly using the landscape for specific activities. This could also explain the presence of lithic artifacts at FLK North 6. To further support this claim, we summarize here some of the features of the small lithic assemblages overlying this archaeological level, belonging to levels 5, 4, and 3 (i.e., FLK North 5, FLK North 4, and FLK North 3) (based on de la Torre's [2006] study).

At FLK North 5, Leakey (1971) documented 151 lithic artifacts, among which "debitage" was predominant and choppers were the most common tool type. Leakey also identified 29 manuports. When Ludwig (1999) reanalyzed the assemblage, he studied only 111 artifacts. De la Torre (2006) identified 163 pieces, including the purported manuports. Table 23 shows that manuports are extremely well-represented, only overrepresented by the possible flake fragments. As was the case at FLK *Zinj* (de la Torre and Mora, 2006), we believe that several of the lava "manuports" could be natural rocks. This would also be supported by the fact that in most of the modified pieces, quartz is the predominant type of raw material. If the so-called manuports are set aside, one feature that is surprising is the high number of cores with respect to flakes. Most of the cores are made on lava, whereas most of the purported "debitage" (namely, small fragments) is made of quartz (Table 24). This is an important contradiction that demands explanation. Quartz knapping can be demonstrated by the presence of cores, refitting pieces, and even one retouched flake. However, the presence of artifacts functionally related to percussion is also important. The quartz anvils make up 3.7 kg out of the 5 kg of raw material represented. Thus, it may seem reasonable to interpret a substantial part of the purported flake fragments as the result of battering activities and, therefore, as being positives detached from anvils as, is the case at FLK North 6.

Table 23. Categories and raw material types represented in the FLK North 5 lithic assemblage (de la Torre, 2006)

	Quartz		Lava		Total	
	n	%	n	%	n	%
Test cores	–	–	1	1.1	1	0.6
Cores	2	2.6	13	14.9	15	9.2
Retouched pieces	1	1.3	–	–	1	0.6
Hammerstones	–	–	22	25.2	22	13
Anvils	9	11.8	–	–	9	5.5
Flakes	2	2.6	3	3.4	5	3.1
Chips	2	2.6	2	2.3	4	2.5
Possible flake fragments	36	47.4	12	13.8	48	29.4
Angular fragments	22	28.9	1	1.1	23	14.1
Battered fragments	–	–	2	2.3	2	1.2
Unmodified stones	2	2.6	31	35.6	33	20.8
Total	76	100	87	100	163	100

Table 24. Lithic assemblage at FLK North 5 excluding the purported lava manuports

	Quartz		Lava		Total	
	n	%	n	%	n	%
Test cores	–	–	1	1.8	1	0.7
Cores	2	2.6	13	23.2	15	11.3
Retouched pieces	1	1.3	–	–	1	0.7
Hammerstones	–	–	22	39.3	22	16.5
Anvils	9	11.8	–	–	9	6.8
Flakes	2	2.6	3	5.4	5	3.7
Chips	2	2.6	2	3.6	4	3
Possible flake fragments	36	47.4	12	21.4	48	36.3
Angular fragments	22	28.9	1	1.8	23	17.4
Battered fragments	–	–	2	3.6	2	1.5
Unmodified stones	2	2.6	–	–	2	1.5
Total	76	100	56	100	132	100

Quartz unmodified pieces are included since quartz is supposed to have been transported into the site but lava is not (see de la Torre and Mora, 2006)

At FLK North 5, hammerstones constitute 39.3% of the lava pieces, being represented by higher numbers than cores and even than combined frequencies of flakes and flake fragments (see Table 24). Given their high number and their size, it can be argued that elements of large dimensions are better represented than smaller ones, and battering activities seem to be well-supported. The lack of smaller pieces (only 3% of chips are < 20 mm) may be the result of postdepositional processes involving some hydraulic disturbance.

According to Leakey (1971: 69), the lithic assemblage of FLK North 4 was composed of 67 flaked pieces and 25 manuports. This small assemblage was deemed even smaller by Ludwig (1999), who identified only 55 pieces, among which were 28 cores and only 23 debitage products. According to Leakey, choppers and polyhedrons were the most abundant artifacts, whereas the "debitage" only constituted 29.8%. In de la Torre's (2006) study, he found 83 pieces, including unmodified lithic material. As in the underlying level, the most striking feature of FLK North 4 is the abundance of unmodified pieces. As can be seen in Table 25, the purported manuports are the most abundant type in the assemblage (28 pieces, or 53.8%). Including the unmodified material, there is a total of 18.4 kg of raw material at this level. Manuports are the most important type, accounting for 8.3 kg; almost the double that of quartz pieces (4.3 kg). Once again, we believe that it is a mistake to consider the vesicular lava manuports as related to the flakes and the used artifacts (de la Torre and Mora, 2006). Therefore, as in the previous case, unmodified lithic material from this level was excluded from our analysis. Table 26 shows that flaking products are underrepresented compared to other sites. This raises the issue of taphonomic bias and differential preservation according to size. As was the case with the previous level, this assemblage is extremely small and very likely biased.

The FLK North 3 lithic assemblage is composed of 171 pieces and 39 manuports, according to Leakey (1971: 72). De la Torre (2006) counted up to 214 pieces, including unmodified material (Table 27). This assemblage weight 28.8 kg total. Once again, the poor quality of the unmodified pieces suggests their natural origin. If we exclude them from the analysis, we notice that the total amount of

Table 25. Categories and raw material types represented in the FLK North 4 lithic assemblage (de la Torre, 2006)

	Quartz		Lava		Total	
	n	%	n	%	n	%
Test cores	–	–	–	–	–	–
Cores	–	–	8	15.4	8	9.6
Retouched pieces	–	–	–	–	–	–
Hammerstones	4	12.9	9	17.3	13	15.6
Anvils	1	3.2	–	–	1	1.2
Flakes	–	–	4	7.7	4	4.8
Chips	5	16.1	–	–	5	6
Possible flake fragments	7	22.6	3	5.8	10	12
Angular fragments	14	45.2	–	–	14	16.9
Battered fragments	–	–	–	–	–	–
Unmodified stones	–	–	28	53.8	28	33.7
Total	31	100	52	100	83	100

Table 26. FLK North 4 lithic assemblage, excluding the unmodified material (de la Torre, 2006)

	Quartz		Lava		Total	
	n	%	n	%	n	%
Test cores	–	–	–	–	–	–
Cores	–	–	8	33.3	8	14.5
Retouched pieces	–	–	–	–	–	–
Hammerstones	4	12.9	9	37.5	13	23.6
Anvils	1	3.2	–	–	1	1.8
Flakes	–	–	4	16.7	4	7.3
Chips	5	16.1	–	–	5	9.1
Possible flake fragments	7	22.6	3	12.5	10	18.2
Angular fragments	14	45.2	–	–	14	25.5
Battered fragments	–	–	–	–	–	–
Total	31	100	24	100	55	100

Table 27. Categories and raw material types represented in the FLK North 3 lithic assemblage (de la Torre, 2006)

	Quartz		Lava		Total	
	n	%	n	%	n	%
Test cores	–	–	5	4.8	5	2.3
Cores	–	–	11	10.6	11	5.1
Retouched pieces	3	2.7	–	–	3	1.4
Hammerstones	2	1.8	28	27	30	14
Anvils	8	7.3	1	1	9	4.2
Flakes	8	7.3	8	7.7	16	7.5
Chips	17	15.5	1	1	18	8.4
Possible flake fragments	52	47.3	7	6.7	59	27.6
Angular fragments	19	17.3	–	–	19	8.9
Battered fragments	–	–	–	–	–	–
Unmodified stones	1	0.9	43	41.3	44	20.6
Total	110	100	104	100	214	100

quartz transported (3.7 kg) is smaller than the amount of lavas showing traces of hominid use (15.5 kg). Flakes and debris are once again underrepresented compared to the abundant presence of cores and anvils.

The data from levels 5, 4, and 3, together with the description of the level 6 assemblage, underscore the lack of flakes and flaking products at FLK North. If we compare these patterns with those of FLK *Zinj*, where debitage constitutes up to 90% of the assemblage (de la Torre, 2006), we see that none of the FLK North levels has a sufficient number of flakes, given the presence of cores. Isaac (1986) named those objects detached from cores as detached pieces and named cores as flaked pieces. Most of the detached pieces at the various levels of FLK North seem to have been caused by battering and not by intentional flaking. As can be seen in Table 28, all the levels in the FLK North site have very low numbers of detached pieces, which in some cases could be explained by postdepositional processes (Schick, 1984; Isaac and Marshall, 1981; Petraglia and Potts, 1994). The representation of detached pieces is particularly marginal in

Table 28. Main characteristics of the lithic assemblages in the FLK North levels (de la Torre, 2006)

	Number of pieces*	Total weight (g)	Detached pieces (%)	Flaked pieces (%)
FLK North Level 6	128	15,948	80.5	19.5
FLK North Level 5	130	15,952	64.7	35.3
FLK North Level 4	55	10,098	60	40
FLK North Level 3	170	18,954	67.9	32.1

* All unmodified material has been excluded

levels 4 and 3. Schick (1984) pointed out that even in the most disturbed experimental assemblages that she analyzed, the proportion of cores with respect to debitage was never higher than 10%.

As we have seen, level 6 has an artifact density similar to levels 5, 4, and 3, classified by Leakey (1971) as levels with diffused materials. The distribution of lithic types at FLK North 6 shows neither qualitative nor quantitative differences with respect to those levels. If levels 5, 4, and 3 are considered background deposits, in which any association between lithics and fauna is merely incidental, the same interpretation should be applied to FLK North 6. The minimal lithic industry in this level seems to be related to battering activities. Therefore, it is not unexpected that anvils and hammerstones are predominant over flakes, which are basically absent from the assemblage (see Table 22). Leakey (1979: 92) had thought that the high number of anvils at the site could be related to the breakage of bones to obtain marrow. However, Leakey herself discarded this idea, given that most elements are fairly complete. Additionally, the small size of the anvils means that they could not have been used to break open the elephant bones. The lack of percussion marks, notches, and the completeness of most limb bones further disprove any functional link between the anvils and the bone assemblage. Therefore, we argue that the inferred battering activities at the site took place as an independent episode from the accumulation of the bone assemblage.

Conclusions

According to evidence drawn from the taphonomic analysis of the fauna at FLK North 6 and the technological study of the 130 stone pieces at the site, there is only one scenario in which hominid activity could have been functionally linked to the faunal assemblage at the site: that would be if hominids had carefully defleshed the carcasses without leaving a single cut mark, discarding complete bones without any interest in marrow, all while using a stone tool kit that they brought into the site, used, and did not discard. We believe this is highly unlikely, since at nearby sites in which this functional link was established (such as FLK *Zinj*), the taphonomic evidence of carcass exploitation by hominids is fairly abundant and the functional aspect of the lithic assemblage supports this assertion, with a large number of intentional flakes present at the site.

A more parsimonious explanation is that FLK North 6 is a palimpsest. Binford (1981) argued that the artifacts from this level were not spatially associated with the elephant remains, but were inversely correlated with them, and that they were instead associated with the background scatter of bovid and suid bones. Such an assertion is not taphonomically supported. No functional link could be established between the stone tools and the faunal assemblage. Ultimately, the present work supports Leakey's claim that the non-elephant bone assemblage is a background accumulation. We believe that it is a natural bone cluster, as is the accumulation of proboscidean bones.

Natural bone clusters in certain *loci,* as documented in Miocene sites and in specific places in Olduvai Bed I (channel) and Olduvai Bed II (wetlands) are similar – even denser – than the non-elephant accumulation at FLK North 6 (Cushing, 2002). Comparisons with Miocene sites and the landscape scatters at Olduvai during Early Pleistocene times are

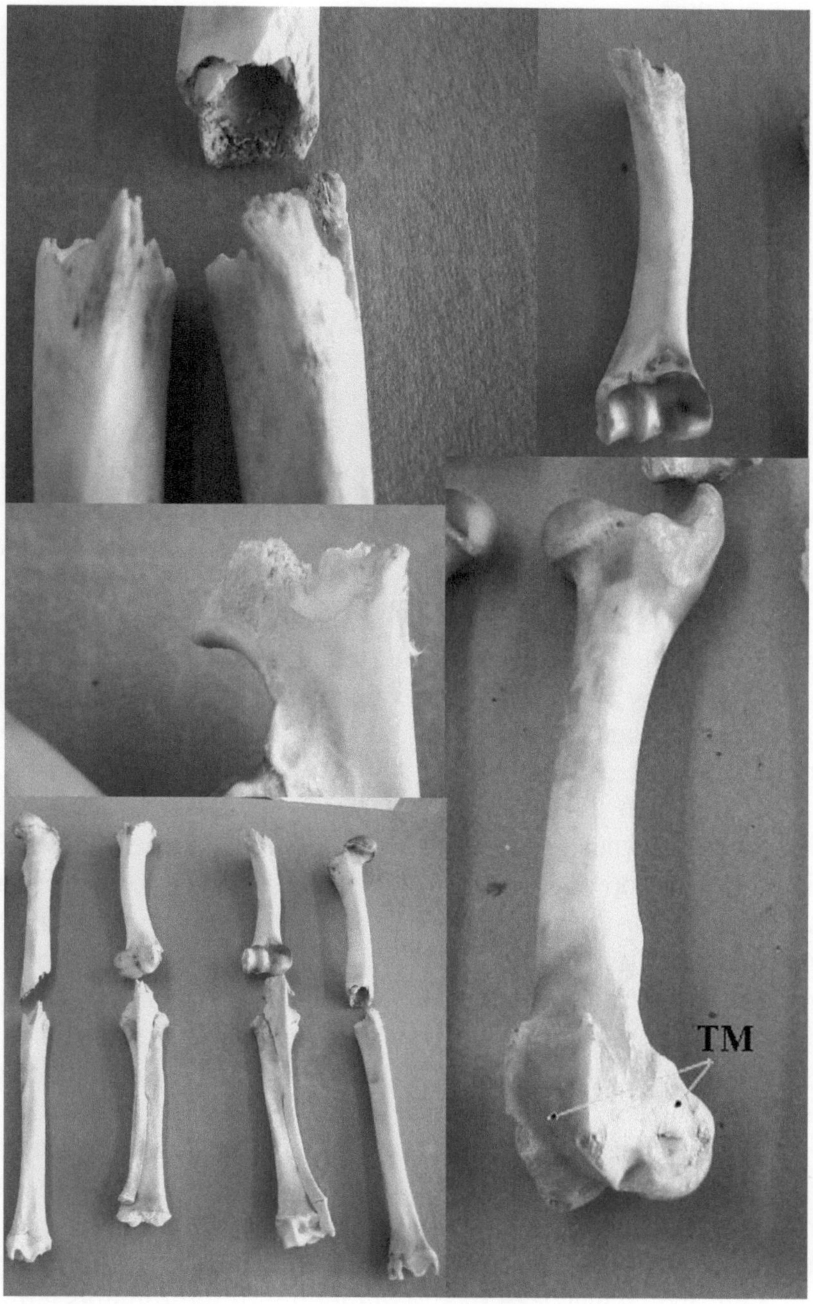

Figure 66. Various types of bone modification by leopards, furrowing on proximal epiphyses from humeri and tibiae and the oleocranon of the ulna, as well as both ends of femora, can be moderate or intensive. Arrows on the distal femur show small tooth marks (punctures) similar to those observed on some specimens from FLK North 6 (Figure 12). The bones shown are from the taphonomic collection created by Brain and stored at the Transvaal Museum.

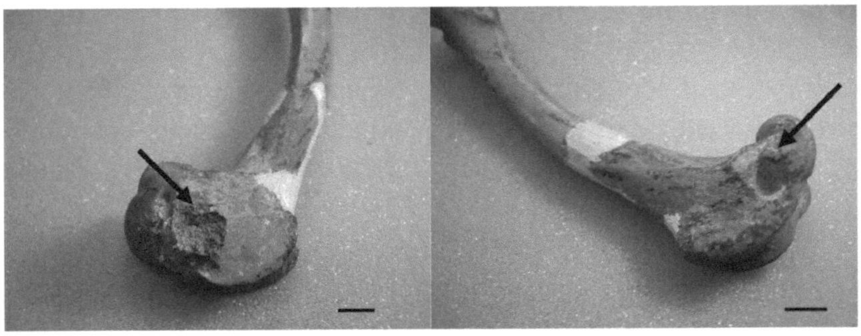

Figure 67. Tooth marks on the medial and lateral sides of the distal epiphysis of a femur from FLK North 6, similar to those created by felids (see femur in Figure 66) (scale = 1 cm).

warranted even if the landscape scatters are time-averaged. The FLK North 6 deposit spans about 50 cm of depth. It is not a single discrete horizon such as FLK *Zinj* (Leakey, 1971). The fossils contained in the thick stratigraphic interval at FLK North 6 could have been deposited in different times during long time intervals. The occurrence of elephant bones from two individuals (one almost complete and another represented by a couple of bones) can be explained by natural causes, since such type of accumulations are documented in modern savannas around alluvial areas like the one in which FLK North 6 would have been found (Haynes, 1991). The complete individual lies on top of some stone tools, suggesting that its deposition took place after the artifacts had been discarded.

The site seems to be the result of natural bone deposition in a favored place, which may be the result of animals frequently visiting the place (as suggested in Fossil-Lagerstätten deposits), or of multiple agents participating in the palimpsest, among which carnivore transport could also be included. As a matter of fact, the low tooth-mark frequency, as well as the presence of complete bones from small adult individuals (not modified by hyenas) bearing punctures on the epiphyses (Figures 66 and 67), show similar kinds of modifications to those made by felids, namely, leopards. Hyaenid bone modification could also be supported by the way that some of the bones (even for middle-sized carcasses) are gnawed (see Figure 59), by the lack of axial bones, as well as by the presence of hyaenid coprolites. However, the mild degree of bone modification is rather suggestive of the intervention of a felid, such as leopard. The small number of carcasses represented by postcranial bones could be similar to the small open-air accumulations generated by felids (Tappen, 1992; Cavallo, 1998). Before or after the formation of the hyaenid- and felid-modified bovid and suid assemblage, the elephant bones were incorporated into the deposit. Therefore, the site shows all the traces of a multiple agent and time-averaged process in which hominids intervened briefly, performing some activities that were not related to the bone assemblage. This kind of site could easily be found in ecotone alluvial habitats in modern savannas. This is suggested by the depositional context of FLK North 6 (silts) in the lake-margin area (Leakey, 1971) where the lake floodplain–alluvial plain ecotone exists, usually characterized by the transition of lacustrine forests to floodplains.

FLK North 6 shows that early site interpretation has to be done carefully and that taphonomic arguments must prevail over ecological ones to establish agents and order of site formation. In this book, we argue that FLK *Zinj* is completely exceptional, since the rest of the Bed I sites share different features. Such a distinction may be due to the fact that the other Bed I sites may have had more complicated taphonomic histories than FLK *Zinj*, which need to be properly unraveled before a sound behavioral interpretation can be provided.

Chapter 8 will further elaborate on this approach and will show that, in contrast with FLK North 6, some intervention of hominids in the accumulation of the bone assemblage at FLK North 1–2 (the largest site in Bed I) can be supported. This raises a very important issue for taphonomists: can we accurately interpret archaeological assemblages that were created through multiple independent agents, given that most of our referential frameworks have been created under the assumption that there is almost always interaction among the different agents who accumulate and modify faunal assemblages?

8. A palimpsest at FLK North 1–2: independent carnivore- and hominid-made bone accumulations

M. DOMÍNGUEZ-RODRIGO AND R. BARBA

Introduction

Current interpretive models of early human behavior are based on referential frameworks that model carnivore and hominid interaction (discussed in Chapter 2). These experimental models have been applied to Olduvai Bed I "living floors" and constitute the basis for "central-place foraging" (Isaac, 1983), "transport resource" (Potts, 1991), and "near-kill location" (O'Connell, 1997) interpretive models (discussed in Chapter 1). All the previous interpretations are mainly focused on zooarchaeological approaches to the analysis of faunal assemblages, based on studies of skeletal part frequencies (Bunn, 1982), and ecological neo-taphonomic studies (Potts, 1988). Also, all "carnivore" experimental scenarios in East African archaeology have been modeled through what hyenas do to bones (Blumenschine, 1988). The remainder of carnivores have been marginally studied regarding their bone-modifying behaviors (but see Brain, 1981; Selvaggio, 1994; Cavallo, 1998).

In Domínguez-Rodrigo *et al.* (in press), it is argued that experiments designed to model passive scavenging by hominids lack relatedness between what was theorized and what was being experimentally modeled. It was previously assumed that carnivores (felids and hyenas) modified bones in a similar fashion. Cavallo (1998) conducted a study that, in part, investigated tooth mark densities in leopard-created bone assemblages. He concluded that a high percentage of tooth marks was expected on prey limb bones, similar to tooth mark rates documented in assemblages modified by hyenas. However, he overlooked the fact that most leopard tooth marks in the assemblage occurred on bone fragments, not on partial or whole bone specimens. In contrast, limb bone specimens with intact medullary cavities display lower tooth mark frequencies. If hominids were scavenging from felid kills, they would disregard broken, resource-depleted bone fragments and instead concentrate on the exploitation of marrow-containing whole bones; tooth mark frequencies in this scenario would therefore be very low (see Domínguez-Rodrigo *et al.* for full discussion). A fundamental aspect of the triple-stage passive scavenging model (discussed in Chapter 2) is that felids, not hyenas, initiated carcass exploitation. It is therefore inappropriate to substitute data on felid-imparted bone damage for hyena-imparted bone damage. This calls for a reconsideration of passive scavenging models and their application to archeofauna.

In this chapter, we raise another important question: could felids, hyenas and hominids have participated in assemblage formation without interdependence? If so, how would archaeologists discriminate which agent did what? Skeletal part profiles, age mortality profiles, and bone surface modification frequencies and

M. Domínguez-Rodrigo, R. Barba, and C.P. Egeland, Deconstructing Olduvai: A Taphonomic Study of the Bed I Sites, 127–163.

distributions, all discussed in Chapter 3, have been used to discriminate between carnivore and hominid contribution to site formation. Although previous observational and experimental datasets are useful interpretive tools for some specific archaeological sites, they are still incomplete, and they have been misapplied in constructing referential frameworks to model hominid-carnivore independent contribution to site formation.

The concept of a palimpsest involves assuming that accumulations are both created *and* modified by more than one agent (Binford, 1978, 1981). So far, all the interpretive frameworks applied to Plio-Pleistocene African sites are based on the assumption that the creation of an assemblage is mainly the result of the actions of single agent, whether carnivore (e.g., Binford, 1981) or hominid (e.g., Bunn and Kroll, 1986, 1988; Potts, 1988; Oliver, 1994). Even the more recent multiple-stage models (e.g., Blumenschine *et al.*, 1994; Selvaggio, 1994; Blumenschine 1995; Capaldo, 1995) posited that bones were mainly accumulated at sites by a single collecting agent, hominids, even though the initial capture and consumption of carcasses was attributed to carnivores. In all these scenarios there is an implicit notion that multiple agents participated in the modification and attrition of bone assemblages but *not* in their accumulation. This idea seriously impacts analytical toolsets (Chapter 3) used by taphonomists for the interpretation of assemblage formation. For example, an assemblage created through transport of carcasses to the same place by independently operating hominids and carnivores could generate skeletal part and carcass size profiles that mimic the conditions expected in an assemblage created by hominids scavenging from carnivore kills or dens, especially if carnivore contribution to the assemblage surpasses that of the hominids.

We lack conceptual and referential frameworks to account for multiple but *independent* agents of bone accumulation at sites. Here we present an archaeological example from levels 1 and 2 (both located in Bed I) at FLK North, the same site discussed in Chapter 7. FLK North 1–2 is situated on the far side of the ridge, which lies to the north of FLK. It is situated beneath Tuff If and has a depth of about 2–3 ft (less than 1 m) according to Leakey (1971), comprising a single horizon of grey-brown silty clay. Levels were artificially subdivided for convenience in excavation, although three different levels were archaeologically detectable (Leakey, 1971).[1] Faunal materials stored at the National Museums of Kenya were labeled lumping together levels 1 and 2, so independent taphonomic histories of each level cannot be distinguished. However, most of the materials seemed to be clustered at the top of the sequence in level 1. Level 3 seemed to be very poor in remains. The main archaeological levels were identified as "two periods of occupations within the bed" (Leakey, 1971: 70). The site was interpreted by Leakey (1971) as a living floor.

The FLK North 1–2 archeofauna was previously studied by Bunn (1982, 1986). He interpreted the site as a central foraging place. He found a substantial amount of hominid-modified bone, although less than at FLK *Zinj*. However, in this chapter we show that the total amount of bone specimens bearing hominid-induced marks is as low as 21 cut-marked fragments and 12 specimens probably bearing percussion marks.[2] We present data that suggest that most of the accumulation was made by carnivores and that hominids, despite the abundant lithic repertoire, seem to have been independently involved only with bones (based on cut-marked specimens) from a maximum of 4 carcasses out of the 45 (as reported by Bunn 1986) represented by bones at the site. Only two appendicular specimens have mixed marks created by both hominids and

[1] A recently excavated test trench revealed that there is no basis for differentiating between levels 1and 3 (Andrews, personal communication).

[2] Most of them not securely identified, since the association of striae and pits is not as straightforward as in experimentally derived marks; see discussion later in this chapter.

carnivores, and in both cases, carnivore modification appears on the near-epiphyseal or epiphyseal sections of the specimens, suggesting secondary access to those fragments after they were discarded by hominids.

Results

TAXA AND SKELETAL PART FREQUENCIES

Bunn (1982) identified diverse taxa at FLK North 1–2 including *Theropithecus* sp., *Canis* sp., *Crocuta* sp., *Ceratotherium simum, Hippopotamus gorgops, Giraffa jumae, Hipparion* sp., *Equus oldowayensis, Syncerus acoelotus, Tragelaphus strepsiceros*. Most of these taxa were represented by scanty remains from one individual. He also documented remains of four suidae carcasses. Remains from three tragelaphini and three antilopini could not be identified to species. However, all these species only made up half of the total MNI documented at the site. A total of 11 carcasses from *Parmularius altidens* and 10 carcasses from *Antidorcas recki* were also documented at the site. In contrast to the previously mentioned taxa, mostly represented by very few bones (very often just teeth), the *Parmularius* and *Antidorcas* samples are represented by several elements of the skeleton. It seems that they resulted from a different taphonomic history from the other set of bones. Our analysis has identified a higher number of individuals of both species than was previously reported by Bunn (1986) (Table 29).

The bone assemblage at FLK North 1–2 is composed of 2,873 specimens larger than 20 mm (Table 30). It is by far the largest assemblage in Olduvai Bed I. The present study has significantly enlarged the sample of bovids, the most abundant taxa, by doubling the number of specimens initially identified (Table 31). An analysis of specimen size distribution shows that small fragments are

Table 29. Minimum number of individuals identified at FLK North 1–2

	MNI
P. altidens	29
A. recki	13
Alcelaphini 3B	1
Tragelaphini 3B	2
Bovini	1
Hippotragini	1
Equid	1
Kolpochoerus sp.	1
Metridiochoerus sp.	1
Suidae 3[a]	1
C. simum	1
*G. jumae**	1
H. gorgops	1
Dinofelis sp.	1
Crocuta crocuta	1
Canis mesomelas	3
Large felid aff. *P. leo*	2
P. pardus	1
Theropithecus sp.	1
Total	63

*Leakey(1971)

Table 30. Number of identified specimens (NISP) for each skeletal element in the faunal assemblage at FLK North 1–2

	Bovid	Suid	Equid	Hippo	Rhino
Horn	21	–	–	–	–
Skull	79	1	–	–	–
Teeth	96	36	–	3	–
Mandible	63	4	–	–	–
Vertebra	138	2	–	–	–
Ribs	263	2	–	–	–
Pelvis	92	4	–	–	–
Scapula	77	2	–	–	–
Humerus	171	2	2	–	–
Radius-ulna	233	10	1	–	–
Carpals-tarsals	187	8	1	1	1
Metacarpal	124	1	–	–	–
Femur	230	–	–	–	–
Tibia-fibula	382	5	3	–	–
Metatarsal	167	3	–	–	–
Patella	14	2	–	–	–
Phalanges	122	1	–	–	–
Other	19	–	1	–	–
ULB*	31	–	–	–	–
ILB*	21	–	–	–	–
LLB*	21	–	1	–	–
Indeterminate	99	–	–	–	–
Total	2,650	83	9	4	1
% total NISP	96.5	3	0.3	0.1	0.04

* ULB, upper limb bones; ILB, intermediate limb bones; and LLB, lower limb bones.

Table 31. Frequencies of specimens identified for each major faunal group, comparing Leakey's (1971) list and the present study

	Leakey (1971)		MDR&RB	
	NISP	%	NISP	%
Bovidae	1351	78.5	2650	90.3
Carnivora	193	11.2	185	6.4
Suidae	118	6.9	83	2.9
Equidae	52	3	9	<1
Hippopotamidae	4	0.2	4	<1
Primates	1	<1		
Rhinocerotidae	1	<1	1	<1

well-represented in all carcass sizes. Specimens smaller than 4 cm make up between 20% (middle-sized carcasses) and 40% (small carcasses) of all fragments represented (Figure 68). Similar proportions can be observed if using a sample of similarly dense fragments such as shafts (Figure 69). In this case, however, the number of shaft specimens smaller than 4 cm is substantially lower than that for the whole sample, and is also lower than that reported for experimental assemblages of hammerstone-broken bone

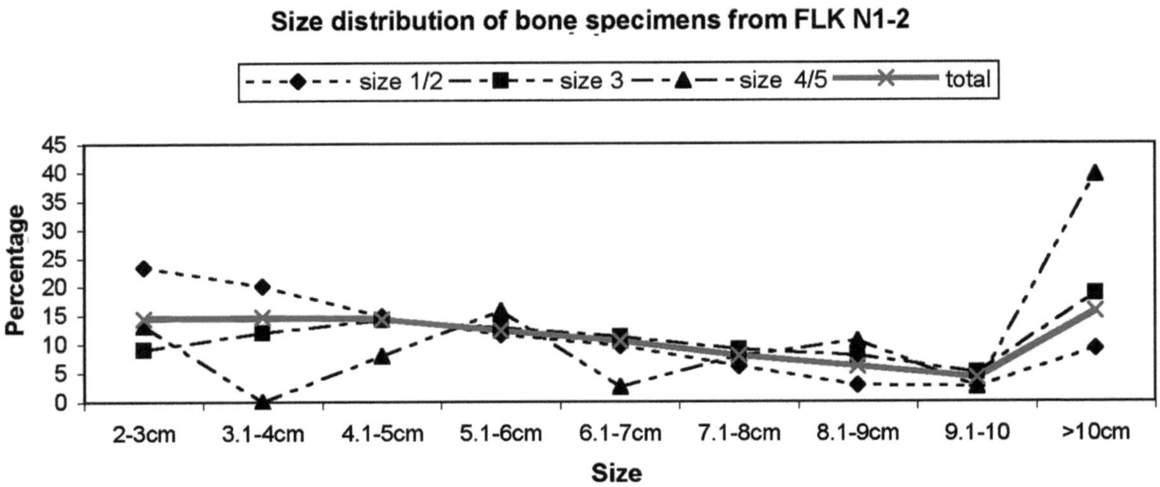

Figure 68. Size distribution (shown as %NISP) of bone specimens at FLK North 1–2 by carcass size for all elements.

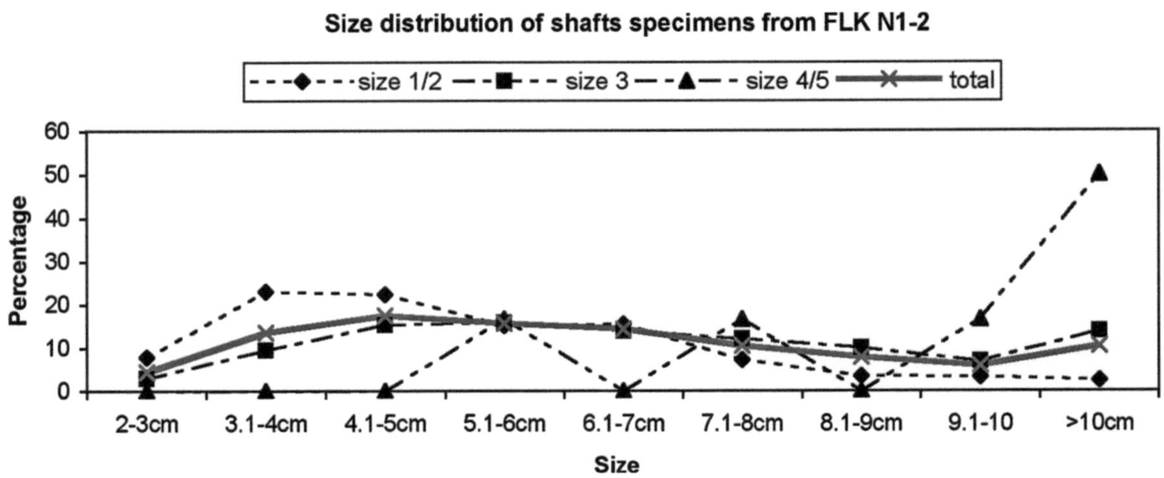

Figure 69. Size distribution (shown as %NISP) of shaft specimens at FLK North 1–2 by carcass size.

(Blumenschine, 1988, 1995). Fragmentation in the assemblage has therefore been less intense than in experimental assemblages created through dynamic loading. Specimen size representation, together with the absence of abrasion or polishing on any bone fragments, suggests that the assemblage remains mostly undisturbed, as suggested by the clayish lithology of the geological context where it was found (Leakey, 1971).

Table 29 shows the taxa identified in our study. FLK North 1–2 is mainly composed of 29 *P. altidens* and 13 *A. recki*. The presence of other carcasses is rather marginal, since most of them appear in very small numbers (MNI = 1) and are represented by very few bones (usually no more than five each). By contrast, both *Parmularius* and *Antidorcas* are represented by bones belonging to most of the skeleton. It therefore seems to be a very specialized accumulation.

Bones from all anatomical elements have been recorded (Table 30). A minimum estimate of 624 elements (not including phalanges and carpo-tarsals) have been identified. Limbs make up 59% of the assemblage. Axial bones (which here includes pelves and scapulae for taphonomic purposes) constitute 34%; cranial bones make up the remaining 7% of the assemblage (Table 32) . Element quantification was carried out including shaft specimens together with long bone epiphyseal ends (Tables 32 and 33). This method is time-consuming but it is significantly more accurate than using limb ends alone, as the contrasting results of MNE estimates using shafts and epiphyses separately show (Table 32).

There is a substantial difference in the element counts reported by Bunn (1986) versus ours. The reason is that he could never display all limb shaft specimens to be properly studied, which would imply using an extensive amount of space, because of former research policies at the National Museums of Kenya. In contrast, we were allowed to proceed this way. Furthermore, Capaldo's unpublished study of the site produced a few more boxes with limb shaft fragments that were previously classified as "indeterminate." We included those in our analysis and displayed all specimens by element to be properly identified and quantified. Considering the MNE and MNI calculated by including shafts, we see an overall high representation of most limb bones, suggesting that bones at the site are the result of complete carcass transport, rather than being isolated elements (Figure 70). Interestingly, meaty long bones (with the exception of femora) are more represented than the denser metapodials.

When focusing on epiphyses, a low frequency of the least dense long limb portions can be observed (Figure 70). Humeri are most represented by distal ends and radii by proximal epiphyses for both carcass sizes. This contrast in epiphysis representation is wider in

Table 32. Minimum number of elements (MNE) for each skeletal element and bone portion (for appendicular bones) in small and large carcasses

	Small	Large	Total
Skull*	13	29	42
Scapula	15	19	34
Pelvis	22	26	48
Ribs	28	47	75
Vertebrae	24	29	53
Humerus			
Epiphyses	17	31	48
ep. + shaft	23	43	66
Radius			
Epiphyses	19	32	51
ep. + shaft	19	36	55
Metacarpal			
Epiphyses	17	22	39
ep. + shaft	18	29	47
Femur			
Epiphyses	8	23	31
ep. + shaft	11	38	49
Tibia			
Epiphyses	22	41	63
ep. + shaft	23	62	85
Metatarsal			
Epiphyses	20	42	62
ep. + shaft	20	48	68

MNE estimates are compared for when epiphyses are used alone, and when epiphyses and shafts are combined.
* including mandible.

Table 33. NISP for each bovid long limb bone portion according to element type at FLK North 1–2

	NISP		
	Size 1/2	Size 3	Size 4/5
Humerus complete	4	3	0
Humerus prox.	9	7	
Humerus prox. + shaft	1	1	
Humerus shaft	29	76	1
Humerus dist.	7	13	
Humerus dist. + shaft	6	14	
Radius complete	6	4	0
Radius prox.	8	26	
Radius prox. + shaft	11	28	
Radius shaft	34	84	
Radius dist.	3	7	
Radius dist. + shaft	3	4	
Metacarpal complete	10	6	0
Metacarpal prox.	7	4	
Metacarpal prox. + shaft	2	9	
Metacarpal shaft	23	42	
Metacarpal dist.	3	5	1
Metacarpal dist. + shaft	4	6	1
Femur complete	0	0	0
Femur prox.	8	5	1
Femur prox. + shaft	0	0	
Femur shaft	70	124	2
Femur dist.	1	11	
Femur dist. + shaft	4	3	
Tibia complete	4	3	1
Tibia prox.	7	12	
Tbia prox. + shaft	4	3	
Tibia shaft	82	191	3
Tibia dist.	10	13	
Tibia dist. + shaft	8	24	
Metatarsal complete	7	4	0
Metatarsal prox.	6	15	
Metatarsal prox. + shaft	9	23	
Metatarsal shaft	24	48	
Metatarsal dist. + shaft	1	9	
Metatarsal dist. + shaft	5	14	
Total	420	841	10

middle-sized carcasses. Femora from small carcasses are slightly better represented by proximal ends, whereas the trend is the opposite for larger animals. As is expected in the case of density-mediated attrition, tibiae are more widely represented when using distal epiphyses. Metapodials do not show marked contrast in epiphysis preservation, with the exception of metatarsals, which show a bias toward proximal ends. The profile obtained is similar to that documented by Brain (1981) in Hottentot bone accumulations that have undergone carnivore ravaging.

Further insight can be gained from looking at the representation of Bunn's (1983b) circumference types, outlined in Chapter 3. Hyenas tend to have a larger proportion of complete circumference shafts (Type 3) in their dens than are found in human-broken bone assemblages, where shaft fragments smaller than 50% of the section (Type 1) are more widely represented. This proportion changes in assemblages where carnivores have secondary access, in which complete shaft circumferences are rarer due to intensive bone destruction caused by marrow consumption. Bone shaft sections can therefore be used as an analytical tool to differentiate low-competition settings (e.g., dens) from high-competition settings (e.g., bone clusters in open spaces). The high number of complete bones at FLK North 6, presented in Chapter 7, is thus suggestive of a very low-competition environment. In FLK North 1–2, the distribution of shaft sections is very distinct from hammerstone-carnivore and carnivore-only scenarios where intense ravaging was modeled (Figure 71). Type 1 shafts are as abundant as Types 2 and 3 together, with a far higher number of complete bones than has been experimentally observed. This suggests a very low-competition setting where bone destruction was reduced. Bones at FLK North 1–2 are more intensively broken than at FLK North 6, but much less so than carnivore-only models. The presence of bones that are complete with others that are very fragmented also suggests different depositional moments with different amounts of carnivore post-ravaging taking place.

In sum, the distribution of shaft sizes, the low frequency of axial bones, the higher preservation of dense long limb portions over less dense epiphyses, and the distribution of shaft circumference types all suggest

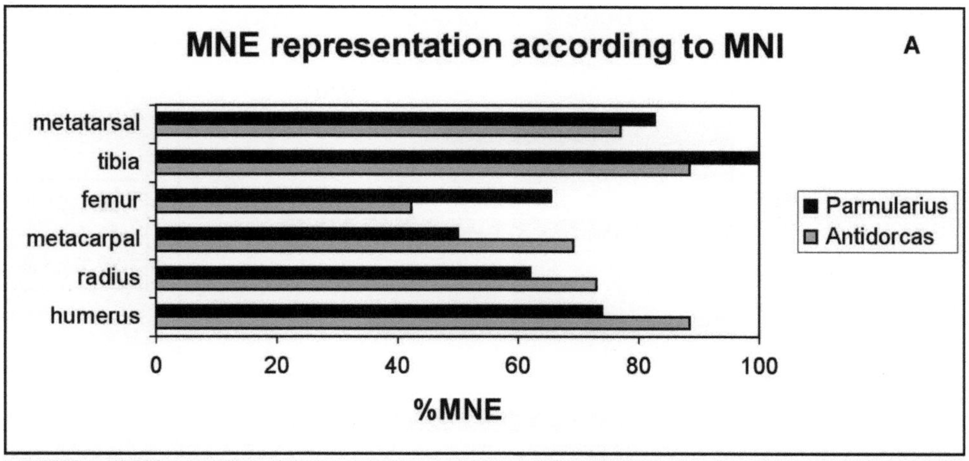

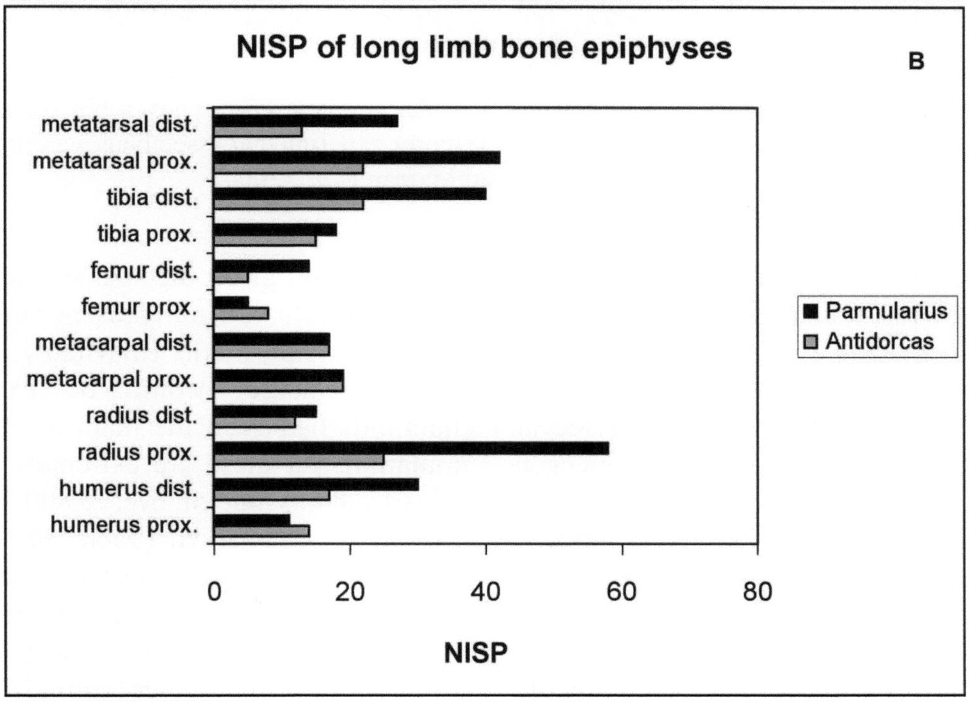

Figure 70. (A) Frequency of elements (%MNE) for each long limb bone according to the expected number of elements given the MNI of the assemblage. (B) Frequency of identifiable specimens (NISP) for each long limb bone end portion.

postdepositional carnivore ravaging, irrespective of the agent of bone accumulation.

BONE BREAKING AND CARCASS DEFLESHING

In order to understand who accumulated the FLK North 1–2 assemblage, we search for traces of the agents involved in their modification, using the criteria outlined in Chapter 3: bone breakage features (breakage planes, notches), and those features resulting from carcass defleshing (tooth marks and cut marks) and marrow extraction (tooth pits and percussion marks) by carnivores and hominids. We can use the experimental referential frameworks discussed in Chapters 2 and 3 to interpret these data, provided that there is interaction between hyenas and hominids.

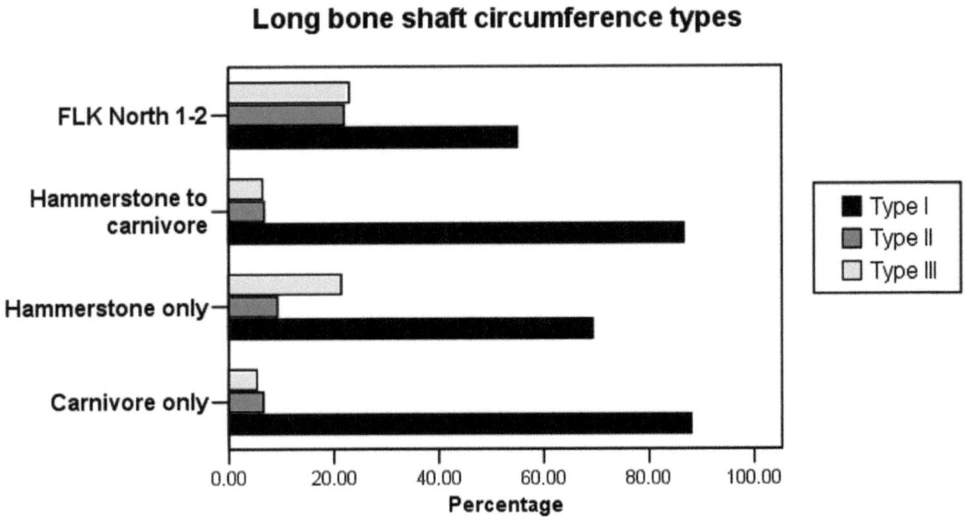

Figure 71. Distribution (in percentages) of Bunn's (1982) circumference types of bovid long bones in experimental assemblages and at FLK North 1–2. Data for experimental assemblages are from Marean and Spencer (1991) and Marean *et al.* (2004).

Bone preservation in the assemblage is very good. Most specimens (85.2%) had good (73.8%) to moderate (11.4%) bone surfaces. Good preservation was defined as complete preservation of the entire bone surface and clear identification of any modification occurring on it. Moderate preservation was defined as more than two thirds of the bone surface of the specimen exhibiting good cortical preservation, on which bone surface modifications could be clearly identified. Bad preservation affected only 14.8% of the long limb bone assemblage, and these badly preserved specimens were discarded from the analysis.

Therefore, given the size and preservation of the collection, a high presence of cut marks and percussion marks (inevitably left during bone breakage) would be expected if hominids had been the main agent of accumulation. Yet these are extremely rare: only 21 specimens out of over 2,800 bone fragments are cut-marked (Table 34). Similarly, very few percussion marks have been found at the site: a total of 12 specimens bearing probable percussion marks have been identified. Midshafts from smaller animals (3/237 = 1.2%) and larger animals (9/507 = 1.7%) bear far fewer percussion marks than are reported

Table 34. Location and behavioral implication of cut marks on each of the cut-marked specimens from small and large carcasses at FLK North 1–2

Small carcasses		Large carcasses	
Element	Behavior	Element	Behavior
1 Proximal femur	Disarticulation	3 Distal femora	Disarticulation
1 Proximal metacarpal	Disarticulation	1 Proximal metatarsal	Disarticulation
1 Distal humerus	Disarticulation	3 Tibia midshafts	Filleting
1 Humerus midshaft	Filleting	2 Radius midshafts	Filleting
		6 Femur shafts	Filleting
		2 Humerus shafts	Filleting

in human (hammerstone)-carnivore experiments (Blumenschine, 1988). The frequencies of percussion marks at FLK North 1–2 are 1.4% for smaller animals and 2% for larger animals if considering only specimens with "good" cortical preservation, instead of 20% on average as would be expected in hammerstone-carnivore experiments (Blumenschine, 1988).

The frequencies reported for FLK North 1–2 are thus outside the range for the human-to-carnivore experimental scenarios. In the small sample of hominid-modified bones, only two specimens have both cut marks and tooth marks, showing some hominid and carnivore interaction. One femur from a *Parmularius* bears disarticulation marks right above the caudal side of condyles and there is very little shaft (only a small portion of the near-epiphysis) remaining. Directly on the immediate breakage plane, a small tooth mark and tooth notch can be seen, clearly indicating that carnivores had broken the bone. This can be better seen on a femur, a metacarpal, and especially on a humerus of *Antidorcas*, where the cut mark appears on the shaft, suggesting defleshing, but the bones were broken by carnivores, as evidenced by a double overlapping notch and an opposing one, resulting from more than one impact point as is often seen in hyena-broken bones (Figure 72).

A recent study by Egeland *et al.* (in preparation; see Table 1 in Chapter 3) on the taphonomy of shafts at the Maasai Mara hyena den shows a high presence of this type of notch (double overlapping), referred to as Type C by Capaldo and Blumenschine (1994). In the Maasai Mara assemblage, 16.7% of shaft fragments had notches. Of that notch

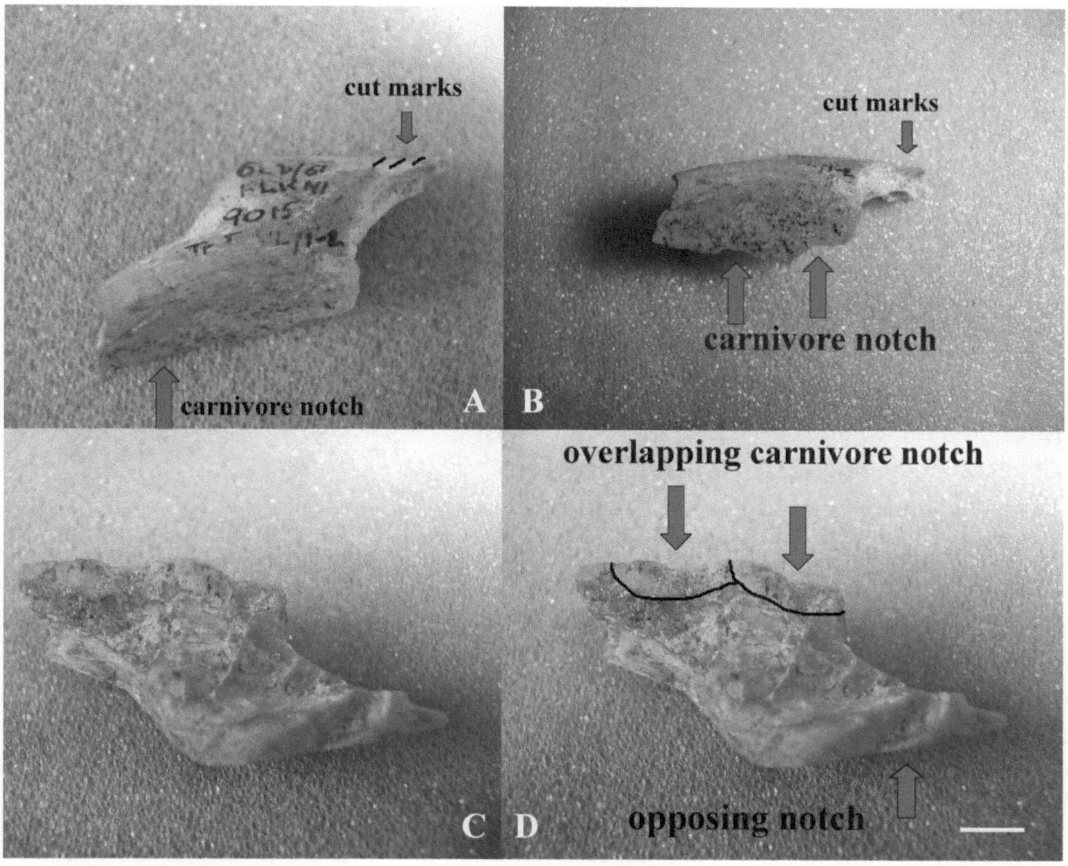

Figure 72. Humerus shaft fragment from a small carcass, showing cut mark location with respect to the location of the carnivore notch (A); the typical outline of the overlapping carnivore notch (B); and the morphology of the overlap of flake scars of the overlapping notch and location of opposing notch (C, D) (scale = 1 cm).

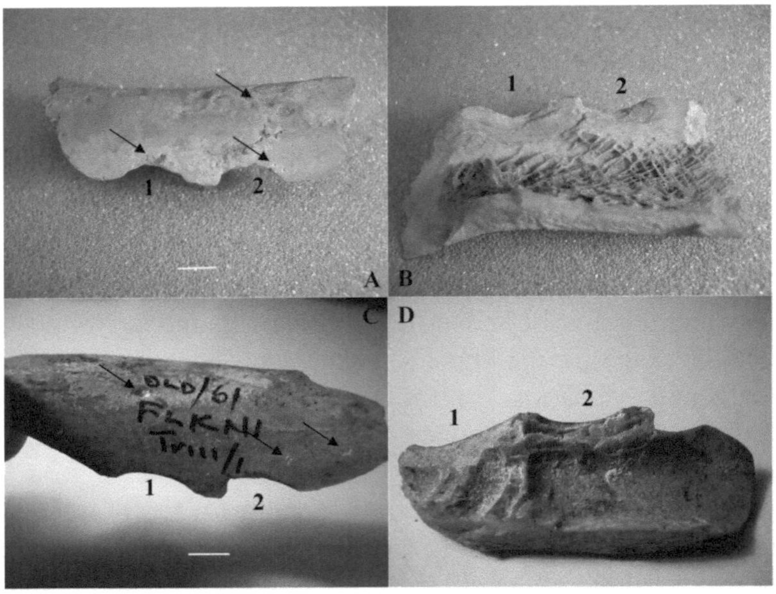

Figure 73. Typical overlapping notches (Type C) from the Maasai Mara hyena den (A,B) and from FLK North 1–2 (C, D), showing morphology and outline from the cortical (A,C) and the medullary (B, D) surfaces (scale = 1 cm).

sample, 34% of the notches were Type C; in the total assemblage, therefore, 5% of specimens had Type C notches. Type C notches were not reported by Capaldo and Blumenschine (1994) to appear in hammerstone-broken samples, but we have observed them in our experimental sample although in lower frequencies. The reason is simple: most hammerstone notches are created through a single impact point. Carnivores, by contrast, sometimes hit bone surfaces with more than one tooth cusp, thereby creating several impact points; breakage therefore follows a winding outline, often in the form of close-set notches (since the carnivore's teeth are close-set), and this also creates overlapping of the negative scars of notches. This is a typical signature of carnivore bone breakage, which often appears accompanied by tooth marks (Figure 73). Opposing notches, like those shown in Figure 74, are also typical of carnivore bone breakage, since upper and lower dentition impact on opposite sites of the bone, which generates the pressure that characterizes static loading. However, hammerstone percussion may also

Figure 74. Typical double opposing notch on a specimen from FLK North 1–2, generated by a carnivore while using its upper and lower tooth cusps to create pressure at opposing impact points (scale = 1 cm).

create some opposing notches, since the side of the bone leaning against the anvil surface reabsorbs the force of the blow, thereby creating a second impact point.

The frequencies of notches at FLK North 1–2 are strikingly similar to those obtained

Table 35. Frequencies of notch types in the Maasai Mara hyena den (Egeland et al. in preparation) and FLK North 1–2 shaft samples

	Maasai Mara	FLK North 1–2
% Notches	132/789 (16.7)	145/744 (19.4)
% Complete notches	45/789 (5.7)	43/744 (5.7)
% Incomplete notches	21/789 (2.6)	29/744 (4)
% Overlapping notches (Type C)	46/789 (5.8)	59/744 (7.9)
% Double-opposing notches	20/789 (2.4)	35/744 (4.7)
% Complete notches/all notches	45/132 (34)	43/145 (29)
% Overlapping notches/all notches	46/132 (34.8)	59/145 (40)
% Double-opposing notches/all notches	21/132 (15.9)	35/145 (24)

in hyena-broken assemblages (Table 35). Complete notches, overlapping, double opposing, and incomplete notches are represented at similar rates. Overlapping notches (Type C) are the most abundant type in the notch sample. The frequencies of both overlapping and double opposing notches are so alike that it looks as if most of the limb bone assemblage was broken by hyenas, based on notches alone.

Measurements of notches at FLK North 1–2 show that most notch breadth:depth ratios and flake scar breadth:depth ratios fall within the range of variation for experimentally modeled carnivore bone breakage (Figure 75). Only seven notches fall outside the range, but they are very close to it and given that the experimental sample size was small, their proximity to the carnivore range could indicate that carnivores generated them, especially since no percussion mark was found on these specimens which could indicate otherwise.

Platform angles were measured in very few specimens, because most of those having a notch usually preserved a large portion of shaft circumference (Bunn's Type 2), which did not allow the use of a goniometer, and we could not use casting materials since these were reserved exclusively for bone surface modifications. Only platform angles on notches from large carcasses could be measured. The small sample (Figure 76) shows strong overlap with both hammerstone and carnivore experimental scenarios, but most of the platform angles are below 120°, which is common for carnivores.

Additional data on bone breakage at the site confirm this interpretation. According to breakage frequencies, the *Antidorcas* bones seem to show a different taphonomic history from those of *Parmularius* (Table 36). About 27% of all long limb elements from *Antidorcas* appeared complete. In contrast, only 7% of limb bones from *Parmularius* are complete. There is a nearly fourfold difference in limb element completeness between these taxa. Most of the breakage is green. Diagenetic fractures seem to have affected mostly metapodials, but their frequencies do not seem to be large enough to invalidate comparisons with referential frameworks for green-broken bone (Figure 77). Apparently, diagenetic breakage of bone, very likely happening after burial, affects denser elements. This probably has implications for the original number of complete elements deposited, which may have been underestimated by our counts. Stylopods, zygopods, and most metapods therefore seem to have been broken fresh, either by static or dynamic loading processes.

The analysis of breakage planes yielded further information in this regard (Figure 78). A total of 490 breakage planes were measured. From bones of medium-sized carcasses, 149

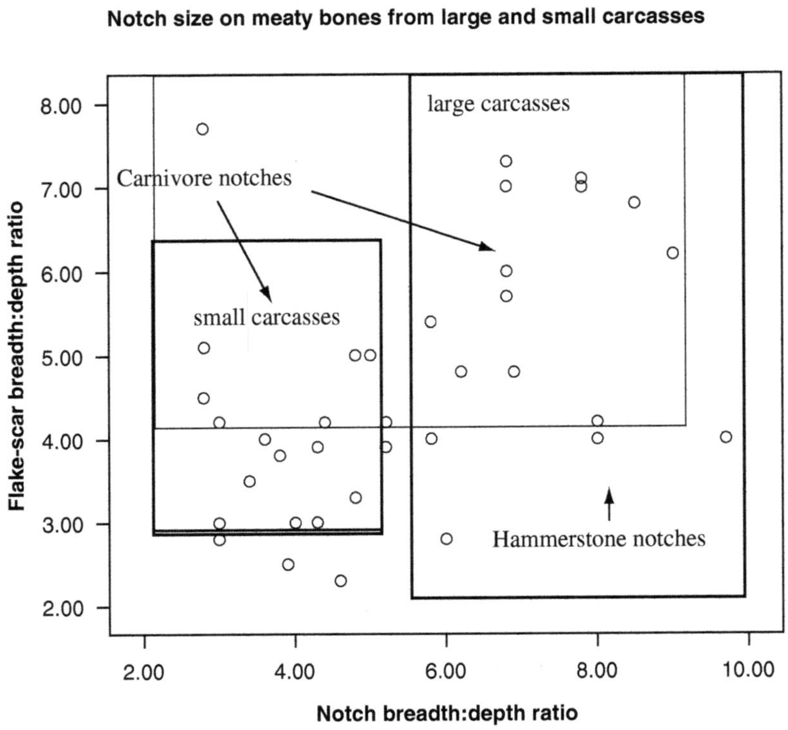

Figure 75. Notch measurements for the FLK North 1–2 specimens, showing notch length:breadth ratio from the cortical surface and flake scar length:breadth ratio from the medullary surface. Boxes represent the ranges of variation (95% confidence intervals) for these ratios as documented in experimental assemblages produced by static (carnivore) and dynamic (hammerstone percussion) loading (Capaldo and Blumenschine 1994). The box with the thinnest outline at the top represents the range of variation for carnivore notches on large carcasses. The box with a moderately thick outline, on the left, represents the range of variation for carnivore notches on small carcasses. The box with the thickest outline, on the right, represents the range of variation for percussion notches on large carcasses. The range of variation for percussion notches on small carcasses falls to the extreme right, outside the graph.

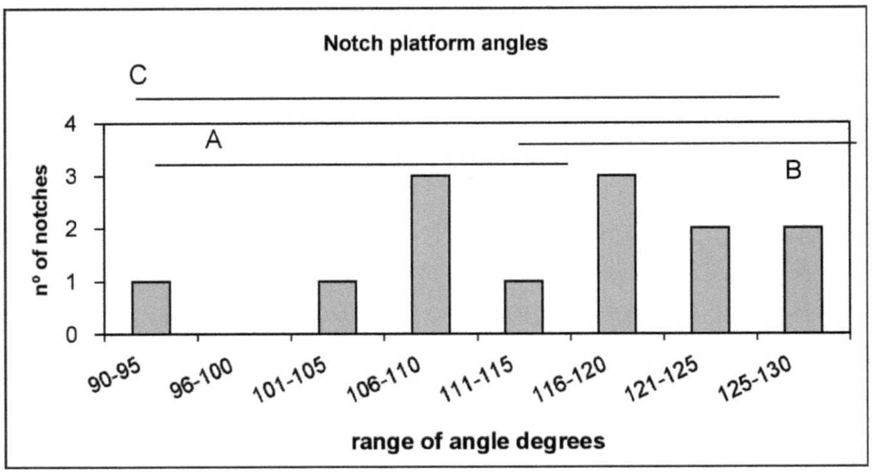

Figure 76. Distribution of notches from FLK North 1–2 according to platform angle. Ranges on top of bars are for carnivore-made notches (a) and hammerstone-broken notches (b) from Capaldo and Blumenschine (1994) and for carnivore-made notches (c) on bones from the Maasai Mara hyena den (Egeland *et al.*, in preparation).

Table 36. Frequencies of complete elements at FLK North 1–2 with respect to the total number of specimens (NISP) and minimum number of elements (MNE) identified

	Complete long limb bones/NISP	
	Size 1–2	Size 3
Humerus	4/57(7)	3/114(2.6)
Radius-ulna	6/54(11.1)	4/116(3.4)
Metacarpal	10/49(20.4)	6/75(8)
Femur	0/81(0)	0/143(0)
Tibia	4/116(2.3)	3/244(1.2)
Metatarsal	7/53(13.2)	4/113(3.5)
Total	31/410(7.6)	20/805(2.5)
	Complete long limb bones/MNE	
Humerus	4/23 (17.4)	3/43 (7)
Radius-ulna	6/19 (31.6)	4/36 (11.1)
Metacarpal	10/18 (55.6)	6/29 (20.7)
Femur	0/11 (0)	0/38 (0)
Tibia	4/23 (17.4)	3/62 (4.8)
Metatarsal	7/20 (35)	4/48 (8.3)
Total	31/114 (27.2)	20/256 (7.8)

oblique planes and 222 longitudinal planes were measured. A total of 68 oblique planes and 51 longitudinal planes from small-sized carcasses were also measured. Longitudinal planes were ambiguous as to the agent of bone breakage: angles <90° in smaller carcasses are more similar to those generated through dynamic loading, whereas angles >90° in larger carcasses clearly suggest a process of static bone breakage (see Chapter 3). With respect to oblique angles, the wide range of variation in smaller carcasses created an overlap with dynamic and static loading scenarios. However, both the mean values and the ranges of variation for oblique angles in larger carcasses fit the ranges of static loading very well, although the overlap might reflect a sample of bones broken by hyenas and, to a much lesser degree, by hominids (Figure 78).

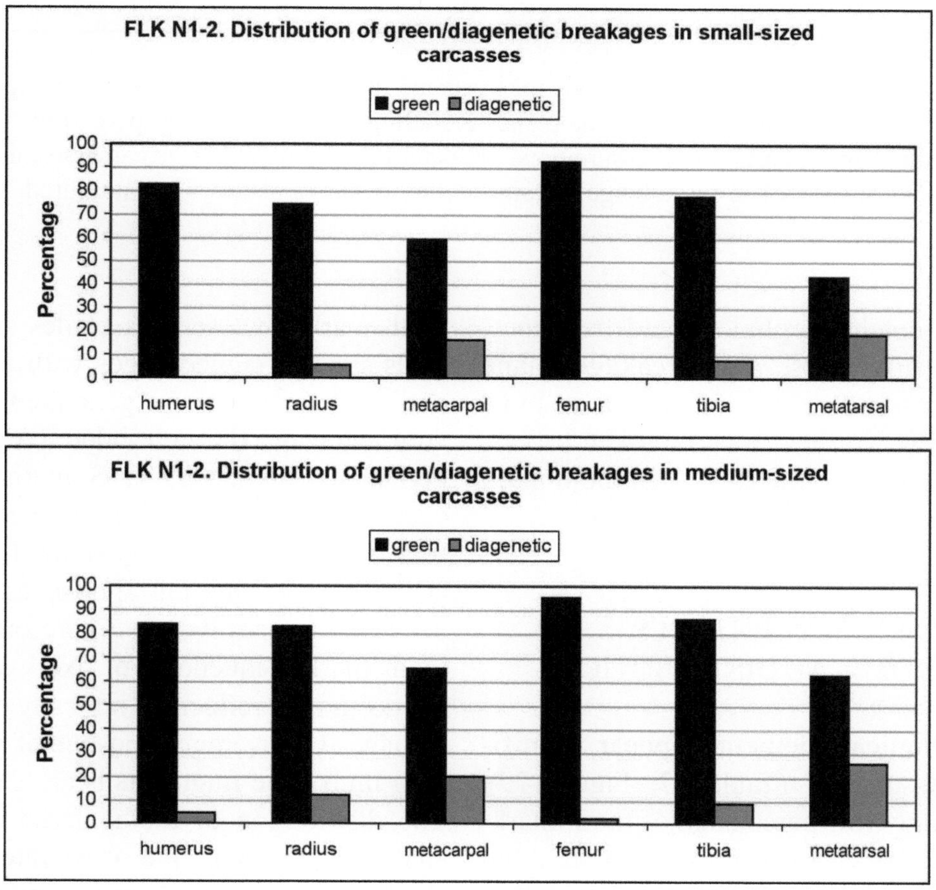

Figure 77. Distribution of frequencies of green fractures and diagenetic fractures for each long limb bone from small and large carcass sizes.

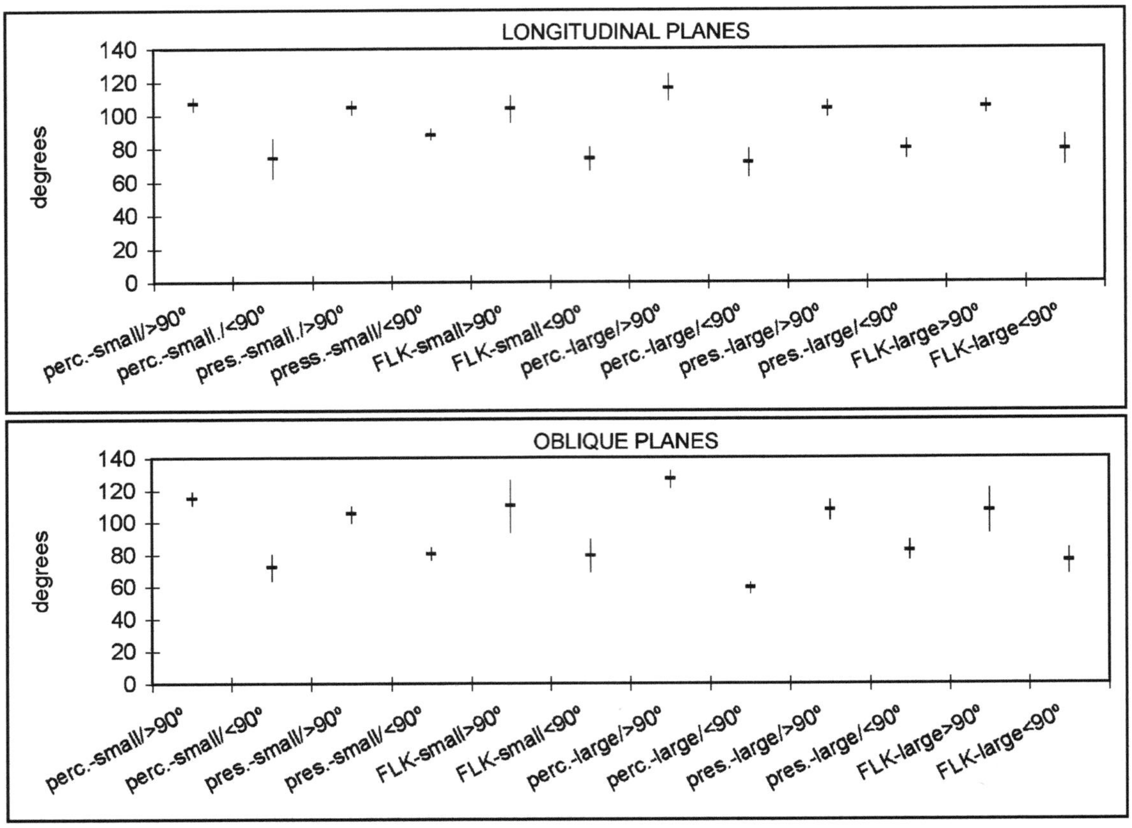

Figure 78. Mean values and 1 SD values for breakage plane angles (in degrees) for experimental assemblages on longitudinal and oblique planes modeling dynamic loading (perc., percussion) and static loading (pres., carnivore pressure/bone breakage) for small and large carcasses, and for the FLK North 1–2 assemblage. A total of 490 breakage planes were measured.

In sum, notch morphology and frequency, notch measurements, and breakage plane measurements seem to indicate that most bones at FLK North 1–2 were broken by carnivores. If that were the case, then tooth mark frequencies should support this assertion.

TOOTH MARK PERCENTAGES AND ANATOMICAL DISTRIBUTION

Most anatomical elements appear significantly tooth-marked (Table 37; Figure 79). The seemingly low percentage of tooth-marking in axial elements is spurious: scapulae and pelves (considered axial elements for taphonomic purposes) showed a low frequency of tooth marks, but scapula blades and innominates showed jagged edges with a crenulated outline typical of those generated by a carnivore chewing through bone (Andrews and Fernández-Jalvo, 1997). Something similar could be seen in vertebrae and ribs. Almost four out of five vertebrae showed this outline at the base of missing apophyses, indicating deletion by carnivores. Therefore, the low tooth mark frequencies of axial elements do not accurately reflect the intensity of carnivore damage. On average, one out of every three long limb bone fragments bears a tooth mark. The distribution of tooth marks is therefore anatomically even and moderate. Table 38 shows tooth mark frequencies and distribution on long limb bone portions. Most tooth marks

Table 37. Tooth mark percentages for each skeletal element in all carcass sizes at FLK North 1–2

	Size 1/2	Size 3	Size 4/5	Total
Vertebra	11/57(19.3)	15/68(22.1)	0/4(0)	26/129(20.2)
Ribs	10/124(8.1)	23/131(17.6)	0/1(0)	33/256(12.9)
Pelvis	8/42(19)	12/45(26.7)	0/2(0)	20/47(42.6)
Scapula	5/32(15.6)	3/42(7.1)	0/0(0)	8/74(10.8)
Humerus	32/56(57.1)	43/113(38.1)	1/1(100)	76/170(44.7)
Radius-ulna	22/67(32.8)	62/163(38)	0/1(0)	84/231(36.4)
Metacarpal	15/49(30.6)	15/71(21.1)	0/2(0)	30/122(24.6)
Femur	27/81(33.3)	46/143(32.2)	2/3(66.7)	75/227(33)
Tibia-fibula	36/120(30)	50/249(20.1)	0/6(0)	86/375(22.9)
Metatarsal	15/53(28.3)	36/113(31.9)	0/0(0)	51/166(30.7)
Phalanges	6/56(10.7)	3/59(5.1)	0/3(0)	9/118(7.6)
ULB*	1/11(9.1)	0/20(0)	0/0(0)	1/31(3.2)
ILB*	0/6(0)	2/14(14.3)	0/0(0)	2/20(10)
LLB*	2/9(22.2)	3/12(25)	0/0(0)	5/19(26.3)
Indeterminate	0/1(0)	4/12(33.3)	0/1(0)	4/14(28.6)
Total	190/764(24.9)	317/1255(25.3)	3/24(12.5)	510/2043(25)

Numerator is total number of tooth-marked specimens, denominator is total number of specimens of each skeletal element, and percentages are in parentheses.
* ULB, is upper limb bones; ILB, intermediate limb bones; and LLB, lower limb bones.

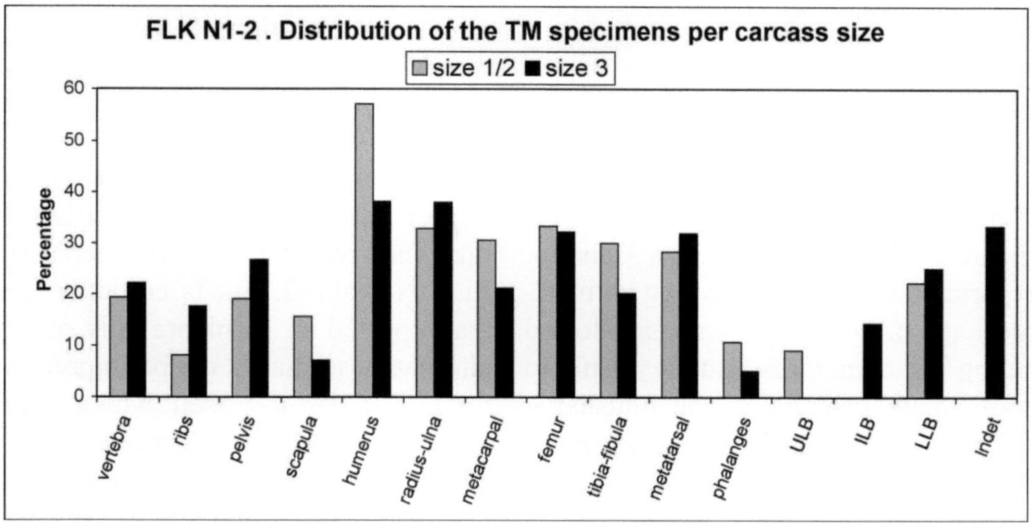

Figure 79. Frequencies of tooth-marking (percentage NISP) in each element from both carcass size groups. ULB, upper limb bone; ILB, intermediate limb bone; LLB, lower limb bone.

cluster on ends, and moderate frequencies are also preserved on midshafts (Figure 80).

Epiphyseal and near-epiphyseal fragments for both carcass sizes appear tooth-marked at rates that are similar to human (hammerstone)-to-carnivore experiments (Figures 81 and 82). This is also the case, surprisingly, for midshaft fragments, which are tooth-marked at rates lower than 35% for both carcass sizes; this is much lower than the >75% expected in carnivore-only models (Blumenschine, 1988, 1995). However, despite being at the upper

Table 38. Frequencies of tooth-marked specimens on long limb bones, by bone portion: epiphyses (epi), near-epiphyses (nep), and midshafts (msh)

	Size 1–2		
	epi	nep	msh
Humerus	16/23(69.6)	2/3(66.7)	11/26(42.3)
Radius-ulna	10/25(40)	2/5(40)	7/29(24.1)
Metacarpal	8/16(50)	0/1(0)	6/22(27.3)
Femur	8/12(66.7)	2/7(28.6)	17/62(27.4)
Tibia	10/30(33.3)	3/7(42.9)	22/75(29.3)
Metatarsal	11/22(50)	0/1(0)	4/23(17.4)
Total	63/128(49.2)	9/24(37.5)	67/237(28.3)
	Size 3		
	epi	nep	msh
Humerus	24/35(68.6)	1/13(7.7)	14/62(22.6)
Radius-ulna	29/65(44.6)	0/3(0)	26/83(31.3)
Metacarpal	3/24(12.5)	1/2(50)	11/39(28.2)
Femur	11/19(57.9)	13/24(54.2)	22/100(22)
Tibia	11/51(21.6)	4/13(30.8)	34/178(19.1)
Metatarsal	22/61(36.1)	1/3(33.3)	13/45(28.9)
Total	100/255(39.2)	20/58(34.5)	120/507(23.7)

Numerator is total number of tooth-marked specimens, denominator is total number of specimens of each skeletal element, and percentages are in parentheses.

limit of hominid-carnivore models, their frequencies are very atypical for such a scenario. Most experimental scenarios dealing with primary breakage of bones by humans followed by ravaging by carnivores usually result in lower tooth mark percentages on midshafts. The reason for such a particular tooth mark frequency at FLK North 1–2 is that we are probably dealing with a palimpsest, in which carcasses were deposited in different moments. Some were consumed by carnivores, namely hyenas, resulting in bone breakage, whereas others were deposited and buried intact.

It should be taken into account that 51 complete elements (14% of total MNE) survived carnivore ravaging and that a large portion of them are devoid of tooth marks. This obviously decreases the overall tooth mark rate. However, this does not affect the rates reported in Figures 81 and 82, since they were only obtained using the fragmented specimens. In this case, we emphasize that we may be dealing again with imperfections of the referential frameworks. High tooth mark frequencies like those reported for carnivore-only scenarios come from experiments where very few bones were abandoned for carnivore ravaging in a single depositional event. We have commented elsewhere (see previous chapters) that such a model may be unrealistic in the light of the abundance of remains from multiple depositional events available at Bed I sites for carnivore post-ravaging. It could be assumed that tooth-marking, the same as bone portion deletion, can be tightly linked to the degree of competition. In low-competition scenarios (in which the number of bones and lack of competition for them can generate less drastic modifications), carnivores, namely hyenas, can alter bones in a different way. The taphonomic study of shafts at the Maasai Mara hyena den, mentioned above, provides more support for this assertion. Shaft specimens from this den were tooth-marked at a rate >50% (Egeland *et al.*, in preparation). This frequency is still higher than that reported for both carcass sizes at FLK North 1–2, but is certainly lower than that reported for carnivore-only scenarios. An alternative to that is the palimpsest nature of the FLK North 1–2 itself, which could be due to the action of more than one tooth-marking agent, with one very conspicuous (like hyenas) and another one less conspicuous (such as felids). Both agents, including one that leaves high frequencies of tooth marks on shafts, and another one that leaves very few (Domínguez-Rodrigo *et al.*, in press), could result in a mixed, intermediate tooth mark frequency like that reported for FLK North 1–2.

Mapping the precise locations of marks on each element can help distinguish the agents and processes involved. A substantial amount of tooth marks are located on midshafts of all six long bone types (Figures 83 and 84),

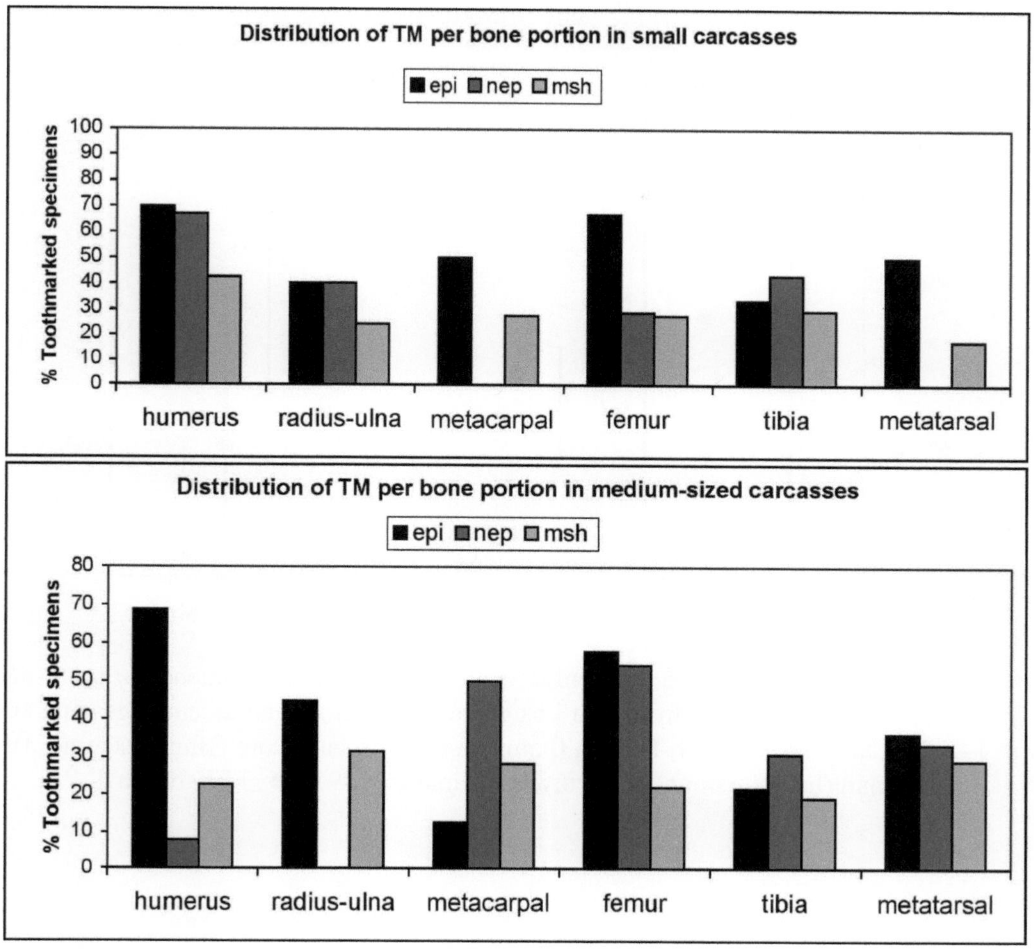

Figure 80. Frequencies of tooth-marking according to long limb bone portion in each element for both carcass size groups. Epi, epiphysis; nep, near-epiphysis; msh, midshaft.

although there is a higher frequency of tooth-marked specimens on epiphyseal and near-epiphyseal fragments relative to midshafts (Figure 80). This clearly shows that carnivores broke most of those elements that did not appear complete. Stylopodial elements (femora and humeri) from *Parmularius* show abundant marks on midshafts because the carcasses from this genus were more heavily ravaged than those of *Antidorcas*. Frequencies of tooth-marked specimens on midshafts of these elements in small carcasses are more reduced (Figure 83). The same can be observed in zygopodial elements (tibia and radioulna), which appear more heavily tooth-marked in larger carcasses (Figure 84). Metapodials show similar intensity in breakage in both cases, the differences being because the *Parmularius* sample is larger. A common pattern for both carcass sizes is that stylopodials and zygopodials exhibit intensive furrowing on the least dense portions. Furrowing on proximal and distal ends in humeri for both carcass sizes is very similar, whereas a contrast can be observed in femora, where furrowing on the distal end in small carcasses is much more moderate (restricted to the caudolateral aspect of the condyle) than in larger ones. Furrowing on ends of tibiae is similar in both carcass sizes, but on radii, distal ends from large carcasses are more intensively ravaged. The overall impression is less

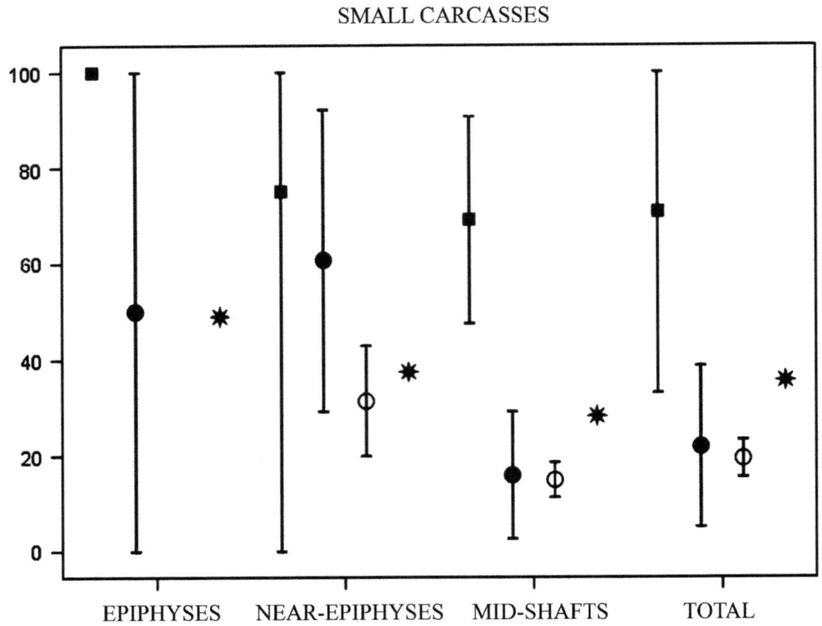

Figure 81. Distribution of the 95% CI (confidence intervals) for the frequency of tooth-marked specimens for each bone portion from small carcasses in experimental assemblages and at FLK North 1–2. ■, Carnivore only; ●, Human (hammerstone)-to-carnivore (Blumenschine, 1995); ○, Human (hammerstone)-to-carnivore (Capaldo, 1997); ✷, FLK North 1–2.

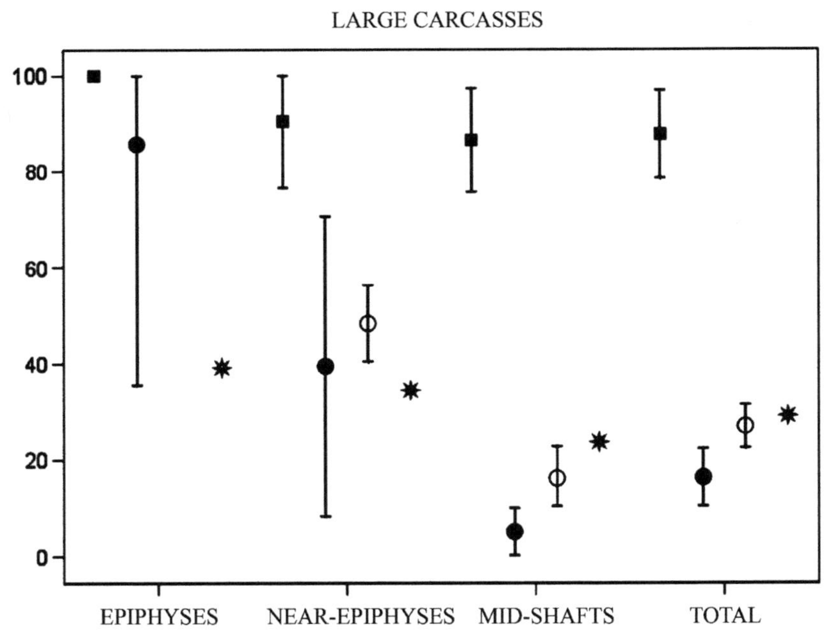

Figure 82. Distribution of the 95% CI (confidence intervals) for the frequency of tooth-marked specimens for each bone portion from middle-sized carcasses in experimental assemblages and at FLK North 1–2. ■, Carnivore only; ●, Human (hammerstone)-to-carnivore (Blumenschine, 1995); ○, Human (hammerstone)-to-carnivore (Capaldo, 1997); ✷, FLK North 1–2.

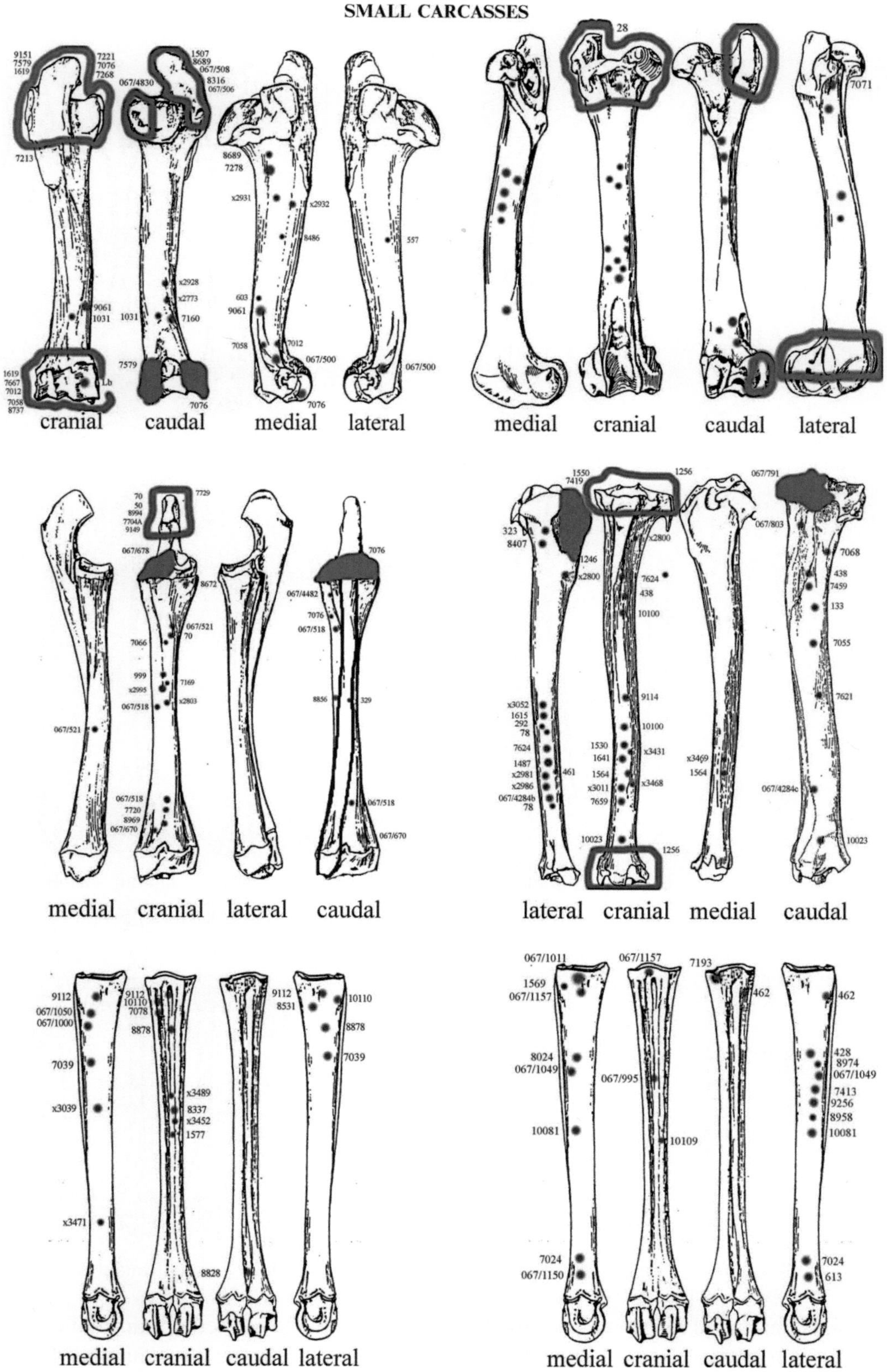

Figure 83. Anatomical distribution of tooth marks on the humerus, femur, radio-ulna, tibia, and metapodials from small carcasses. The letters and numbers indicate specimen number from the FLK North 1–2 assemblage. Specimens bearing marks which could not be anatomically located were not incorporated into this figure. Outlines and shaded areas show furrowing. Bones are redrawn from Pales and Lambert (1971).

LARGE CARCASSES

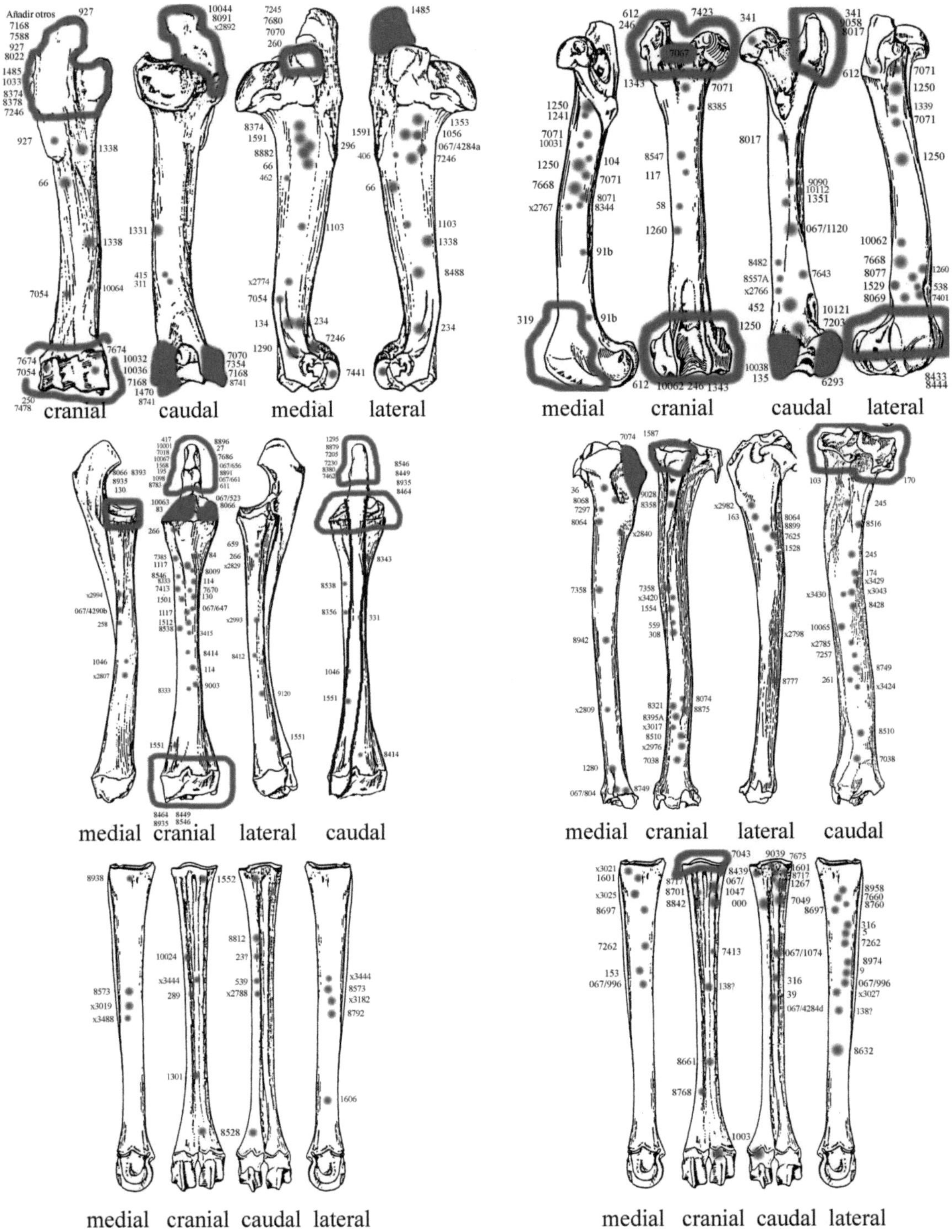

Figure 84. Anatomical distribution of tooth marks on the humerus, femur, radio-ulna, tibia, and metapodials from large carcasses. The letters and numbers indicate specimen number from the FLK North 1–2 assemblage. Specimens bearing marks which could not be anatomically located were not incorporated into this figure. Outlines and shaded areas show furrowing. Bones are redrawn from Pales and Lambert (1971).

ravaging and more bone portion completeness in small carcasses compared to the larger ones.

It can be concluded that most of the breakage so far described can be attributed to hyenas, especially in *Parmularius* bones, given their density. The frequent breakage of dense elements like zygopodials and metapodials through the midshaft clearly supports this interpretation. Alternatively, canids like wild dogs could also have broken those bones, but no experimental research with these carnivores is available.

This receives further support from the analysis of tooth pit size (Figure 85). Recently, Andrews and Fernández-Jalvo (1997), Selvaggio and Wilder (2001), and Domínguez-Rodrigo and Piqueras (2003) have suggested that diverse measurement methods applied to tooth pits could yield positive results in the identification of specific carnivore taxa (Selvaggio and Wilder; Andrews and Fernández-Jalvo) or groups of carnivores (Domínguez-Rodrigo and Piqueras), which might have modified modern and fossil bone assemblages. Selvaggio and Wilder used a method based on the length to breadth ratio, whereas Domínguez-Rodrigo and Piqueras used each variable independently, clearly establishing some significant differences in tooth pit sizes among carnivores (discussed in Chapter 3). This study discovered that pits on cancellous bones could be more easily attributed to carnivore type, whereas pits on dense diaphyseal cortical surfaces were harder to differentiate.

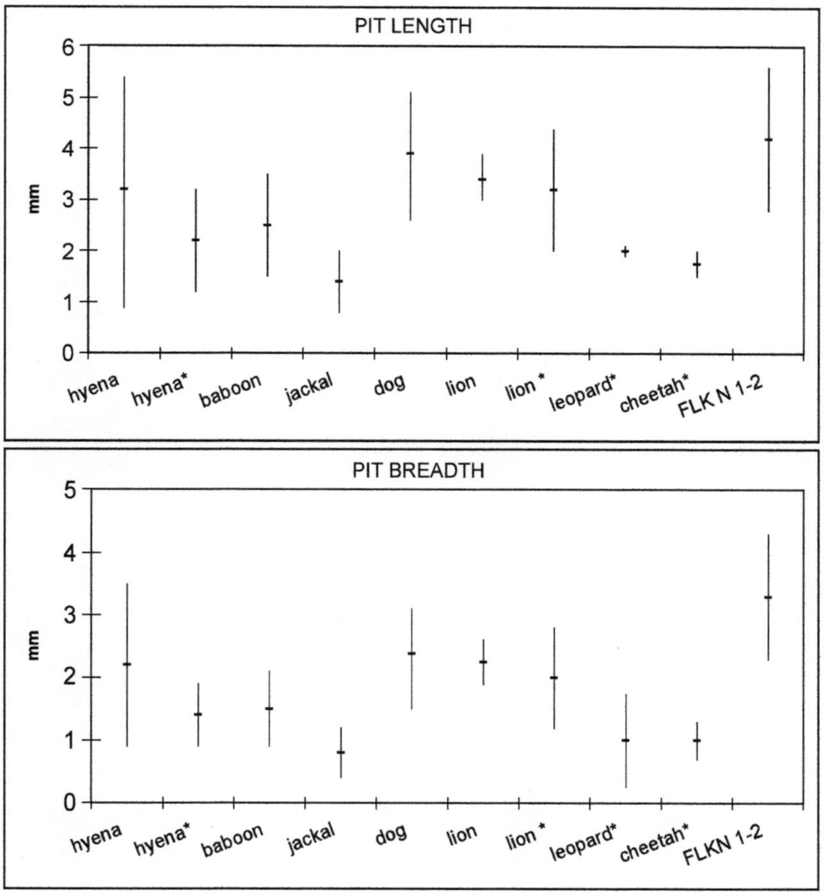

Figure 85. Distribution of tooth pit dimensions from FLK North 1–2 on shafts from large carcasses, and mean values plus 1 SD ranges for experimental assemblages measuring tooth marks made by different carnivore types (see data in Domínguez-Rodrigo and Piqueras, 2003). *Data from Selvaggio (1994).

Despite this, pit marks on cortical areas could be divided into two groups by size: the first group includes marks under 2 mm long and 1.5 mm broad, belonging to small carnivores, such as jackals and to middle-sized felids (cheetah and leopard). Marks above those measurements could be confidently attributed to other larger carnivores. In the second group, those marks over 4 mm long and 2 mm broad could be attributed to hyenas and lions, the only carnivores documented to create pits of this size.

A sample of conspicuous tooth pits from FLK North 1–2 were randomly selected for measuring (Figure 85). Results show that pit size (both in length and breadth) is outside the ranges documented for cheetahs and leopards, falling into the variation range of lions and hyenas. Given that carcass sizes at FLK North 1–2 are smaller than most of the animals that lions prey upon, it is safe to suggest that such large tooth pits could only have been made by hyenas. This would support the tooth marks and notches observed at the site and the intensive bone breakage, not documented among lions.

Bones of both *Antidorcas* and *Parmularius* show two different kinds of preservation. On the one hand, there are carcasses whose bones have undergone intense ravaging by hyenas (Figures 86 and 87), and others are so complete that it seems that hyenas never were responsible for their accumulation and modification (Figure 88). Bones ravaged by hyenas show their typical signatures, deletion of cancellous epiphyses and midshaft breakage in several stages, until the shaft is reduced to a near-epiphyseal section of the element (Figures 86 and 87). This behavior accounts for the tooth mark frequency and distribution observed and for all notches documented.

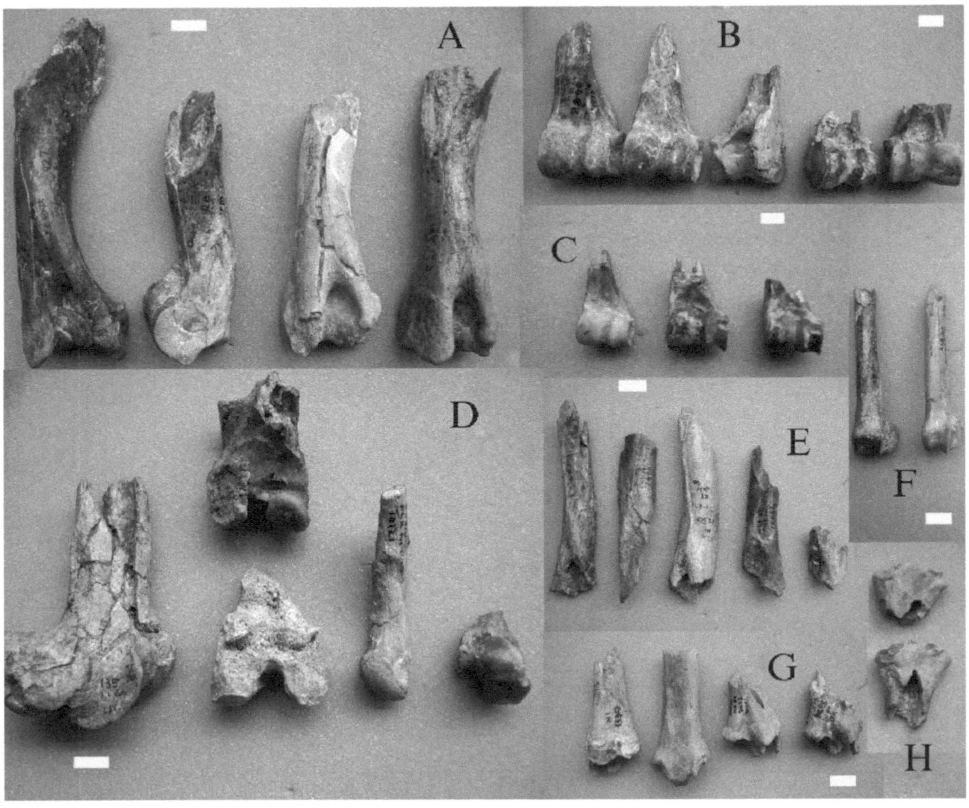

Figure 86. A small sample of fragmented bones at FLK North 1–2. Humeri (A, B), femora (D, E), and tibiae (G) from large carcasses, and humeri (C), femora (F), and radii (H) from small carcasses (scale = 1 cm).

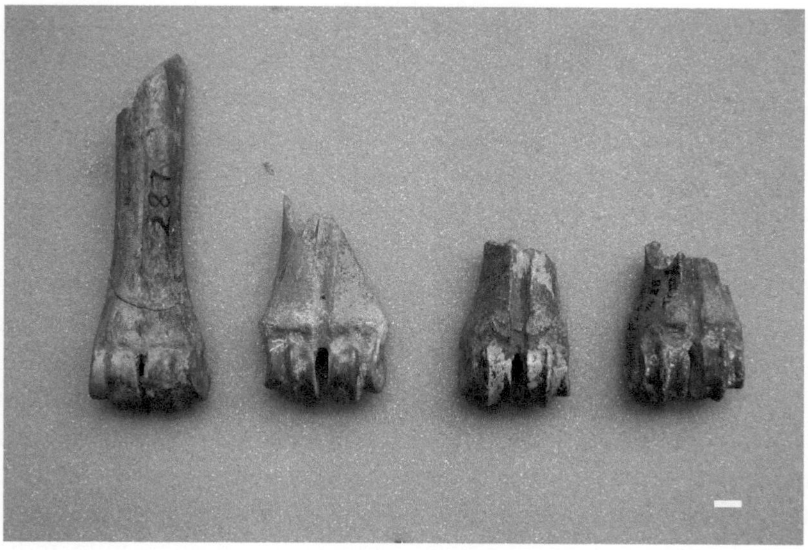

Figure 87. Distal ends of green-fractured metatarsals from large carcasses (scale = 1 cm).

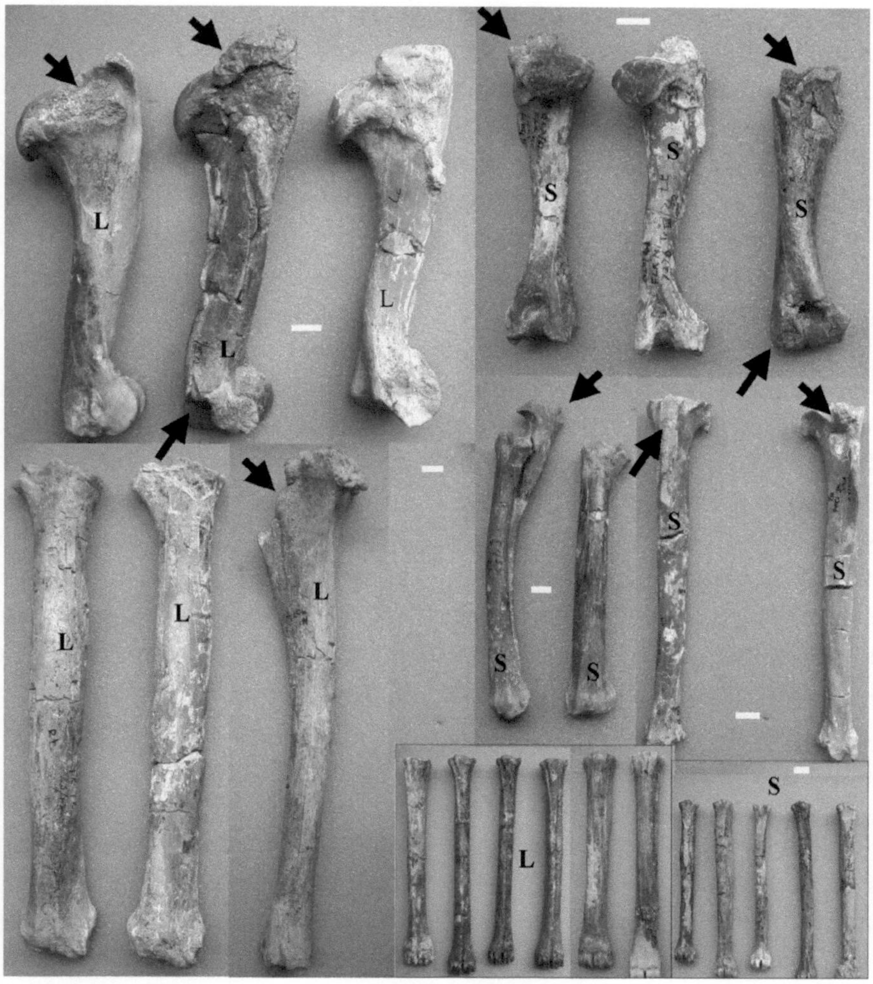

Figure 88. A small sample of the complete bones from FLK North 1–2. The three larger humeri on the left are from large (L) carcasses, whereas the three on the right are from smaller (S) carcasses. The two radii and one tibia on the left are from large carcasses. The smaller two radii and two tibiae on the right are from small carcasses. At the bottom right, the first six metapodials are from larger carcasses and the three to their right are from small carcasses. Arrows indicate furrowing (scale = 1 cm).

The intriguing part is the modifications observed in complete elements, which are more abundant in small carcasses, where one out of three limb bones was retrieved complete. In complete elements, some show a pattern in furrowing on certain ends (Figure 88). Humeri in both carcass sizes may appear slightly furrowed on their proximal ends (more heavily furrowed in small animals), affecting either trochleae or part of the articular surfaces. A very common pattern of furrowing (affecting between one third and one fifth of the distal ends of humeri, depending on carcass size) is documented on the distal caudal epicondylar area in both small and large carcasses. This is connected to the furrowing observed in the oleocranea of ulnae, also very common (almost four out of every ten). Complete tibiae also show some minor furrowing on the proximal epiphysis or, depending on carcass size, complete deletion of only the proximal epiphysis. This pattern contrasts with that observed in bones modified by hyenas, especially with bones from small carcasses. For humeri, for instance, hyenas exhibit a pattern of distal epiphyseal deletion which begins with the lateral condyle, affecting both the cranial and caudal sides equally (Kerbis, 1990). A pattern like the one described for the complete carcass elements from FLK North 1–2 has been reported for felids (Domínguez-Rodrigo *et al.*, in press).

Of the extant large-bodied carnivores, felids are the most specialized flesh eaters, with teeth developed almost exclusively for meat-slicing. There are only very weakly expressed mechanical adaptations for bone breaking in the felid dentition. Behavioral observations of carcass consumption by large cats (e.g., Turner and Anton, 1997) support the morphological indications that felids will likely avoid the intentional contact of teeth with bone. This, in turn, leads to the prediction that felids will impart fewer tooth marks on bone assemblages than will hyenas (e.g., Brain, 1981). Selvaggio (1994) was one of the first zooarchaeologists to provide data in support of this hypothesis by quantifying tooth mark densities (number of marks per bone specimen) in experimental assemblages. For felid-derived assemblages, 75% of the specimens have tooth mark densities < 21 (Selvaggio, 1994). More specifically, up to 96% of small animal bone assemblages displayed tooth mark densities ≤ 20. These values are significantly lower than carnivore-only experimental assemblages, in which hyenas had been responsible for bone modifications and in which frequencies > 70 were usually reported. These results indicate clearly that felids tooth-mark limb bones at much lower rates than hyenas. This contrast is even more acute when observing complete bones remaining unbroken by felids after carcass defleshing. For such specimens, 50% displayed not a single tooth mark (Selvaggio, 1994). This frequency becomes much smaller once the complete element is broken into several specimens (Domínguez-Rodrigo *et al.*, in press). Despite this, felids can conspicuously modify long limb bones. They may furrow both epiphyses of femora and proximal ends of humeri very heavily together with scapulae. They also impart serious damage on the caudal side of the distal humeral epicondyles, leaving the frontal facet of epicondyles intact. This process also affects ulnae, whose oleocranea may disappear. Proximal tibiae also can undergo severe chewing, leaving most of the shaft intact. In sum, felid modification of bones is primarily seen in three areas: caudal distal humerus, proximal ulna, and proximal tibia (see Figure 89).

The fact that these modifications are frequently found in several complete elements from both carcass sizes supports the interpretation that those elements belong to carcasses modified by felids as a result of defleshing and which were also very likely transported by them to the site. The significant numbers of tibiae and especially humeri modified like this (between 20% and 33%, similar to those

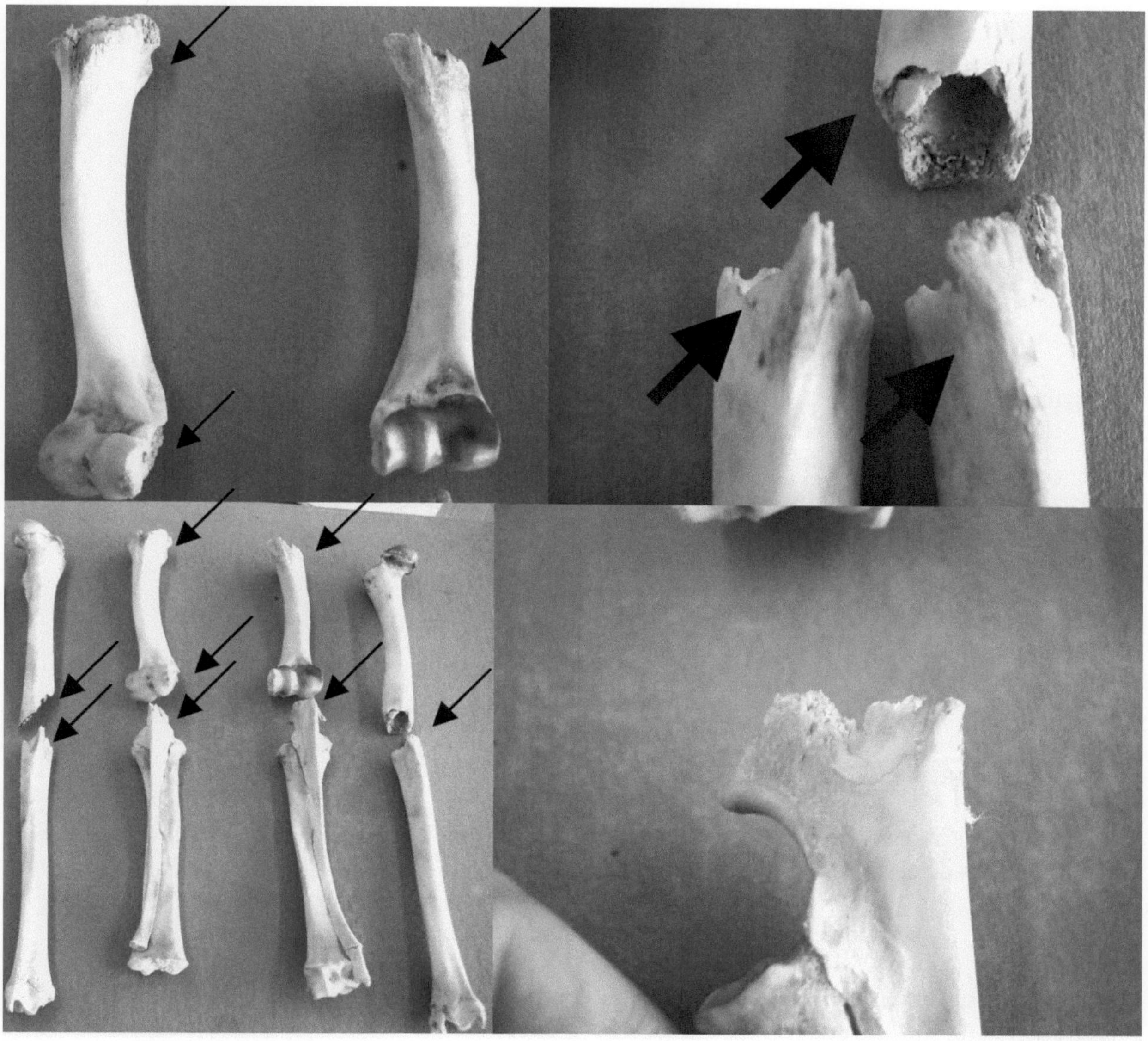

Figure 89. Typical damage to bones inflicted by felids. Images are from leopard-modified elements from the Valencia Farm experiments (Brain, 1981). Arrows show furrowing. This is especially visible on proximal and distal ends of femora and humeri, and on proximal ends of the tibiae (upper right) and ulna (lower right).

documented in felid-modified assemblages) in the remaining broken assemblage would indicate that other carcasses might also have been transported by felids to the site and were later ravaged by hyenas. Humeri with the caudal facet of the epicondylar area furrowed appear in the form of specimens with epiphyses plus long and short shafts, or with the shaft completely reduced to a small portion of the near-epiphyseal area, clearly indicative of moderate to intensive postdepositional ravaging. However, if there is a single anatomical section where felid bone modification is observed in detail (given the relative absence of hyena intervention in the surviving elements), that is on the axial skeleton: more specifically, on vertebrae. Most vertebrae that have been preserved bear small tooth marks and have their centra intact. There are even articulated sections of axial

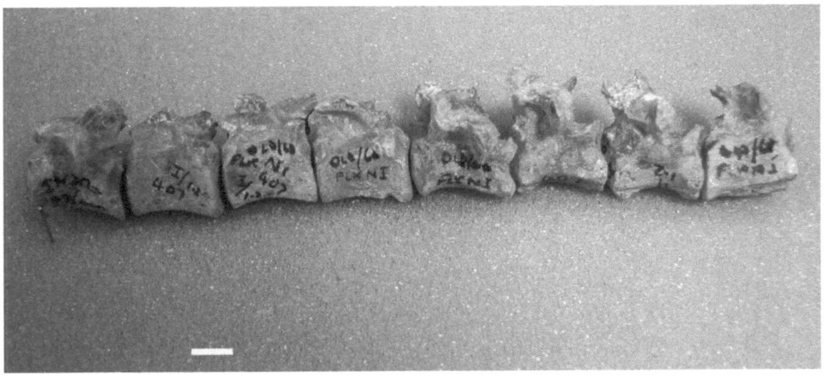

Figure 90. Partial articulated axial skeleton of an *A. recki* from FLK North 1–2, showing intact vertebral centra and deletion of apophyses (scale = 1 cm).

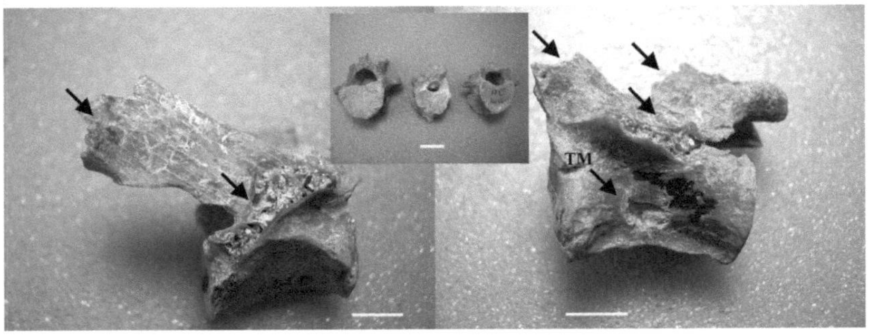

Figure 91. Typical damage documented on vertebrae from small carcasses at FLK North 1–2. Arrows show apophysis deletion and fragmentation. The vertebra to the right has a tooth mark on the centrum (arrow). The small photograph in the middle shows more examples of typical damage to vertebrae from larger carcasses in the assemblage (scale = 1 cm).

skeletons from *Antidorcas* (Figure 90). The only sections that have been deleted are the apophyses (Figure 91). Since these are small carcasses, and none of these elements for these carcass sizes have been documented to survive hyena ravaging or are present at hyena dens, the only candidate for their modification is a felid. Cavallo (1998) and Domínguez-Rodrigo (1999a) described the way that the vertebral spine of bovids is modified by felids. Irrespective of their size, they follow the same pattern, which preferentially destroys apophyses over vertebral centra.

This interpretation could tentatively offer the first taphonomic window to a felid-hyaenid behavioral model of bone accumulation and modification.

Discussion

FLK North 1–2 is the result of two seemingly independent agents of bone transport: felids (who accumulated most of the bones), and hominids (responsible for a smaller amount); and two agents of bone breakage: hyenas (as the principal agent), and hominids (as the marginal agent). The few probable percussion marks (given the excellent preservation of bone surfaces) in the FLK North 1–2 assemblage does not enable us to defend the hypothesis that hominids broke most bones at the site. Measurements of carnivore notches, frequencies of double opposing and especially overlapping notches, measurements of breakage planes, percentages of tooth marks by element type and bone portion, and skeletal part profiles and long bone portion representation are more indicative of carnivores as the main bone-breaking agent at the site. If there are stages in the heuristics of interpretation, the most solid one, that is the one that can easily be defended in a scientific way and devoid of speculation, is that FLK North 1–2 is a bone accumulation made mostly by carnivores and contributed to by hominids in a more marginal way.

Recently, Blumenschine *et al.* (in press) have claimed that almost 40% of the midshaft fragments from FLK North 1–2 bear percussion marks and 10% bear at least one tooth mark. They suggest this is a perfect example of a hominid-to-carnivore model. They do so using two lines of evidence (percussion marks and tooth marks), obviating other taphonomic data related to bone completeness and fragmentation. Blumenschine *et al.* (in press) would have to explain why their analysis of the site yields higher frequencies of percussion marks than their human (hammerstone)-to-carnivore experiments (where all bones were broken), especially when hominids would have abandoned more than 15% of their long limb elements unbroken at FLK North 1–2. Furthermore, if the site is a hominid-to-carnivore assemblage, why are shaft circumference sections so different from human-to-carnivore experimental assemblages? There seem to be too many complete shafts, given such a high frequency of "percussion" marks at the site.

One reason why we differ from Blumenschine *et al.* (in press) in the identification of percussion marks may lie in the fact that over 24% of the long limb bone assemblage at FLK North 1–2 shows traces of micro-abrasion and trampling. This indicates that the assemblage must have been intermittently exposed for enough time for various agents to trample the bones. Indirectly, it also shows that the spot where the bones accumulated must have been attractive to these agents. However, this micro-abrasion may be partially responsible for the overestimates of percussion marks in Blumenschine *et al.*'s (in press) study. Several specimens bear traces of striae with random orientations. This cannot be mistaken for a percussion mark. However, some marks appear in the form of patches containing striae in the same direction (almost always in specimens also bearing clear abrasion marks not caused by stone tools), which sometimes differ from percussion marks in that the striae are not densely packed, as is the case with percussion marks as reported by Blumenschine and Selvaggio (1988, 1991), but are rather spread out; however, in other cases, the striae fields are extremely similar to those produced by hammerstone percussion (Figure 92). The presence of micro- and macro-abrasion caused by trampling also prevented us from incorporating very few specimens with some striae in the cut-marked bone sample.

Following Turner (1983), Pickering and Egeland (2006) divided percussion marks into two classes: pits (associated with striae) and striae fields. Turner (1983) had called the striae fields "anvil scratches," assuming that they resulted from reabsorbed impact on the bone surface lying against the anvil. However, Pickering and Egeland (2006) demonstrated that a large portion of these striae fields are also the result of direct impact by a hammerstone. However, more interesting is the fact that striae fields with very similar morphology to striae fields documented in percussed bones have also been documented in experimentally trampled bones (Figure 92). We observed several such marks in a small, previously unpublished study of bones trampled on sandy silt in the lacustrine floodplain of Lake Natron, which is lithologically similar to the geological contexts where the Olduvai sites occur (Figure 92). In previously published trampling experiments, similarities between long striations and cut marks had been noted, but nothing had been mentioned of the similarities between much smaller and more subtle marks

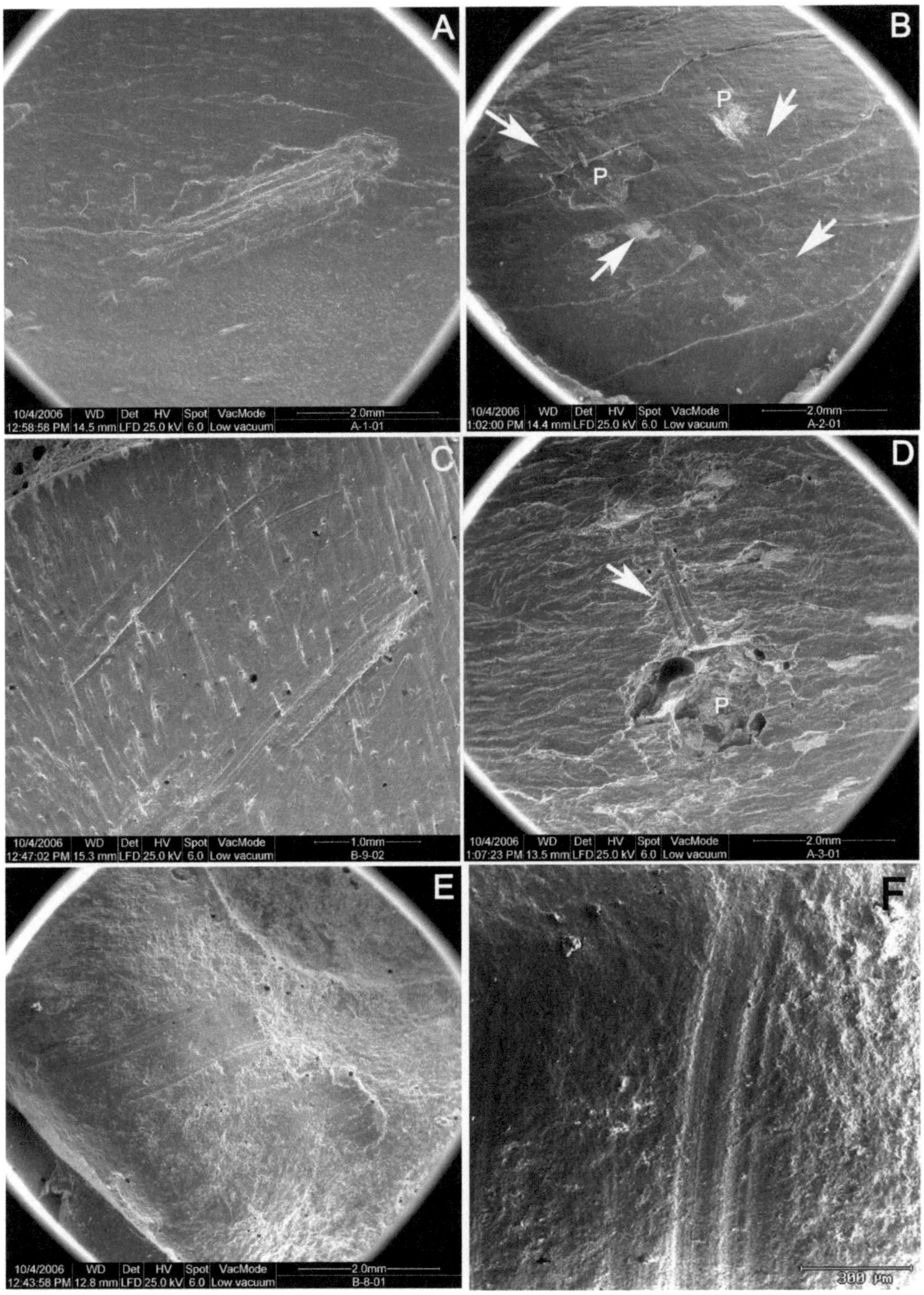

Figure 92. Several examples (at a magnification of 40X) of striae fields created by trampling. (A) striae fields and a shallow fossa at the end from a complete (unbroken) humerus retrieved from the Eyasi hyena den; (B) two shallow pits (P) with associated striation fields emanating across them (arrows) from a trampling experiment; (C) striae fields from a shaft specimen broken and trampled by hyenas found at the Eyasi hyena den; (D) tooth pit (P) and associated striae field emanating from it (arrow) caused by abrasion of the tooth on a rib fragment found at the Eyasi hyena den; (E) striae fields created in trampling experiment; (F) striae fields found on a bone specimen from the Eyasi hyena den (image taken in a different SEM from images A–E). The specimens from the Eyasi hyena den are currently under study (Prendergast and Domínguez-Rodrigo, in preparation). Compare these images with the percussion-made striae fields shown in Figure 2 in Pickering and Egeland (2006).

and the striae fields known as percussion marks (Andrews and Cook, 1985). The striae fields similar to percussion striae fields observed in the experimentally trampled bones needs some magnification to be well-identified. The micro-striae fields could be perfectly mistaken for percussion marks (Figure 92). For this reason, we suggest that in case of documented natural abrasion (e.g., trampling) of any given assemblage, the use of pits is more reliable than the use of isolated striae fields (see below), especially if they can only be documented with aid of lenses. Striae fields without pits are a common signature of trampling, as we have recently also documented in a hyena den near Lake Eyasi (work in progress) (Figure 92).

Fortunately, this issue of equifinality can be overcome. Pickering and Egeland (2006) reported that most percussion marks created by them in an assemblage consisting of 74 bones (which became 1,283 fragments after breakage) were pits (>80%) over striae fields. This is especially true in shaft portions where 85% of the percussion marks documented were pits. Therefore, in any trampled, hominid-discarded bone assemblage, if focusing on pits alone we should be able to identify at least 80% of the original number of percussion marks. In FLK North 1–2, there are only 12 specimens bearing pits with associated micro-striations (even after using magnification) out of *all* the specimens. However, there were more specimens showing microscopic striae fields, most frequently associated with abrasion marks in the remainder surface of the bone. These microscopic striae fields were documented in almost all skeletal elements: pelves, ribs, scapulae, skull, and limbs, sometimes even on complete bones. If they were percussion marks, then the reason for pounding some of those bones with hammerstones is quite unclear. The scarcity of pits suggests that most of the documented striae fields are just natural micro-abrasion features. The high incidence of percussion marks identified in Blumenschine *et al.* (in press) is probably the result of misidentifying these natural microscopic striae fields. This would explain why in the "percussion mark" data reported by Blumenschine *et al.* (in press), epiphyses fall completely outside the range of variation for human (hammerstone)-to-carnivore experiments, and the total count of percussed specimens barely make it to the lower range of their human (hammerstone)-to-carnivore experimental sample.

Since Blumenschine *et al.* (in press) did not publish the list of specimens with marks from FLK North 1–2, it is virtually impossible to explain in more detail the reasons for disagreement in the identification of marks, as is done with tooth marks at FLK *Zinj* (Chapter 5). However, Capaldo – the analyst of the site in Blumenschine *et al.* (in press) – left some notes with some specimens on which he had identified marks (Figure 93). These notes show that natural micro-abrasion is not the only cause for the widely divergent mark frequencies reported by each team. Some specimens bearing tooth marks have been misidentified as percussion marks by this researcher. The extent to which these misidentifications have contributed to their inflated frequencies can only be estimated when a published list of specimens bearing marks is available.

Given that traces of butchery or marrow extraction by hominids are far fewer than those marks left by carnivores, we support the more parsimonious view that carnivores had a larger role in the modification of bones at FLK North 1–2 than hominids had. The next stage of interpretation is more inferential. It would be very easy to think of hyenas as the only carnivore involved in bone accumulation, given that today it is the only one that could create an accumulation similar to that of FLK North 1–2. However, the site lacks juvenile hyaenids, coprolites, and digested bone (only 6 specimens found) as would be expected in a den (Brain, 1981; Brugal *et al.*, 1997). Furthermore, there

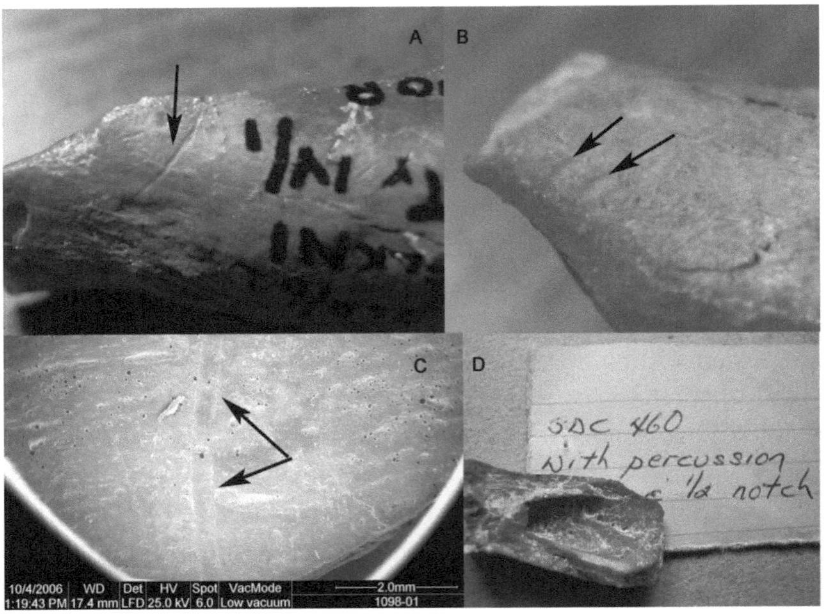

Figure 93. This specimen (phalanx) (A–D) was classified as 8608 together with six other phalanges in the Leakey stored collection. Capaldo singled it out as an example of a hammerstone-broken phalanx bearing traces of percussion (Image D shows Capaldo's labeling of the specimen as SDC 460). Marks occur both near the "notch" (A) and on the opposite side of the notch (B) and are in fact tooth scoring (A) and tooth pitting (B) resulting from a carnivore breaking the bone, rather than hammerstone-imparted marks. This is further supported by the SEM analysis (35X) of the score which clearly shows that it bears all the diagnostic features of a tooth score (C).

are too many complete bones, especially from small animals such as *Antidorcas*, to support the idea that the entire assemblage was exposed to hyena ravaging. The tooth mark frequencies are also too low for a hyena den (Brain, 1981; Bunn, 1982; Fosse *et al*., 1998; Marra *et al*., 2004; Villa *et al*., 2004). Finally, the species profile would be extremely odd for a hyena. The collector of carcasses at FLK North 1–2 was a specialist focusing on two species: *A. recki* and *P. altidens*, which are both relatively small animals: Bunn's (1982) sizes 1 and 3A, respectively.

Typical felid bone damage has been documented at FLK North 1–2 in complete elements, presumably unmodified by hyenas, which offers a different explanation for site formation. According to data from some of the most abundant limb bones (tibiae and metatarsals) and from teeth, up to 6 individuals from *Parmularius* and 5 individuals from *Antidorcas* were subadults. This increases the sample of small individuals in the assemblage. Using biometric measurements, we matched left and right limb elements in each carcass size, which suggests that the bones entered the site as complete carcasses rather than isolated elements or limbs. This is supported by the fact that while axial bones are scarce according to the number of carcasses represented, they are very abundant in the assemblage: one out of three bones is an axial bone. This high presence of ribs and vertebrae, especially after hyena ravaging (and thereby deletion) of part of the assemblage, further suggests that the original number of axial elements before hyena intervention was much higher and, therefore, that carcasses at the site were transported complete. Hyenas do not transport complete carcasses to their dens, but felids do transport complete carcasses (see Cavallo [1998] for leopards, Schaller [1972] for lions, and Domínguez-Rodrigo [1994b] for cheetahs. Repeated transport of small-size carcasses smaller than *Parmularius* to the same spot by leopards has been documented by Cavallo (1998) in the Serengeti. The taxonomic range of prey accumulated by leopards in the Serengeti (Cavallo, 1998) as well as those reported from other studies (Scott, 1987) suggest that the most frequent predatory behavior on mammals by leopards is restricted to very few species, usually no more than three.

About 40% of the carcasses represented at FLK North 1–2 (*Antidorcas* and juvenile *Parmularius*) could easily have been transported and stored into trees by leopards. The remaining carcasses could have been transported either by leopards (without storing them in trees) (Ruiter and Berger, 2000) or by any other felid slightly bigger than a leopard, but smaller than a lion. The best candidate is *Dinofelis*. Remains of *Dinofelis* have been found at the site. The similar physical structure of *Dinofelis* to leopards would have enabled them to exhibit a similar adaptive pattern. However, *Dinofelis* would probably have had a wider predatory range, given their slightly larger size (Lewis, 1997). An interesting insight into the felid responsible for the accumulation of *Parmularius* carcasses is gained by looking at a tooth-marked specimen. A proximal humerus from a *Parmularius* exhibited a tooth puncture that could not have been produced when the bone was defleshed because there was no tooth mark on the opposite side of the neck from the articular surface where the puncture is located. The force necessary to produce such a deep puncture would have required some tooth-marking to occur on the opposite side. Given the absence of such an opposing puncture, we suggest that the puncture was made when the carcass was still fleshed. We tried to match carnivore canines with the puncture, and the best fit was obtained with a lower canine from *Dinofelis* (Figure 94).

Figure 94. Fitting of a canine from *Dinofelis* into the tooth puncture located on one of the *Parmularius* humeri (scale = 1 cm).

Whether one felid (leopard) or two (leopard and *Dinofelis*) are responsible for the bone accumulation at FLK North 1–2, the impressive number of carcasses is clearly the result of time-averaging over a deposit slightly less than 1-m deep. All the carcasses were accumulated over extensive periods of time in which leopards and/or *Dinofelis* might have used the space, and during which hyenas might have been present for part of these depositional events. Hyenas also could have varied in their ravaging behavior at each visit, thus creating the splendid palimpsest that is FLK North 1–2.

Leakey (1971: 61) noted that the site (including all the trenches that were opened separately) comprised two circular areas: "[I]n Levels 1–2 there were three areas where remains were particularly concentrated. Two of these were circular, between 8 and 10 ft. [about 2.5–3 m] in diameter, and lay near to one another." These clusters could correspond very well in size to two different depositional spaces surrounding two trees. Cavallo (1998; personal communication) identified similar clusters of bones around trees where leopards stored carcasses, which fell to the ground after either consumption or drying up. It is obvious that the carcasses of *Antidorcas* are less intensely ravaged than those of *Parmularius*. If the *Parmularius* carcasses were too heavy to be stored in trees and had to be processed on the ground, their remains would have been more likely to have been ravaged by hyenas than those from smaller carcasses, which would be more easily transported into trees and thereby temporarily protected from other carnivores. This would explain the smaller frequencies of tooth marks and higher percentages of complete bones in *Antidorcas*.

An interesting buttress for this explanation comes from examining all the carnivore species represented at FLK North 1–2. This level has the highest number of carnivore specimens and individuals of all the sites from Bed I. A total of 185 remains have been identified, of which 126 are Viverridae. The remaining 59 fragments belong to five different types of larger carnivores: one hyena (two hemimandibles, one pelvis, one distal tibia, one metatarsal, and one ulna), one *Dinofelis* sp. (two incisors and one metatarsal), three jackals (44 skeletal elements), one *Panthera pardus* (four phalanges and one metatarsal), and two large felids (two femora), which were previously identified as *Homotherium* sp., but this taxonomic attribution is dubious (this is not reflected in Lewis [1997] probably because of nonsecure identification). Both femora are very similar to those of *Panthera leo*.

It is logical to think that carnivore remains are more likely to occur in *loci* where individuals spend more time. This suggests that more than one tree-climbing felid was present at the site for long time intervals and also more than one *Panthera* cf. *leo*, which could be due to these large felids bringing carcasses to the same spot. This would explain the presence of carcasses belonging to large animals: equids, one Alcelaphini size 3B (very likely *Connochaetes*), two Tragelaphini size 3B (one being *T. strepsiceros*), one Hippotragini and one buffalo. The presence of leopard and *Dinofelis* bones suggests that these two types of carnivores must have spent enough time in this spot to die there.

The taphonomy of micro-mammals at FLK North 1–2 shows that the environment was very likely an open woodland, with a significant number of trees (Fernández-Jalvo *et al.*, 1998). The small mammals, which were discovered together with the clusters of macro-mammal remains, were interpreted as deriving from eagle owls, which habitually roost in trees and drop their pellets in a rough circle around the base of the tree, and this suggests that the bone clusters were indeed located near or under trees.

Taking into account that the bone accumulation at FLK North 1–2 is almost exclusively the action of carnivores, very likely felids, what role did hominids play in site formation? Bones appeared associated with 1,211 stone tools (de la Torre, 2006); the complete Oldowan tool kit is present (see Chapter 7). De la Torre's (2006) analysis of the lithic industry concluded that flakes were less than 7% ($n = 84$) of the assemblage and that hammerstones and anvils (and their fragments) are more abundant (10%). Considering raw material weight, both hammerstones and anvils constitute 41 kg. Fifty two kilograms of the total 87 kg of raw material brought into the site show clear evidence of battering activities. The presence of 25 quartz anvils further supports the interpretation that battering was the most common activity performed by hominids at the site. The total amount of hammerstone-broken elements at FLK North 1–2 cannot be reconstructed over more than nine limb bones. There are almost three times as many anvils as there are broken elements. The scarce (and in several cases doubtful) presence of hominid-broken bones at the site (especially given the large number of bones present) therefore suggests that these battering activities were not linked to carcass exploitation activities. From a techno-functional point of view, the activities carried out by hominids at the site and the subsequent deposition of lithic materials on the one hand, and the accumulation of most of the fauna on the other, have independent depositional histories. In the light of these taphonomic data, we argue that exploitation of plant and nut resources should be considered to explain the evidence of battering activities.

An experiment on small and middle-sized carcass butchery suggests that a minimum of about 15–20 complete flakes are needed to efficiently butcher a complete carcass (work in progress). This gives us a minimum estimate of flakes per carcass that could be archaeologically tested. Had all carcasses at FLK North 1–2 been defleshed by hominids and not felids, a total of 840 flakes should have been present at the site (as well as a high percentage of cut marks). Instead, only 84 flakes have been found at the site, which limits its butchery to a maximum of five carcasses. The hominid-modified bones discovered at the site account for a minimum of four carcasses, one small and three large. However, the sample is insufficient to fully explain the contribution by hominids to the site. There are two options: hominids contributed, independently from carnivores, with four carcasses to the site or they interacted with carnivores, processing carcasses from their kills. Even if the size of the cut-marked bone sample is too small, here are the reasons why we believe the former option was more likely:

1. Cut mark anatomical distribution was the result of skinning, dismembering, and defleshing activities that would not be expected (since they would produce a negative energy return) if hominids had access to carcasses after felids consumed them.
2. Even if we consider that hominids may have scavenged tree-stored carcasses from leopards at their earliest stages of consumption, anatomical cut mark location argues otherwise. Out of the 21 cut-marked specimens, 9 are from femora, the most cut-marked element, which are usually devoid of flesh after the first stage of carcass consumption by felids (Cavallo, 1998; see Chapter 9).

The specimens that have combined tooth and cut marks reveal order of bone modification. One distal femur from *Parmularius* shows a carnivore notch on the shaft with tooth mark. Another femur from *Antidorcas* shows a crenulated edge of the shaft that is also suggestive of carnivore breakage as the oblique angle of the breakage plane indicates. Finally, a humerus and a metacarpal of *Antidorcas* show a double opposing and overlapping Type C notch typical of carnivore bone breakage (see Figure 72). This means that hominids did not break open those elements, although they did deflesh or disarticulate them, which would not be expected if they were scavenging limited resources from felid kills. We believe that given the lack of convincing evidence otherwise, hominids occupied the site when carnivores were not active in it, because it would have provided a low-competition environment. That would explain why felids would also use it to transport their carcasses. Hominid battering activities – as was the case at FLK North 6 (Chapter 7) and will be the case at other sites (following chapters) –were related only in a very marginal way to bone breakage. The interpretation provided for this level is that it is a good example of palimpsest in which some agents were interdependent (felids-hyaenids) but others were independent (hominids). How can archaeologists distinguish these activities from a taphonomic point of view?

Fortunately, important actualistic work on humans and carnivores provides the fundamental datasets by which a model of an archaeological palimpsest, like the one inferred for the FLK North 1–2, can be constructed. In other words, it is possible to build a composite palimpsest assemblage from available experimental assemblages created independently and exclusively by different agents. We begin simply by considering a single carcass modified by hominids and a single carcass modified by spotted hyenas. Spotted hyenas, unlike lions, often collect bones in shelter sites, such as caves. Because of their general attractiveness to multiple bone-collecting agents, caves often serve as natural venues for the creation of palimpsests, with independent (usually serial) taphonomic contributions from hominids and carnivores. Thus, our simplified model of an archaeological palimpsest is restricted to a two-actor model, but may be usefully expanded later.

The extensive experimental work of Capaldo (1995), using fairly complete carcasses, is probably the best source that demonstrates differences in bone fragmentation by hyenas versus hominids. First, the average survivorship of appendicular fragments after hyena carcass ravaging is ~14 fragments per complete carcass, while a human-processed carcass yields on average 86 appendicular fragments. Significantly, a large majority of fragments created by hyenas and humans are midshaft specimens. Further, bone consumption and fragmentation by hyenas deletes *all* original appendicular specimens in nearly half of the hyena experiments conducted by Capaldo (1995, Appendix 4). Capaldo (1995) reported a ratio of number identified specimens (NISP) to minimum number of elements (MNE) of 1.2 specimens per complete bone recovered after 30 experiments. In Capaldo's (1995) human-created assemblages, the NISP:MNE recovery ratio is much higher, 7.6. The implication for a theoretical palimpsest is that limb bone fragments of a single carcass processed by hominids at an accumulation site will be more highly represented than the limb bone fragments of another single carcass processed by hyenas at the same site.

With regard to bone surface modifications, Capaldo (1995) reports that ~70% of the appendicular specimens recovered in experiments in which hyenas were the primary carcass-ravaging agents display tooth marks. This percentage is only slightly lower than the 84% tooth-marked value reported by Blumenschine (1988, 1995) in his hyena-first experiments. In addition, both researchers demonstrate that secondary access by hyenas to human-fractured bones results in low percentages of tooth-marked appendicular specimens (20% in Capaldo's experiments and 19% in Blumenschine's). More specifically, the percentages of tooth-marked midshaft portions are only 15% in Capaldo's human-to-hyena experiments and 10% in those of Blumenschine.

When these data are combined, we obtain a theoretical assemblage of 100 total limb bone specimens, of which most are midshafts: 86 from carcass processed by humans only, plus 14 from the carcass processed by hyenas only. Of these 100 specimens, 20% are expected to bear tooth marks; 9 (or 10%) of the 86 limb bone specimens generated by hominid-first carcass processing will be tooth-marked secondarily by hyena scavengers, while 11 (or 80%) of the hyena-first fragments will be tooth-marked.

If hyenas were to deposit the limb bones of yet another carcass to the assemblage, tooth mark frequencies would increase to 27% (31 out of 114 specimens). In other words, the addition of one carcass by hyenas would increase the previous tooth mark percentage by 7%. Following the addition of a sixth carcass, tooth mark frequency increases at a rate of ~2% per carcass. Only after the addition of a ninth carcass does the tooth mark frequency for all carcasses exceed 50%.

We note here that our simple model does not consider the contribution of hominid-derived carcasses, subsequently ignored (and thus unmodified) by secondary hyena scavengers. This complicating factor would certainly change the theoretical implications developed above. In addition, the relatively substantial hyena involvement that is modeled above need not be the only way in which the relative contribution of hominid carcass accumulation would be diluted in a palimpsest. Fragmentation of bones by hyenas in many modern dens (e.g., Maguire *et al.*, 1980; Bunn, 1982, 1983b; Skinner *et al.*, 1986; Skinner and van Aarde, 1991; Lam, 1992; Pickering, 2002; Domínguez-Rodrigo, personal observations), is often less intense, with the generation of far fewer fragments per element, than observed in the open-air settings of Capaldo's and Blumenschine's experiments. If this is the case for an archaeological palimpsest, fewer hyena-contributed carcasses could account for a higher frequency of tooth-marked specimens.

The important message here is that although tooth mark frequencies will be very high in palimpsest assemblages to which hyenas have made a significant contribution, these frequencies *do not address even indirectly* the timing of hominid access to other carcasses which contribute to those assemblages. This conclusion further supports earlier contentions that only hominid-imparted bone surface modifications hold the potential to construct realistic inferences of hominid involvement with animal carcasses. Our opinion on the high analytical utility and explanatory power of cut mark placement, both on intra-skeletal and intra-bone scales, is well-published and asserts that hominids often gained early access to the carcasses of large animals (Domínguez-Rodrigo, 1997a, 1997b; Domínguez-Rodrigo, 1999a, 1999b, 2002; Domínguez-Rodrigo *et al.*, 2002; Domínguez-Rodrigo and Pickering, 2003; Pickering *et al.*, 2003), agreeing with the conclusions of some previous analysts (e.g., Bunn, 1982; Bunn and Kroll, 1986). As a supplement to that work, here we focus on cut mark and hammerstone percussion mark frequencies, and the way in which these frequencies can provide information on the relative contribution of early hominids to faunal palimpsests.

As was discussed in Chapters 2 and 3, a number of modern cut-marked assemblages are available, derived from both ethnoarchaeological and experimental studies. Across these datasets, a consistent range of 10–30% of all specimens recovered after hammerstone bone breakage are cut-marked (e.g., Bunn, 1982, in identifiable bones only; Domínguez-Rodrigo 1997a, 1997b, in the H1/S1/2d butchery experiment only; Lupo and O'Connell, 2002). Turning to hammerstone percussion marks, we return to Blumenschine and Selvaggio's (1991) human (hammerstone)-to-carnivore (hyena) experiments discussed in Chapter 3. Percussion mark frequencies on limb bone specimens (irrespective of bone portion) varied by carcass body size (following Brain's [1974, 1981] size classes): size class 1 and 2 limb bone assemblages showed percussion mark frequencies of ~30%; size class 3, 21.5%; size class 4, 53.8%. Additionally, their percussion mark to tooth mark ratios ranged from 1.5:1 (in size class 3 carcasses) to 2:1 (in size class 2 carcasses). In other words, when hyenas have secondary access to bones after hominids, percussion marks are nearly twice as frequent as tooth marks. We believe that the *relative* contribution of hominids to the bone assemblage formation can be calculated from these frequencies.. If cut mark and percussion mark frequencies are significantly lower than in experimental assemblages where humans were the primary agents of carcass processing, this probably indicates a relatively minor hominid contribution to the fossil assemblage.

As discussed at the beginning of this chapter, cut mark frequencies at FLK North 1–2 are far lower than expected in a hominid-carnivore model: 21 specimens out of 2,650 bovid specimens (0.7%). Particular patterns in tooth-marking, discussed earlier, may be due to two different agents (felids and hyaenids) mixing their bone "signatures" (since felids tooth mark bones are very low rates, whereas hyaenids show higher frequencies), while at another point in time, hominids contributed a few independently broken bones. We reemphasize that this is just a first (and simplified) step to model an accurate ESA palimpsest. Much more work is needed. However, the important point remains that when reconstructing the formation of fossil faunas, zooarchaeologists need to consider much more diligently a probable prehistoric reality: hominids and carnivores operated not just inter-dependently (as currently modeled in passive scavenging scenarios), *but also independently*.

Conclusions

Carnivores (principally) and hominids (marginally) contributed independently to the formation of the FLK North 1–2 bone assemblage. The site, in the form of a vertically dispersed archaeological deposit, is time-averaged and indicates very extensive periods of time between depositional events. It is obvious that the site attracted both carnivores and hominids at different times. If both agents transported resources to it, it must have been because it provided the necessary safety for carcass consumption. Therefore, it must have been a very low-competition setting, as suggested by the moderate destruction of bone and high survival of elements from very small carcasses (see contrast in Capaldo, 1995). Previous interpretations of some Bed I sites as located in the middle of lacustrine floodplains, which are usually devoid of vegetation (Blumenschine and Masao, 1991, 1995) are unsubstantiated. A carnivore or a hominid would not repeatedly bring carcasses to the same spot unless that spot affords some protection.

The fauna accumulated at the site indicates a mosaic of environments nearby. However, most of the individuals are Antilopini and Alcelaphini. This indicates a predominance of open spaces in the vicinity of the site. Savanna carnivores, with the exception of hyenas, do not transport their prey over long distances. Domínguez-Rodrigo (1994b) differentiated between three types of carnivore strategies. *In situ* consumption of prey is the most widespread carnivore strategy. This does not lead to any bone accumulation unless it is carried out systematically on favoured *loci* where serial predation takes place. This could explain the background non-proboscidean cluster at FLK North 6, discussed in Chapter 7; that is, the same site as FLK North 1–2, four archaeological levels below. When under competition, carnivores transport their prey to a peripheral area. In most of the hunting episodes by lions and cheetahs observed by Domínguez-Rodrigo (1994b), when peripheral transport took place this usually did not exceed 300 m. Brain (1981) also considered longer transport of prey, exceptional. This means that if carnivores in Early Pleistocene savannas were behaving similarly, the abundance of *Antidorcas* and *Parmularius* at FLK North 1–2 must have been obtained within a radius of ~300 m around the site. The open floodplain between the site and the lakeshore seems to have been the place of carcass obtainment.

De la Torre and Mora (2005) claim that manuports in Bed I are natural ecofacts. If so, they can only be found in the transition between the alluvial plain that only rarely is flooded, and the floodplain that is seasonally or regularly flooded. In terms of vegetation, that would often correspond to the transition between the lacustrine forest (more often, transitional woodland) and the open edaphic grasslands of the floodplain. This is usually a low-competition setting. Hominids may have been part of the predatory guild at that time (see Chapter 2), and therefore it should come as no surprise that they were overlapping in space with other carnivores. Even if FLK North 1–2 is a vertical deposit, the large abundance of remains would be better understood if natural background scatters at Olduvai, either along the landscape or in *loci* where preferential bone accumulations were naturally created, could be better known. Fortunately, Bed I has preserved some of these spots (see Chapter 11).

FLK North 1–2 suggests that the highly specialized taxonomic range of animals present was accumulated over a long time span. Multiple agents have been shown to have independently contributed to assemblage formation. No signs of a possible link between hominids and the felid-accumulated remains were detected. Such an interaction might be more easily visible in a less time-averaged faunal assemblage. More vertically restricted assemblages can be found in the underlying levels of this site (FLK North 3 and parts of FLK North 4), which further support the assertions produced in this chapter. These underlying levels can also be used to test a version of the passive scavenging hypothesis (hominids scavenging leopard tree-stored kills) once again. Some authors have suggested that this would be the only passive scavenging behavior enabling hominids to acquire significant amounts of meat (Cavallo and Blumenschine, 1989). If hominids had invested energy in scavenging carcasses from leopard-sized felids, we should expect some taphonomically detectable proof of this interaction to be preserved in the archaeological record. The next chapter leads us through a taphonomic analysis of FLK North 3 and 4 as a test of this hypothesis, while continuing the study of independent-agent palimpsests.

9. A taphonomic study of FLK North 3 and 4: a felid–hyaenid and hominid palimpsest

M. DOMÍNGUEZ-RODRIGO, R. BARBA, AND E. ORGANISTA

Introduction

We have seen in the previous chapter that the faunal accumulation at FLK North 1–2, in Olduvai Bed I, was not in fact hominid-made, but rather was the result of carnivores accumulating carcasses in specific *loci* on the landscape that were also used by hominids, presumably when carnivores were not around. The taxonomic and taphonomic signatures from those assemblages suggest that felids were probably the main bone accumulators and hyenas the main bone modifiers.

Why were carnivores and hominids selecting the same destinations for transporting and accumulating carcasses? It has been argued that the Bed I sites were located in the middle of a lacustrine floodplain (Blumenschine and Masao, 1991). However, in such ecological contexts, no trees or refuge areas exist and competition is high. Why, then, would a carnivore choose to accumulate carcasses in such an area, if the sole goal of doing so is to avoid competition? Why would a hominid do the same? Is it possible that sites were instead located on the distal end of a lacustrine floodplain, in the transition toward the alluvial plain where the lacustrine forest must have existed? This scenario would better agree with C3 isotopic signatures detected by Sikes' (1994) analysis of Lowermost Bed II paleosols. This would also explain the location of the ecofacts which were previously interpreted as manuports. As discussed in the previous chapter, de la Torre and Mora (2005) have shown that most "manuports" are similar to those natural cobbles that high-energy water flow may bring into the edge of lacustrine floodplains during the rainy season. Alternative and at the same time complementary to this interpretation is the fact that the portion of the floodplain where the site appears could have been located near a drainage area. River channels meandering across the flood plain often change location, and when they overflow their banks they deposit a graded sequence of sands and silts, thinning away from the channels. This produces sandy levees alongside channels, and if these are built up sufficiently they may rise above lesser flood levels and support trees. Under certain circumstances, they may even support areas of forest, as for instance on the Tana River floodplain (Andrews *et al.*, 1975).

Cavallo's (1998) observations of Serengeti leopards, lions, hyenas, baboons, and vultures suggest that if sites were in the edges of floodplains, they could have been occupied for sustained periods by carnivores and hominids in a manner analogous to the foraging behavior of modern savanna-dwelling baboons. This and other studies suggest that predation risk to hominids at sites would have been negligible due to their reliance on refuge trees and that – based on modern savanna ecology and taphonomic evidence – competition for carcasses in the habitats where sites were formed could

M. Domínguez-Rodrigo, R. Barba, and C.P. Egeland, Deconstructing Olduvai: A Taphonomic Study of the Bed I Sites, 165–189.

have been also low (Blumenschine, 1986, 1995; Capaldo, 1995; Domínguez-Rodrigo, 2001).

The assemblages from most Olduvai Bed I sites exhibit extensive, high-density clusters of bones that are unique when compared to similar bone accumulations seen at carnivore den or kill sites in modern East African savanna-woodland environments such as the Serengeti (Potts, 1984, 1988). Furthermore, they contain an equally unique diversity of bovids, suids, equids, and carnivores: mixed species and multiple kills in the same locations. The abundant small animal remains at these sites fit well with the pattern of high bone survival for small carcasses, as documented by Blumenschine (1989) and Domínguez-Rodrigo (1996) in riparian woodlands of the Serengeti and of Galana and Kulalu, respectively. In each of these woodlands predator pressure is low.

In the previous chapter, it was argued that felids, leopards or *Dinofelis*, could have been responsible for a large part of the fauna accumulated in Levels 1 and 2 at FLK North. Cavallo (1998) has repeatedly observed leopards accumulating carcasses in the same trees. In 1990, he recorded a total of 14 kills during 22 days of continuous observations, for an average of 1 kill every 1.57 days. Cavallo (1998) also recorded four instances of multiple kills consisting of a total of eight carcasses of size 1 and size 2 animals in the same or adjacent trees. The bones from these carcasses would have eventually fallen on the ground and would have created a significant accumulation, if postdepositional disturbance due to ravaging was minimal.

It has been argued that hominids might have scavenged from similar tree-stored kills (Cavallo and Blumenschine, 1989). This could explain the association of stone tools with what is being labeled as a felid-accumulated and -defleshed carcass assemblage. Blumenschine's (1986) carcass consumption sequence showed that the graded nutrient yields of flesh- and marrow-bearing units of a carcass (i.e., parts ranked from highest to lowest in terms of net yield per unit of handling time) appear to "condition" the selection of and order in which the parts are consumed. Importantly, he observed that hindquarters are consumed first by lions and hyenas, followed by forequarters. After consuming head flesh, these carnivores then move back to consume first the hindlimb marrow then the forelimb marrow. Finally, with hyenas only, the head contents are eaten.

Cavallo (1998) documented that leopards, like lions, consume the flesh of prey from back to front. An important variation, however, is the leopard's early consumption of facial flesh and nasal marrow and other parts of the head. Although leopard consumption of the lower viscera, pelvic and femur flesh, and femur and humerus marrow correspond with Blumenschine's sequence, there is a significant deviation in the delayed consumption of forelimb flesh and associated upper limb marrow-yielding bones by leopards. That is, whereas it is highly unlikely to scavenge flesh from a lion kill (Domínguez-Rodrigo, 1999a), it could theoretically be possible to do so from a leopard tree-stored kill. Cavallo (1998) also found that tree-stored kills of small antelopes remained potentially scavengeable resources for 2–3 days, as opposed to Blumenschine's (1986) recording of 1 h for similarly sized prey that were killed and consumed on the ground by lions and hyenas.

However, if hominids were accessing carcasses at the late stage of leopard consumption as represented in Cavallo's collection, bones would be defleshed and no cut marks would appear on their surfaces. In contrast, the analysis of the distribution of cut marks on small-sized carcasses at FLK *Zinj* proves otherwise (Chapter 6). The high incidence of cut marks together with the even anatomical distribution on the appendicular section (Bunn and Kroll, 1986) can only be obtained when replicating primary access to fleshed carcasses (Domínguez-Rodrigo, 1997a; Domínguez-Rodrigo and Barba, 2005). This clearly suggests that if hominids were scavenging small-sized carcasses from leopards, they must have done it regularly from the earliest stages of carcass consumption by these felids. This means that if most flesh was still present, leopards would have left even lower tooth-mark frequencies than those reported for the last stage of carcass consumption by leopards in Cavallo's collection.

If hominids had access to carcasses after the first stage of leopard consumption, leopards would have already deleted all pelvic flesh including upper hindlimbs, viscera, and facial and mandibular flesh. Therefore, no cut marks would be expected on the ventral side of ribs (suggestive of evisceration), nor on the midshafts of femora (devoid of the scraps of flesh that may survive carnivores' initial consumption), nor on the facial part of the skull. Likewise, these sections are more likely to bear tooth marks from leopards.

Hominid access to tree-stored carcasses in an intermediate stage of carcass consumption by leopards would result in the total absence of cut marks on middle and distal sections of ribs (since the ribcage flesh has disappeared), and also a lack of hominid-inflicted modifications (cut marks and percussion marks) on the tibia, since it may be defleshed and demarrowed. Later access would only allow access to marrow, and therefore, no cut marks would be expected (Figure 95). Some of these anatomical areas are not very likely to be preserved in the

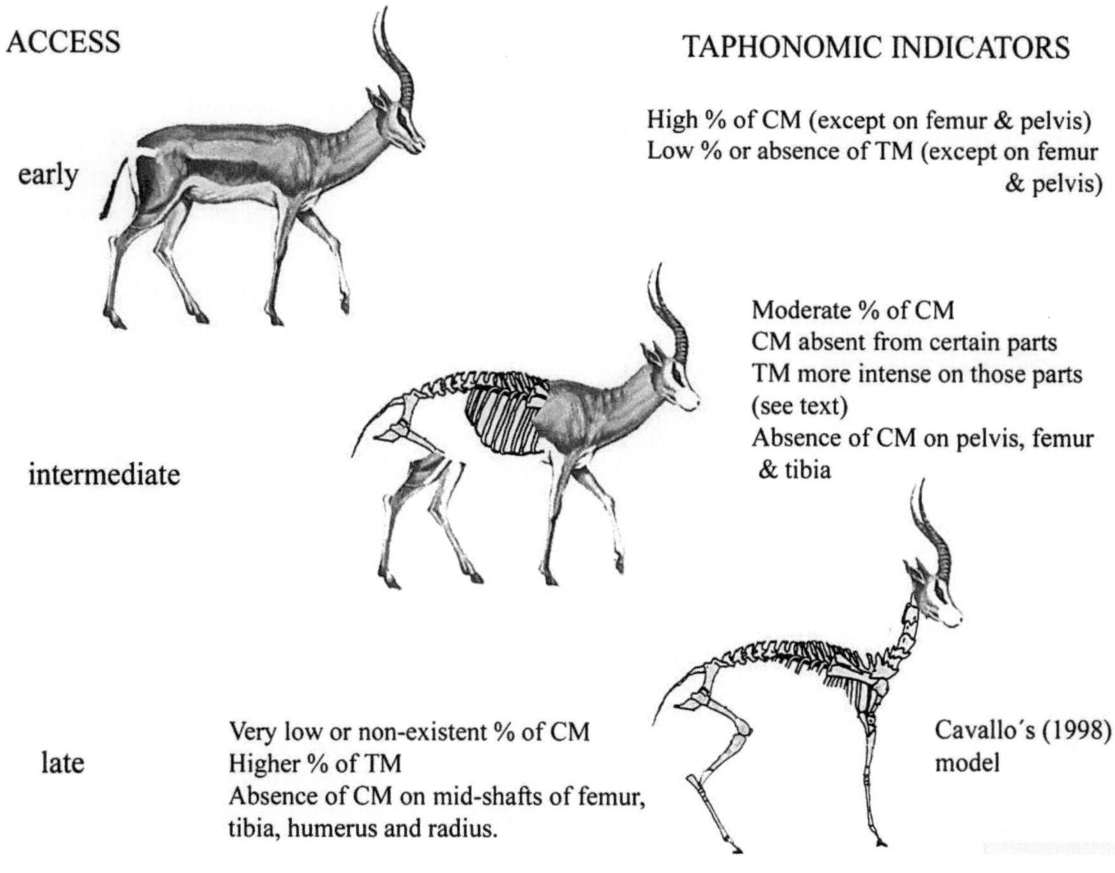

Figure 95. Three theoretical stages of hominid access to tree-stored carcasses by leopards, and the anatomical distribution of flesh resources available (Cavallo, 1998). Taphonomic attributes indicate location and frequency of expected bone surface modifications (tooth marks and cut marks) imparted by both hominids and leopards.

archaeological record given their preferential deletion by postdepositional ravaging carnivores (Marean *et al.*, 1992; Capaldo, 1995). However, early and intermediate access to carcasses scavenged from leopards should be reflected in very low or nonexistent frequencies of cut marks on the midshafts of femora and tibiae, respectively. The front limb is consumed late in the sequence and hominid-created marks on its bones would not be informative about the timing of access to the carcass by hominids.

The faunal assemblages from Levels 3 and 4 at FLK North are ideal to test the passive scavenging hypothesis. Whereas it has been shown that there are no grounds for a passive scavenging model from middle-sized carcasses from felid kills at FLK *Zinj* (see Chapters 5 and 6), the leopard-kill model for small carcasses has not been properly addressed in other sites. If hominids obtained resources from leopard kills, it would be expected that they would have exploited the remaining flesh and marrow. This would be reflected in the presence of cut marks and percussion marks and in low frequencies of tooth marks on midshaft sections, which would indicate that carnivores did not break open any bone. FLK North 3 and 4 are an ideal testing scenario since their bones show very good cortical preservation and most bone surface modifications can be easily identified. If both levels represent sites where hominids interacted with leopards, two scenarios are feasible: either the sites were the actual leopard kill sites, or the sites were central places or refuges to which hominids brought carcass remains scavenged from felid kills. One would expect to find the traces of such interaction in cut mark and percussion mark frequencies and distributions. Likewise, one would assume that hominids would have thoroughly exploited carcasses that were small and limited in resources, if there were previous consumption of parts by leopards. However, if there were taphonomic confirmation that carnivores were the main bone-breaking agents at the site, that result would contradict this scenario.

FLK North 3 and 4 have been selected to test this hypothesis, given the abundance of small carcasses, and to gather more information about felid and hyaenid bone accumulations and modifications. Both levels contain clear indications that carcasses were brought into the site by carnivores, and that they were consumed by both felids and hyaenids. Hominid presence at the site has been once again confirmed by the presence of stone tools (see Chapter 7) in a deposit that is thinner than that of FLK North 1–2 and, therefore, presumably spans less depositional time and could show more resolution.

According to Leakey (1971), FLK North 3 and 4 were living floors. Level 3 is situated in a gray-brown silty clay, while Level 4 was found in a dark chocolate-brown silty clay, varying in thickness from 1.5 ft to 4 in. (about 46–10 cm). In certain areas, Leakey (1971: 62) describes this level as "lensed out entirely, so that Level 3 rested directly on Level 5." Archaeological materials in FLK North 4 were scattered over the excavated area without any recognizable pattern. Some parts of FLK North 3 were difficult to differentiate from the overlying FLK North 1–2. In our analysis, we have excluded those specimens that were classified as belonging to either of the three levels and have focused on those that were clearly separated by level.

Results

SKELETAL ELEMENT REPRESENTATION

Before applying any referential framework to the interpretation of these two archaeological levels, it is necessary to ascertain the degree to which these levels underwent postdepositional modification by nonbiotic agents, particularly hydraulic sedimentary processes. Not a single specimen was found to show any traces of abrasion or polishing due to water transport. It is safe to conclude that neither assemblage was transported by water into the location where it was found. The clay sedimentary context also supports this assertion. Bone size representation at FLK North 3 indicates better preservation of small specimens from small carcasses than of small specimens from larger ones, when all elements are considered (Figure 96). When comparing similarly dense bone portions, this difference disappears. However, specimens smaller than 30 mm are slightly less represented than expected. Overall, bone size distribution suggests a high degree of assemblage integrity, with only minor sedimentary disturbance which may account for the disappearance of a small fraction of the smaller specimens.

Postdepositional disturbance seems more significant at FLK North 4, where there is a sharp contrast in the representation of small fragments from small and large carcass sizes (Figure 96). This is also reflected when considering similarly dense bone portions, like midshafts, although the contrast in this case is smaller. This could tentatively be indicative of a postdepositional disturbance that differentially affected larger carcasses, perhaps suggesting different deposition times for each carcass type, and therefore, alternating use of the space by bone-accumulating agents. When considering midshafts alone, very few fragments smaller than 30 mm are seen. This is suggestive of minor to moderate postdepositional disturbance by sedimentary processes, which are more significant than at FLK North 3.

In both archaeological levels, the taxonomic range represented seems to be dominated by *Parmularius altidens* and *Antidorcas recki*, as was the case with FLK North 1–2, discussed in Chapter 8 (Table 39). Preliminarily, this might suggest similar agents operating in all these levels of FLK North. Like the overlying levels, both FLK North 3 and 4 assemblages are

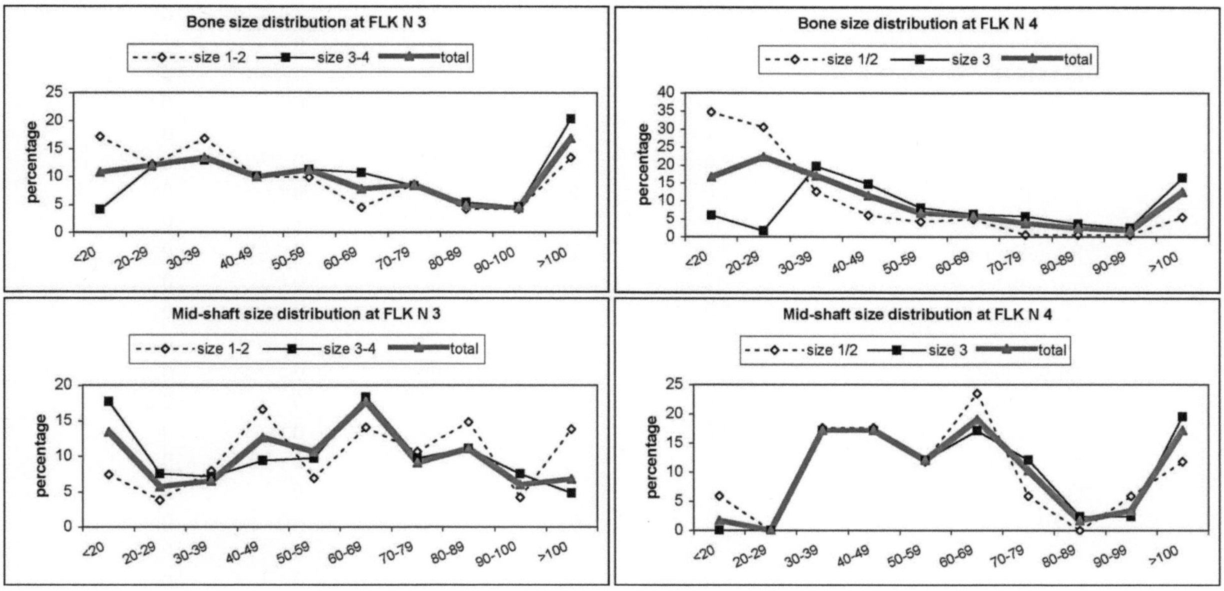

Figure 96. Distribution of frequencies for each specimen size range at FLK North 3 and 4, given for all bones (top two graphs) or for midshafts alone (bottom two graphs) for both carcass sizes. Measurements are in mm.

Table 39. Minimum number of individuals (MNI) identified at FLK North 3 and 4

MNI	FLK North 3	FLK North 4
P. altidens	8	8
A. recki	7	3
Tragelaphus strepsiceros	1	1
Connochaetes gnu	1	0
Alcelaphini size 4	0	1
Redunca redunca	0	1
Hippotragini	1	1
Metridiochoerus sp.	1	0
Kolpochoerus sp.	2	0
Suidae size 3	0	4
Giraffa jumae	1	1
Canis mesomelas	4	1
Crocuta crocuta	0	1
Panthera pardus	0	1
Total	21	23

dominated by elements from bovids (Tables 40 and 41). The greatest number of specimens comes from the skull and limbs (Table 40). When considering appendicular bone portion distribution (Table 42; Figure 97), both assemblages show identical patterns: distal epiphyses of humeri are several times more represented than are proximal ends, which are under 10% of what was expected. The pattern is similar for tibiae but reversed for radii, where proximal ends outnumber distal ends. Metapodials are well-represented and proximal ends are more abundant than distal ends. Overall, the pattern is similar to that reported for FLK North 1–2 and suggests a significant degree of density-mediated attrition, probably due to postdepositional ravaging.

Table 40. Number of identified specimens (NISP) for each skeletal element at FLK North 3 and 4

	FLK North 3			FLK North 4		
	Bovid	Suid	Giraffid	Bovid	Suid	Giraffid
Horn	13	0	0	3	0	0
Skull	37	3	0	15	1	0
Teeth	65	22	5	85	20	1
Mandible	27	1	0	21	5	0
Vertebra	67	1	0	29	0	0
Ribs	80	1	2	17	3	0
Pelvis	49	0	0	12	0	0
Scapula	26	0	0	13	0	0
Humerus	36	0	2	24	1	0
Radius-ulna	71	3	1	25	0	0
Carpals-tarsals	52	1	2	37	0	0
Metacarpal	45	0	0	21	0	0
Femur	53	0	0	23	0	0
Tibia–fibula	91	0	0	40	0	0
Metatarsal	32	0	1	25	0	0
Patella	4	0	0	4	0	0
Phalanges	30	2	0	130	14	0
Other	6	0	3	30	0	0
ULB*[C6]	3	0	0	0	0	0
ILB*	13	0	0	2	0	0
LLB*	7	0	0	1	0	0
Indeterminate	32	0	0	11	0	0
Total	839	34	16	568	44	1

* ULB, Upper Limb Bones; ILB, Intermediate Limb Bones; LLB, Lower Limb Bones.

Table 41. Number of specimens from each faunal group identified by Leakey and by Domínguez-Rodrigo (MDR), Organista (EO) and Barba (RB)

	FLK North 3				FLK North 4			
	Leakey (1971)		MDR & EO		Leakey (1971)		MDR & RB	
	NISP	%	NISP	%	NISP	%	NISP	%
Bovidae	579	71.9	839	93	447	65.3	568	90.4
Carnivora	117	14.6	13	1.4	148	21.6	15	2.4
Suidae	61	7.6	34	3.7	48	7	44	7
Equidae	26	3.2	0	0	11	1.6	0	0
Giraffidae	16	2	16	1.7	30	4.4	1	0.2
Hippopotamidae	–	–	0	0	1	0.1	0	0
Total	799	100	902	100	685	100	628	100

Table 42. NISP of each bovid long limb bone portion by element type at FLK North 3 and 4

	FLK North 3			FLK North 4	
	Size 1/2	Size 3–4	Size 5	Size 1/2	Size 3–4
Humerus complete	2	2	0	0	0
Humerus prox.	1	1	0	1	1
Humerus prox. + shaft	0	0	0	0	0
Humerus shaft	4	7	1	4	8
Humerus dist.	4	2	0	4	4
Humerus dist. + shaft	7	10	0	1	1
Radius complete	2	1	0	2	0
Radius prox.	4	2	0	0	4
Radius prox. + shaft	3	5	0	0	4
Radius shaft	6	25	0	5	1
Radius dist.	0	0	0	0	0
Radius dist. + shaft	1	2	0	0	2
Metacarpal complete	5	1	0	0	3
Metacarpal prox.	0	0	0	0	3
Metacarpal prox. + shaft	11	7	0	0	2
Metacarpal shaft	3	12	0	1	3
Metacarpal dist.	1	2	1	2	5
Metacarpal dist. + shaft	5	3	1	0	2
Femur complete	0	0	0	0	0
Femur prox.	3	0	1	4	3
Femur prox. + shaft	4	1	0	0	0
Femur shaft	14	22	2	1	10
Femur dist.	5	3	0	3	2
Femur dist. + shaft	0	1	0	0	0
Tibia complete	0	0	1	0	0
Tibia prox.	3	0	0	3	1
Tibia prox. + shaft	0	3	0	0	0
Tibia shaft	22	28	3	4	11
Tibia dist.	0	1	0	0	4
Tibia dist. + shaft	8	12	0	0	3
Metatarsal complete	2	6	0	0	3
Metatarsal prox.	0	0	0	1	1
Metatarsal prox. + shaft	5	3	0	2	8
Metatarsal shaft	3	9	0	0	6
Metatarsal dist.	0	0	0	0	3
Metatarsal dist. + shaft	1	4	0	0	1
Total	129	175	10	38	99

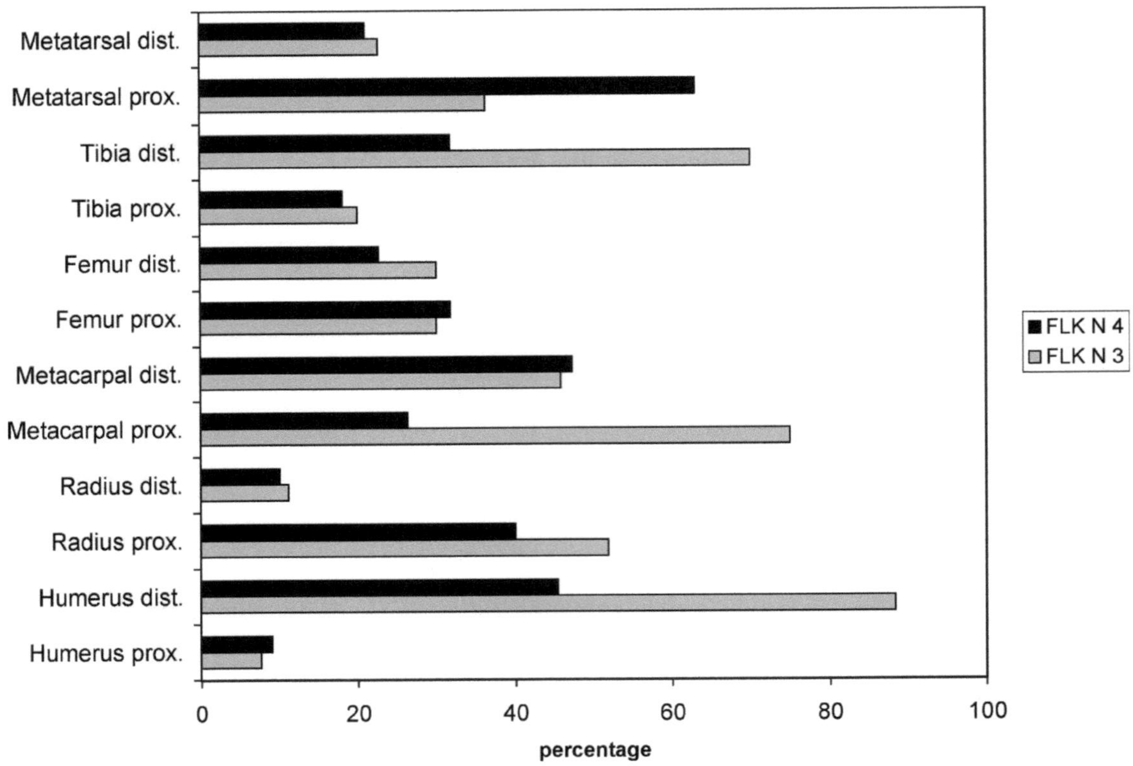

Figure 97. Representation of long limb bone portions (%MNE) at FLK North 3 and 4, relative to the frequencies expected according to carcass MNI. Only carcasses from *Parmularius* and *Antidorcas* have been considered. Frequencies have been calculated using 15 MNI for FLK North 3 and 11 MNI for FLK North 4.

Estimates of complete element representation (MNE) show a predominance of long limb bones and skull elements (Table 43). Axials are underrepresented; they are more abundant at FLK North 3, but the number of carcasses accumulated there is also higher. Long limb bone representation seems to be slightly higher when shafts are included in element counts, rather than considering epiphyses alone. Despite the low relative frequency of axials, when we compare element presence at each site by grouping axial bones (including pelves and scapulae for taphonomic purposes), skulls, and long limb bones, we find that axials comprise 45%, limbs 49%, and skulls 6% of the main elements at FLK North 3 (Table 43). For FLK North 4, the frequencies are: axials 22%, long limbs 68%, and skulls 10%. Axial bones are thus better-represented at FLK North 3.

A total of 15% of all long limb bones at FLK North 3, and 10% at FLK North 4, are complete (Table 44). Complete bones from large carcasses at both sites are represented in similar frequencies. However, complete bones from small carcasses are represented twice as often at FLK North 3 compared to FLK North 4. This, together with the higher frequency of axials at FLK North 3, suggests a more intense ravaging process at FLK North 4.

FRAGMENTATION AND SHAFT CIRCUMFERENCES

Green fractures are predominant in both assemblages. Only a small fraction of specimens show diagenetic breakage, which is more abundant in bones from larger carcasses (Figure 98).

Table 43. Minimum number of elements (MNE) for each skeletal element and bone portion (for appendicular bones) in small and large carcasses. MNE estimates are given using epiphyses alone and using epiphyses and shafts combined

		FLK North 3			FLK North 4		
		Small	Large	Total	Small	Large	Total
Skull*		7	8	15	3	8	11
Scapula		10	10	20	2	6	8
Pelvis		13	9	22	4	5	9
Ribs		4	30	34	6	2	8
Vertebrae		16	39	55	3	17	20
Humerus							
	Epiphyses	11	13	24	5	5	10
	ep. + shaft	13	15	28	8	6	12
Radius							
	Epiphyses	9	8	16	2	7	9
	ep. + shaft	12	15	27	6	7	13
Metacarpal							
	Epiphyses	14	8	22	1	12	13
	ep. + shaft	14	8	22	2	13	15
Femur							
	Epiphyses	7	3	10	5	4	9
	ep. + shaft	10	8	18	5	4	9
Tibia							
	Epiphyses	8	12	20	2	7	9
	ep. + shaft	13	17	30	3	7	10
Metatarsal							
	Epiphyses	7	9	16	3	13	16
	ep. + shaft	7	9	16	3	15	18

*Including mandible

Table 44. Percentages of complete bones at FLK North 3 and 4. Numerator is the total number of complete elements, denominator is the total estimated number of elements. Percentages are in parentheses

	FLK North 3			FLK North 4		
	Size 1–2	Size 3–4	Total	Size 1–2	Size 3–4	Total
Humerus	2/13 (15.3)	2/15 (13.3)	4/28 (14.2)	0/8(0)	0/6(0)	0/14 (0)
Radius	2/12 (16.6)	1/15 (6.6)	3/27 (11.1)	2/6(33.3)	0/7(0)	2/13 (15.3)
Metacarpal	5/14 (35.7)	1/8 (12.5)	6/22 (27.2)	0/2(0)	3/13(23.1)	3/15 (20)
Femur	0/10 (0)	0/8 (0)	0/18 (0)	0/5(0)	0/4(0)	0/9 (0)
Tibia	0/13 (0)	0/17 (0)	0/30 (0)	0/3(0)	0/7(0)	0/10 (0)
Metatarsal	2/7 (28.5)	6/9 (66.6)	8/16 (50)	0/3(0)	3/15(20)	3/18 (16.6)
Total	11/69 (16)	10/72 (13.8)	21/141 (14.8)	2/27(7.4)	6/52(11.5)	8/79 (10.1)

Metapodials seem to be the most affected by diagenetic breakage. Therefore, bone size distribution seems to have been minimally affected by diagenetic breakage. Long limb bone shaft circumference, following Bunn's (1982) three types (discussed in Chapter 3), reveals relative frequencies which contrast with those seen in experimental scenarios (Figure 99). There is a high percentage of Types 2 and 3, which may be an artifact of the reduced number of specimens smaller than 30 mm (Figure 96). It is very likely that postdepositional disturbance at

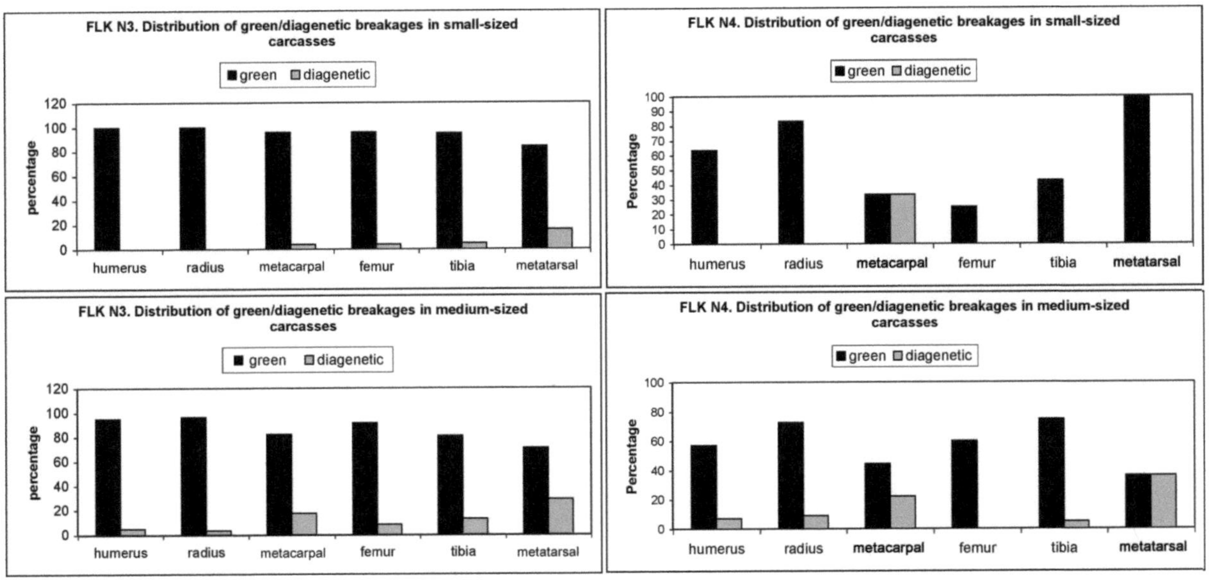

Figure 98. Frequencies of green and diagenetic breakages on bones from small and large carcasses at FLK North 3 and 4.

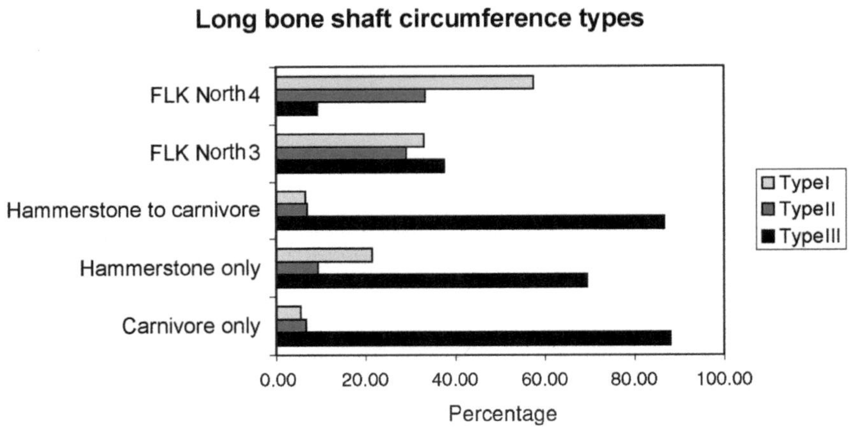

Figure 99. Distribution of long bone shaft circumference types (Bunn, 1982) in experimental assemblages and at FLK North 3 and 4. Data for experimental assemblages are from Marean and Spencer (1991) and Marean *et al.* (2004).

FLK North 4 can account for such a high frequency of Types 2 and 3 and the reduced representation of Type 1. However, the frequencies documented at FLK North 3 cannot be accounted for in the same terms, since bone size distribution suggested minimal post-depositional disturbance by sedimentary processes. In this level, the higher frequency of Types 2 and 3 (especially Type 3) should be attributed to carnivores. In experimental scenarios, hyenas were used to model frequencies of shaft circumference types (Bunn, 1982). It has been noted elsewhere that the same was done for carnivore–hominid–carnivore modeling of tooth-mark frequencies (see Chapter 5). However, it has been shown that felids do not create the same frequencies as do hyenas, given that most of the limb bones that felids abandon from consumed carcasses are Type 3 (Cavallo, 1998 Domínguez-Rodrigo *et al.*, in press). This indicates that the high frequency of Types 2 and 3 is not solely due to the action of hyenas. If hyenas had access to all the bones in both assemblages, the resulting distribution of circumference types should match any of the experimental scenarios shown in Figure 99. The overall percentages of Type 3 shaft specimens in hyena experiments are substantially low. However, the high incidence of this type at FLK North 3 and FLK North 4 (although in the latter, postdepositional disturbance cannot be ruled out) could very likely have resulted from a multiple-event depositional process, in which carcasses were accumulated by felids (initially producing a high frequency of Type 3 specimens), and in which hyenas intervened on and off and broke bones (forming Types 1 and 2), thereby creating the frequencies of the three types shown in Figure 99. This hypothesis needs further support from other taphonomic approaches, which we will address below.

Studies of breakage planes and angle measurements have been shown to help differentiate between dynamic versus static loading (Alcántara *et al.*, 2006; see discussion in Chapter 3). Transversal planes at FLK North 3 and 4 were too small to be discriminating. Oblique planes from bones of small carcasses showed much overlap with experimental scenarios modeling both dynamic and static loading, and are therefore also uninformative. Oblique planes from bones of large carcasses with angles <90° are also ambiguous, but those >90° are more in accordance with modeled expectations for static loading (Figure 100). Longitudinal planes also showed ambiguity in

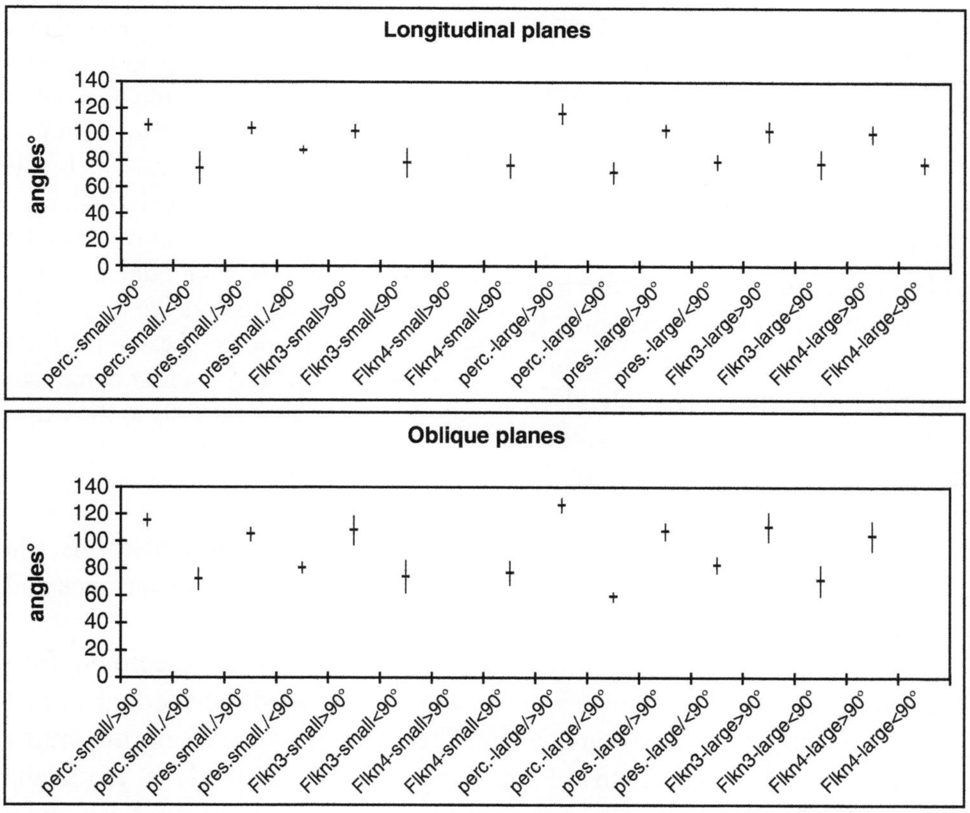

Figure 100. Mean values and 1 SD values for experimental assemblages (Alcántara *et al.*, 2006) on longitudinal and oblique planes modeling dynamic loading (hammerstone percussion, perc.) and static loading (carnivore pressure, pres.) for small and large carcasses, and for the FLK North 3 and 4 breakage planes.

the angles of breakage planes measured on bones from small carcasses. However, longitudinal planes from larger carcasses with angles >90° also supported static loading.

The study of notches is also relevant to unravel the agent of bone breakage (see Chapter 3). The sample size of shaft specimens at FLK North 4 is small and none of the notches observed were complete. FLK North 3 showed a higher number of complete notches, although the number is still small compared to that of other sites, like FLK North 1–2. There are a total of 41 notches documented at FLK North 3. From these, there are 10 that are double opposing, 11 that are Type C overlapping notches (Capaldo and Blumenschine, 1994), 11 single incomplete, and 9 complete notches. Most of the complete notches occurred in specimens that had a circumference greater than 50% and therefore, the use of goniometers for measuring platform angle was impossible, so angle measurements were excluded from this study. However, the two ratios (notch breadth:depth and flake scar breadth:depth; see Chapter 3) could be derived from the nine complete notches observed at FLK North 3. All the measurements are within the experimental range of variation for static loading (Figure 101). That is, they support breakage of bones by carnivores. Furthermore, the high frequency of double opposing and overlapping notches at FLK North 3–4.7% and 5.2% of all midshaft specimens, respectively – is only slightly lower than that reported for hyena-ravaged midshaft assemblages (Egeland *et al.*, in preparation).

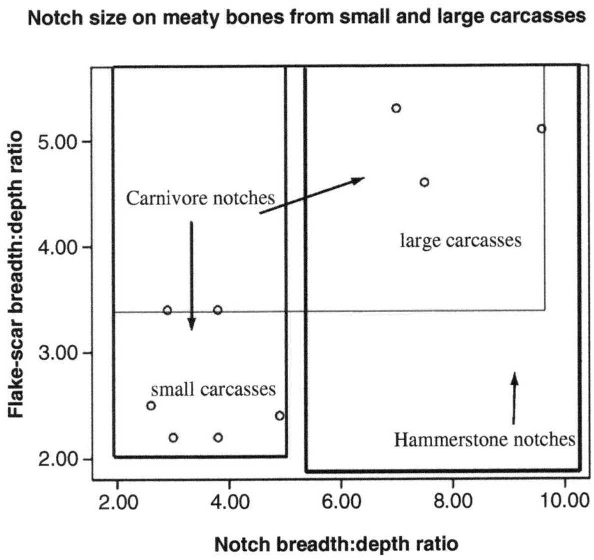

Figure 101. Notch measurements for the FLK North 3–4 specimens, showing notch length:breadth ratio from the cortical surface and flake scar length:breadth ratio from the medullary surface. Boxes represent the ranges of variation (95% confidence intervals) for these ratios as documented in experimental assemblages produced by static (carnivore) and dynamic (hammerstone percussion) loading (Capaldo and Blumenschine, 1994). The box with the thinnest outline at the top represents the range of variation for carnivore notches on large carcasses. The box with a moderately thick outline, on the left, represents the range of variation for carnivore notches on small carcasses. The box with the thickest outline, on the right, represents the range of variation for percussion notches on large carcasses. The range of variation for percussion notches on small carcasses falls to the extreme right, outside the graph.

BONE SURFACE MODIFICATIONS

Bone surface preservation at FLK North 3 and 4 is very good. Cortical surfaces are less affected by biochemical processes than at FLK *Zinj*. The surfaces on bones from both assemblages can be described as almost pristine for most specimens. Therefore, any hominid- or carnivore-made modifications should have been recorded on those surfaces and should be easily identifiable.

Only one radius specimen from a size 3 animal appeared cut-marked at FLK North 3. The cut mark appears on the cranial midshaft and can be defined as a cut-marked "hot zone," where scraps of flesh do not usually survive after felids' primary consumption of carcasses (Chapter 6). At FLK North 4, only one pelvic specimen and two phalanges from an antilopini bore hominid-made marks. The pelvic specimen had some cut marks on the acetabular neck. One cut mark was also identified on one phalangeal shaft. The other phalange showed percussion marks and a notch probably caused through dynamic loading. That is the only percussion mark that has been recorded in both assemblages, which suggests that most if not all of the remaining bone breakage is attributable to nonhominid agents. A total of 3 cut-marked specimens out of more than 1,500 bone specimens (0.1%) underscores the marginal contribution of hominids to both assemblages.

By contrast, tooth-marked specimens are fairly abundant. Bones from small and large carcasses are tooth-marked at rates that range from 13 to 33%, depending on carcass size and assemblage (Table 45; Figure 102). About one out of every five specimens belonging to vertebrae and ribs appears tooth-marked.

Table 45. Tooth-mark frequencies by skeletal element and carcass size at FLK North 3 and 4. Numerator is total number of tooth-marked specimens, denominator is total number of specimens for each skeletal element, and percentages are in parentheses

	FLK North 3			FLK North 4		
	Size 1/2	Size 3/4	Total	Size 1/2	Size 3/4	Total
Vertebra	2/17 (11.7)	13/51(25.4)	15/68 (22)	1/3(33.3)	5/26(19.2)	6/29(20.7)
Ribs	0/11(0)	14/70 (20)	14/81 (17.2)	2/10(20)	2/7(28.6)	4/17(23.5)
Pelvis	11/26 (42.3)	11/23 (47.8)	22/49 (44.8)	4/6(66.7)	3/6(50)	7/12(58.3)
Scapula	3/14 (21.4)	4/12 (33.3)	7/26 (26.9)	2/5(40)	4/8(50)	6/13(46.2)
Humerus	10/15 (66.6)	15/21 (71.4)	25/36 (69.4)	1/10(10)	6/14(42.9)	7/24(29.2)
Radius-ulna	7/15 (52)	19/59 (32.2)	26/74 (35)	5/7(71.4)	2/18(11.1)	7/25(28)
Metacarpal	6/23 (26)	7/22 (31.8)	13/45 (28.8)	1/3(33.3)	1/18(5.6)	2/21(9.5)
Femur	10/25 (40)	13/18 (72.2)	23/53 (43.3)	1/8(12.5)	2/15(13.3)	3/23(13)
Tibia–fibula	7/32 (21.8)	23/59 (38.9)	30/91 (33)	3/15(20)	3/25(12)	6/40(15)
Metatarsal	1/10 (10)	6/22 (27.2)	7/32 (21.8)	3/3(100)	10/22(45.5)	13/25(52)
Phalanges	0/14 (0)	0/18 (0)	0/32 (0)	0/101(0)	4/29(13.8)	4/130(3.1)
Total	57/202 (28.2)	125/375 (33.3)	182/587 (31)	23/171(13.5)	42/188(22.3)	65/359(18.1)

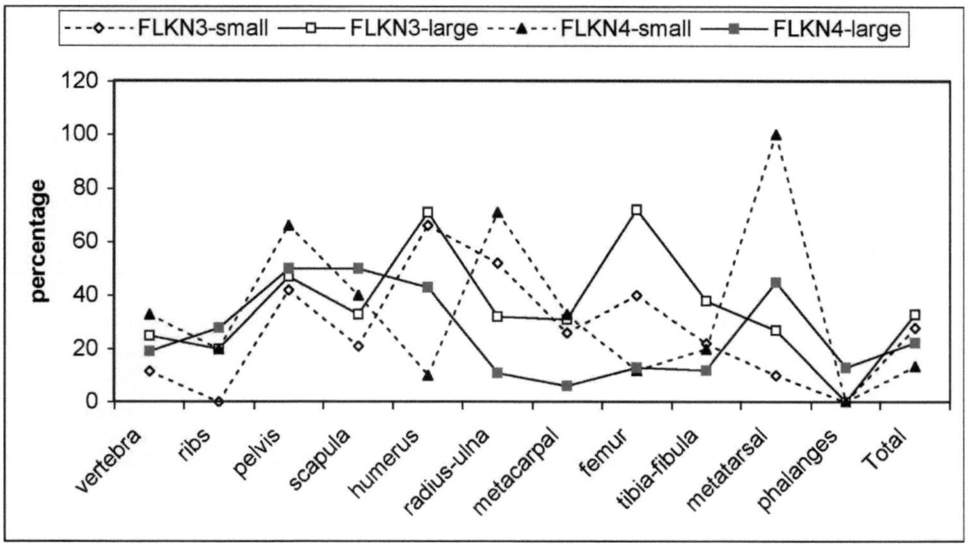

Figure 102. Tooth mark frequencies (as %NISP) on skeletal elements from small and large carcasses at FLK North 3 and 4.

A higher rate of tooth-marking was observed for scapulae and pelves, where 20–50% of specimens appear tooth-marked; that is, almost one out of every three elements on average. At FLK North 3, upper limb bones are highly tooth-marked, with frequencies > 40–60% in smaller animals and > 70% in larger animals, suggesting intensive exploitation of these anatomical sections by carnivores. Intermediate long bones are tooth-marked at significantly lower rates and metapodials appear as the least tooth-marked appendicular elements at the site. This trend is only observed on bones from large carcasses at FLK North 4. Long limb bones from smaller carcasses are less tooth-marked.

The significant contribution of carnivores (very likely hyenas) to bone breakage, as reflected by the tooth-mark frequencies on most skeletal elements, is further supported by the breakdown of tooth-mark frequencies by long bone portion (Table 46; Figure 103). Epiphyses at FLK North 3 and 4 appear tooth-marked at low rates. In smaller animals, the frequencies fall within those reported for human (hammerstone)-to-carnivore experiments. In larger carcasses they are even lower, as is the case in larger carcasses from FLK North 4. Near-epiphyses also show frequencies similar to those of human-to-carnivore experiments, but their meaning is questionable at best given the small sample size. We classified as near-epiphyseal specimens those that showed at least one third of the medullary surface with cancellous tissue. This contrasts with criteria used by Blumenschine and colleagues, since Capaldo's unpublished classifications as "near-epiphyseal" fragments, as identified by him at both sites, very often showed less than one tenth of cancellous tissue covering the medullary surface. Tooth marks on midshaft specimens were close to those for hominid–carnivore models, but frequencies for larger carcasses at both sites were outside the limits of variation showed by all experimental scenarios, suggesting that tooth-marking at the site did not follow the typical carnivore–hominid sequence that was previously modeled.

If considering tooth marks alone, one could argue that the close fit with human-to-carnivore experiments suggests primary access to carcasses by hominids. If so, where are the cut marks and especially the percussion marks resulting from such behavior? A more careful look at tooth mark distribution according to

Table 46. Frequencies of tooth-marked specimens on long limb bones at FLK North 3 and 4, by bone portion: epiphyses (epi), near-epiphyses (nep) and midshafts (msh). Numerator is total number of tooth-marked specimens, denominator is total number of specimens for each skeletal element, and percentages are in parentheses

	Size 1/2			
	epi	nep	msh	total
FLK North 3	23/62 (37)	1/3 (33.3)	15/60 (25)	39/125 (31)
FLK North 4	5/21(23.8)	2/9(22.2)	6/13(46.2)	13/43 (30)
	Size 3/4			
	epi	nep	msh	total
FLK North 3	33/67 (49.2)	2/4 (50)	45/155 (29)	80/226 (35.3)
FLK North 4	11/59(18.6)	2/8(25)	8/32(25)	21/99 (21.2)

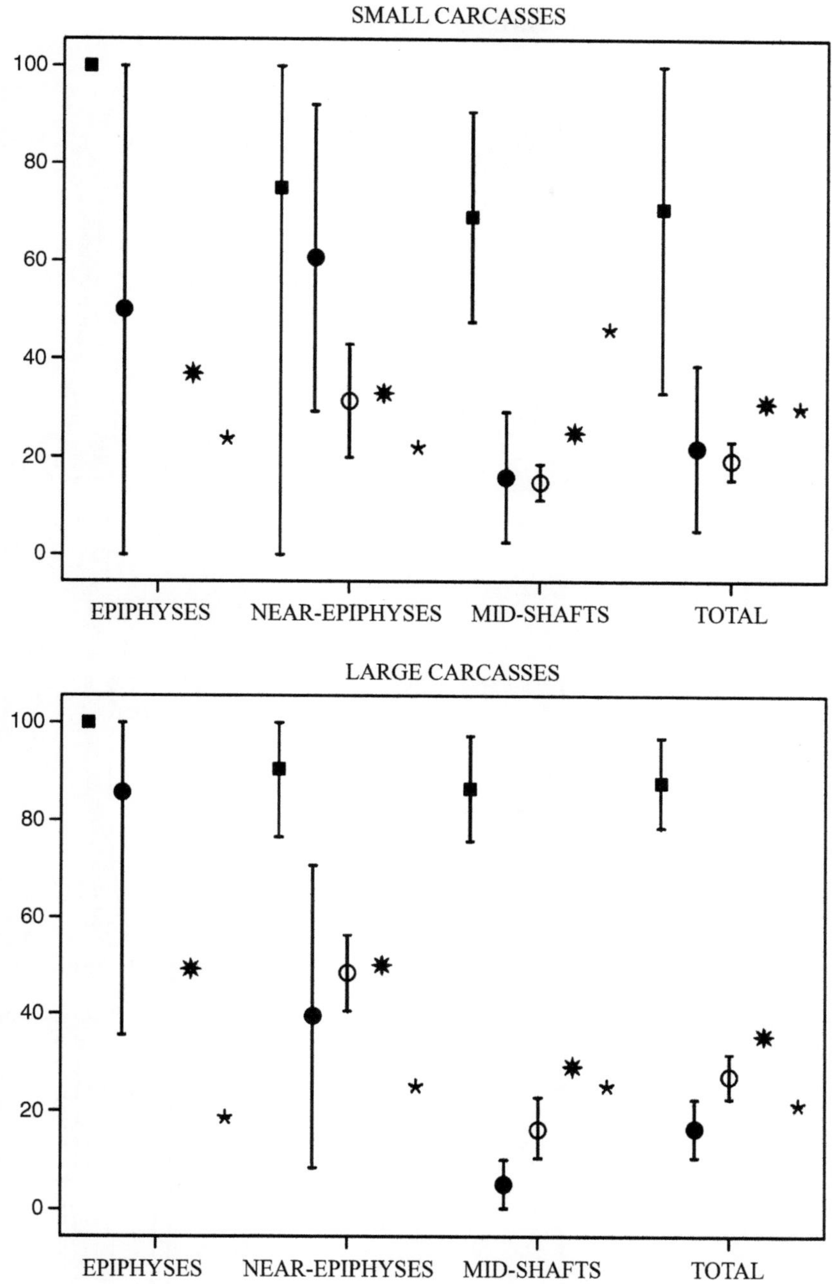

Figure 103. Distribution of the 95% CI (confidence interval) analysis of tooth-marked specimens for each bone portion from small carcasses in experimental assemblages and from FLK North 3–4. ■, Carnivore only; ●, Human (hammerstone)-to-carnivore (Blumenschine, 1995); ○, Human (hammerstone)-to-carnivore (Capaldo, 1997); ✷, FLK North 3; ★, FLK North 4.

bone type and section shows that most bones were probably broken by carnivores (Figures 104–107). Tooth marks occur on midshafts as well as on ends, where furrowing is frequent. Most tooth-marking and furrowing cluster on upper limb bones in small carcasses. By contrast, metapodials from small animals appear tooth-marked very marginally. In larger carcasses, tooth marks are more evenly distributed. The distribution of tooth marks on all long bones suggests that carnivores (very likely hyenas) were the main bone-breaking

SMALL CARCASSES

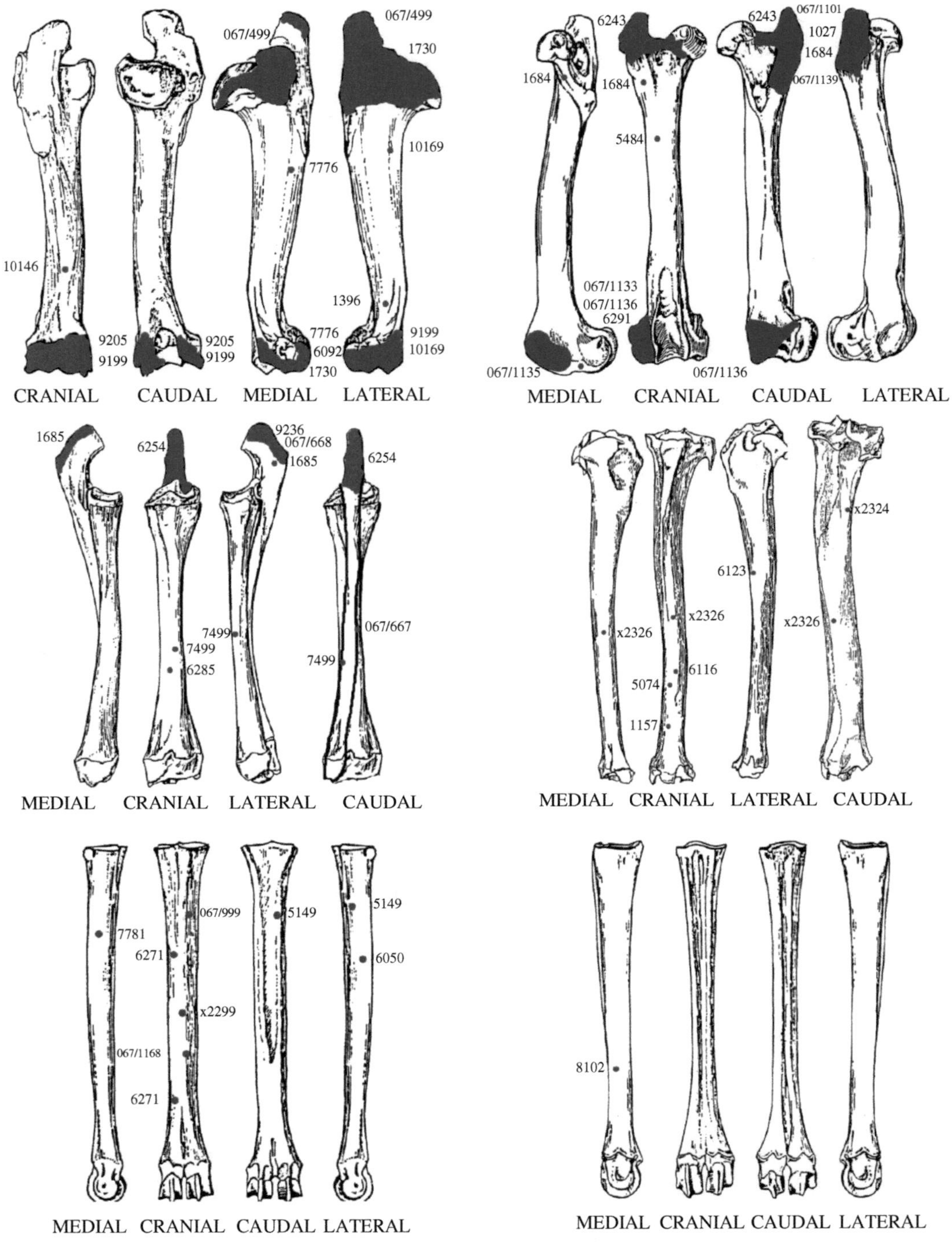

Figure 104. Distribution of tooth marks on long bones from small carcasses. Letters and numbers indicate specimen number from the FLK North 3 assemblage. Shaded areas show intensive furrowing. Specimens bearing marks which could not be anatomically located were not incorporated into this figure. Bones are redrawn from Pales and Lambert (1971).

LARGE CARCASSES

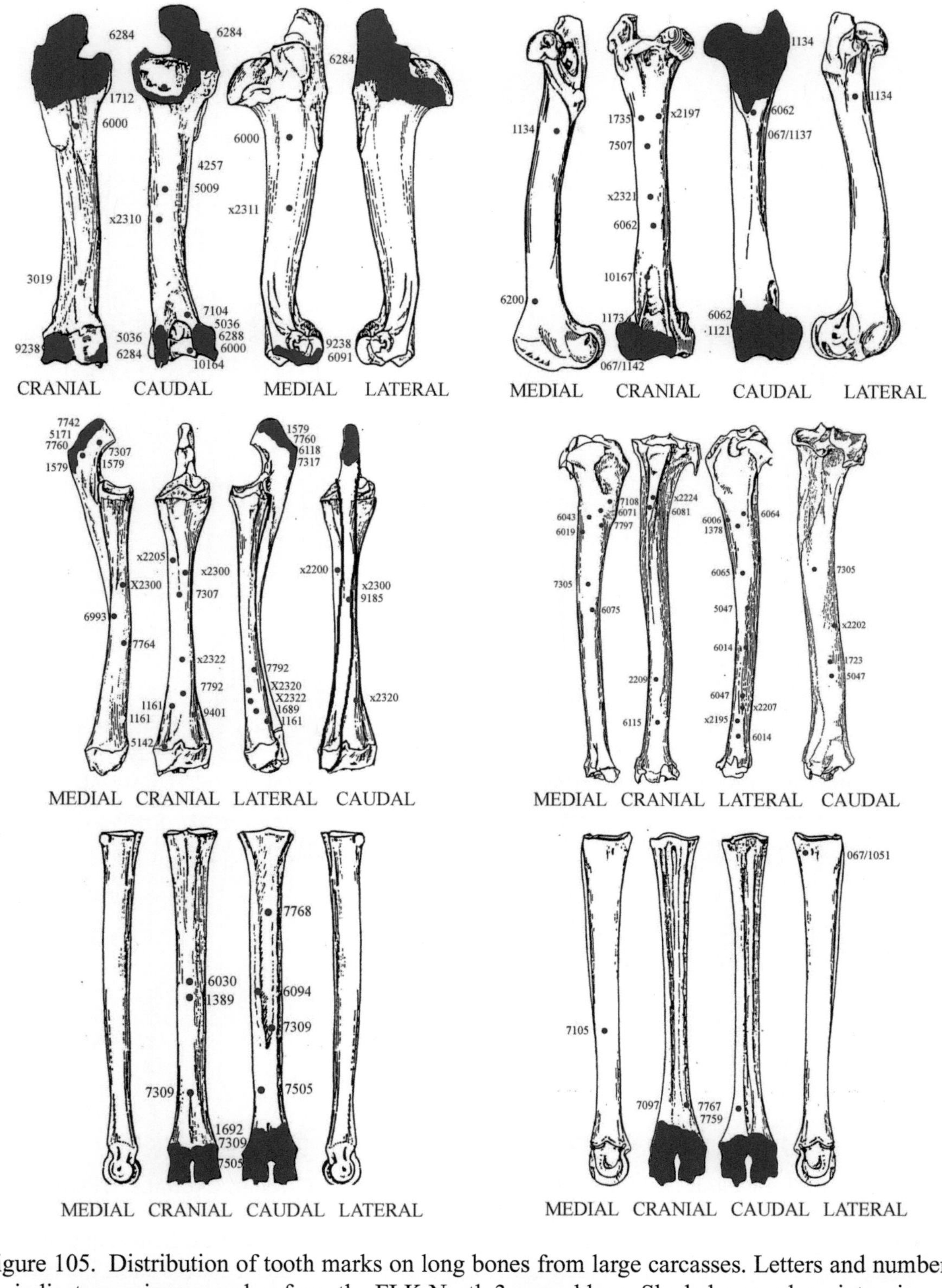

Figure 105. Distribution of tooth marks on long bones from large carcasses. Letters and numbers indicate specimen number from the FLK North 3 assemblage. Shaded areas show intensive furrowing. Specimens bearing marks which could not be anatomically located were not incorporated into this figure. Bones are redrawn from Pales and Lambert (1971).

SMALL CARCASSES

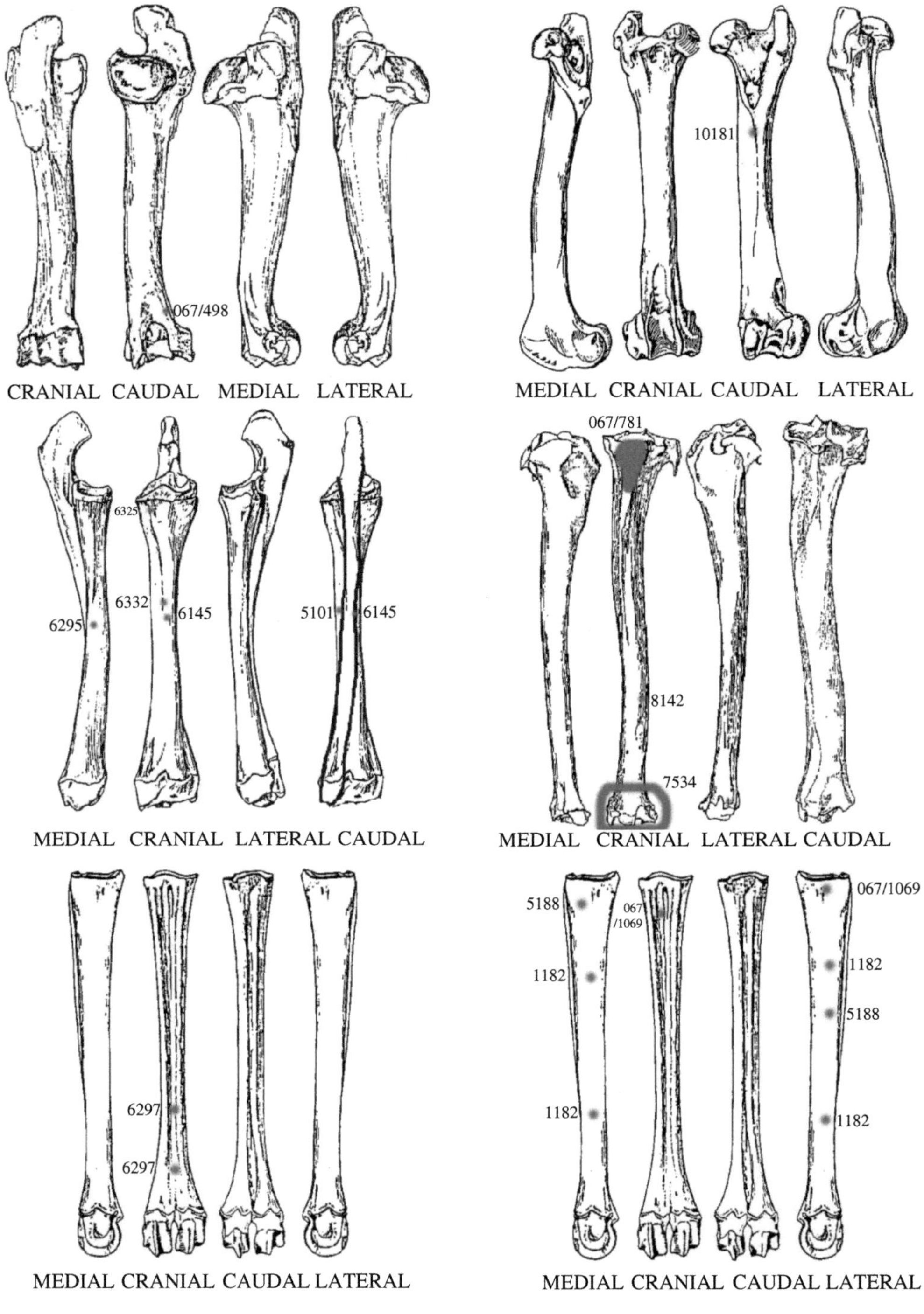

Figure 106. Distribution of tooth marks on long bones from small carcasses. Letters and numbers indicate specimen number from the FLK North 4 assemblage. Shaded areas show intensive furrowing. Specimens bearing marks which could not be anatomically located were not incorporated into this figure. Bones are redrawn from Pales and Lambert (1971).

LARGE CARCASSES

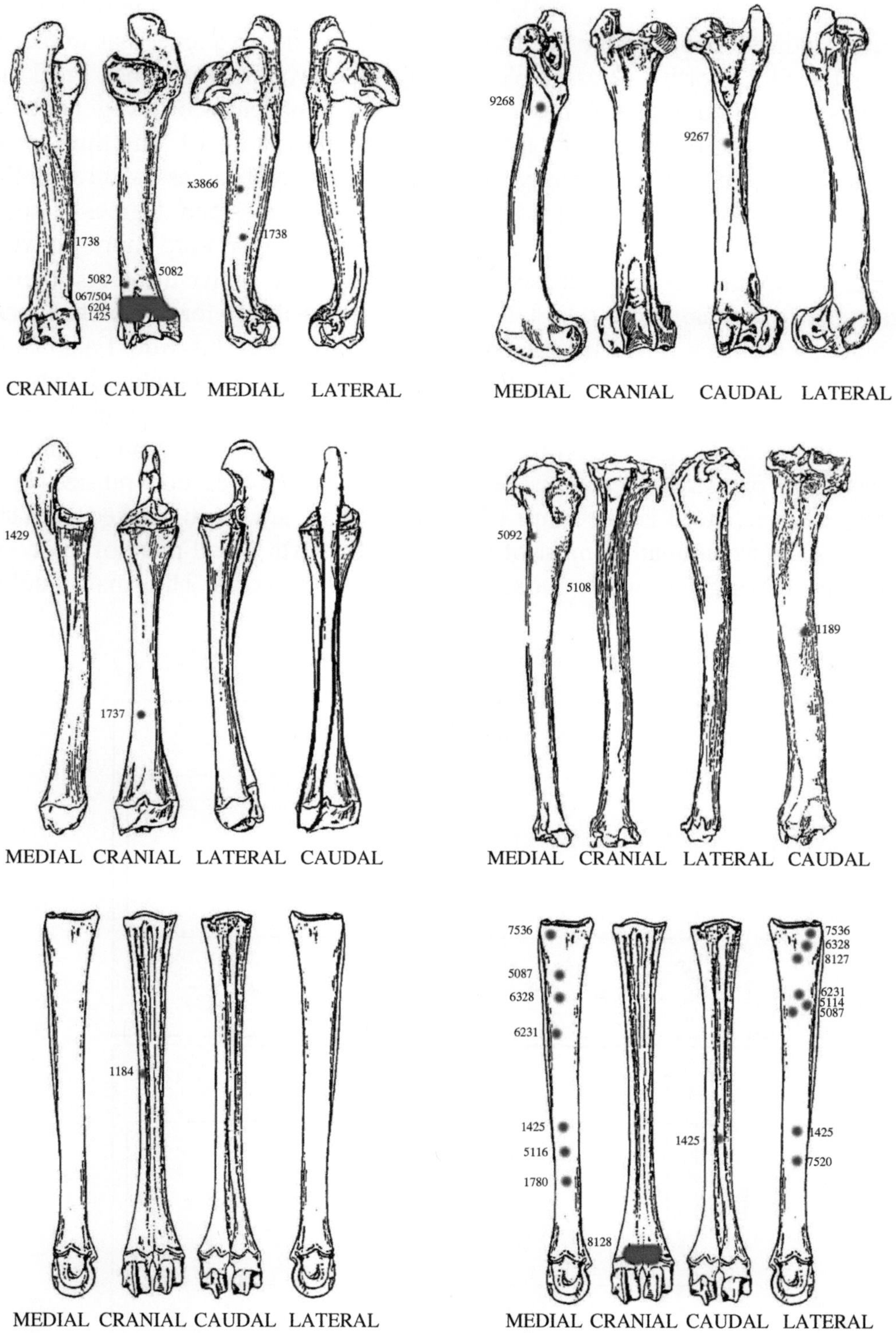

Figure 107. Distribution of tooth marks on long bones from large carcasses. Letters and numbers indicate specimen number from the FLK North 4 assemblage. Shaded areas show intensive furrowing. Specimens bearing marks which could not be anatomically located were not incorporated into this figure. Bones are redrawn from Pales and Lambert (1971).

agent (and the only one we can ascertain, given the absence of percussion marks). The frequencies reported are similar to those obtained for the FLK North 1–2 assemblage, where it was suggested that the moderate frequency of tooth-marked specimens (about 30%) was the result of hyenas partially ravaging the accumulation originally made by felids.

Frequencies of tooth marks in all these sites are substantially lower than those reported for experimental assemblages where hyenas have access to all bones. They are also at least three times higher than those reported for felid-first modified bone assemblages (Domínguez-Rodrigo *et al.*, in press). It therefore seems that the frequency reported is the result of a mix between felids accumulating most of the assemblage and leaving few tooth marks on bones, and hyenas partially ravaging the assemblage, increasing the frequency of tooth marks, but not to a level that would be typical of a scenario where hyenas break all bones. The incompleteness of ravaging could also be explained in terms of long time spans for the accumulation of carcasses, through fluctuating trophic dynamics and degrees of competition, in which hyenas were more active at some times and less active at others, thereby creating the possibility for complete bones to enter the lithosphere. This would also explain why the percentage of complete bones, especially from small carcasses, is so high.

Further support for our argument comes from the analysis of tooth pit size (Figure 108). There were few tooth-marked shaft specimens at FLK North 4 and most of the conspicuous marks were scores. The small number of pits

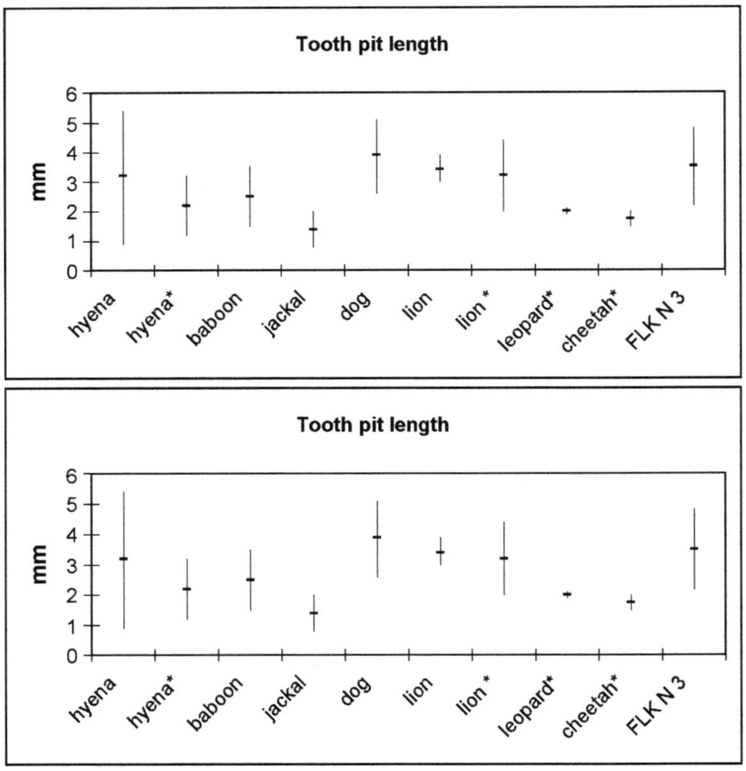

Figure 108. Distribution of tooth pit dimensions from FLK North 3 on shafts from large carcasses and mean values plus 1 SD ranges for experimental assemblages measuring tooth marks by different carnivore types (see data in Domínguez-Rodrigo and Piqueras, 2003). (* Data from Selvaggio [1994].)

available forced us to focus on the larger FLK North 3 sample. A random selection of conspicuous tooth pits (n = 50) from midshafts from this level were measured. Their sizes are outside the ranges reported for medium-sized felids and fit those of hyenas best (see Chapter 3 for background). This suggests hyena bone breaking at the site, which, together with breakage plane angles and notch analysis, supports the interpretation that hyenas were the main demarrowing agent.

However, as was the case at FLK North 1–2, felids seem to have played some significant role in the accumulation of carcasses at both archaeological levels. Domínguez-Rodrigo *et al.* (in press) showed that one typical way in which felids modify humeri is the partial or total deletion of the proximal end, depending on carcass size, and furrowing of the caudal epicondyles of distal ends (in proportions usually lower than 30%). Kerbis (1990) argued that hyenas showed a pattern in their modification of distal humeri, first deleting the cranio-caudal condyle, and then proceeding towards the medial side of the epiphysis. Figure 109 shows several humeri from FLK North 3. The humeri in the background (D) belong to *Antidorcas*. These elements could easily disappear under the effect of hyena ravaging; they rarely survive in their dens. These humeri show mild stages of ravaging, all of them lacking the proximal end; this is systematically documented in humeri from carcasses consumed by felids. Figure 109C shows a humerus from a modern carcass consumed by a leopard, with furrowing on the proximal end. Proximal ends of humeri from larger carcasses are more prone to survival (Figures 109A and 109B). Both in small and large carcasses, most distal ends from humeri show intact cranial aspects of condyles, but the caudal epicondylar areas show furrowing (see arrows in Figure 109 for multiple examples). This pattern is more in accordance with what felids do, rather than with hyena ravaging. This also explains why the oleocranea of ulnae, even from small animals, appear furrowed (Figure 110). Scapulae also show fairly intact necks and glenoid fossae, with occasional tooth-marking in the form of isolated pits or punctures like those reported for leopards (Figure 111).

Further evidence comes from our study of surviving axial elements. Most vertebrae show furrowing of apophyses but intact centra (bodies) even in small-sized carcasses (Figure 112). Together, these observations argue in favor of felids as the main carcass accumulators at both archaeological levels. The taphonomic signatures documented in these assemblages are very similar to those of FLK North 1–2 (Chapter 8). Given that FLK North 1–2 overlies FLK North 3 and 4, and that level distinction was established on arbitrary criteria (Leakey, 1971), it is very likely that FLK North 3 and 4 belong to the same depositional process along an extremely large time span, in which felids (probably more than one type), hyaenids and water flows (at least for FLK North 4) alternated in the configuration of the palimpsest.

Conclusions

Carnivores were responsible for the accumulations of bones at FLK North 3 and 4. This is supported by tooth-marking on long bone midshafts, by the types and frequencies of notches on limb bones, by notch ratios and breakage plane angles. This interpretation is also supported by the fact that despite good cortical preservation, not a single percussion mark was observed on long limb bone specimens. With respect to which carnivores were responsible for the accumulation, we make similar claims for felids as we did for the FLK North 1–2 assemblage. The reasons are

1. The FLK North 3 and 4 bone assemblages are dominated by two taxa: *P. altidens* and *A. recki*. This indicates a specialized carcass collector. Hyenas typically have a

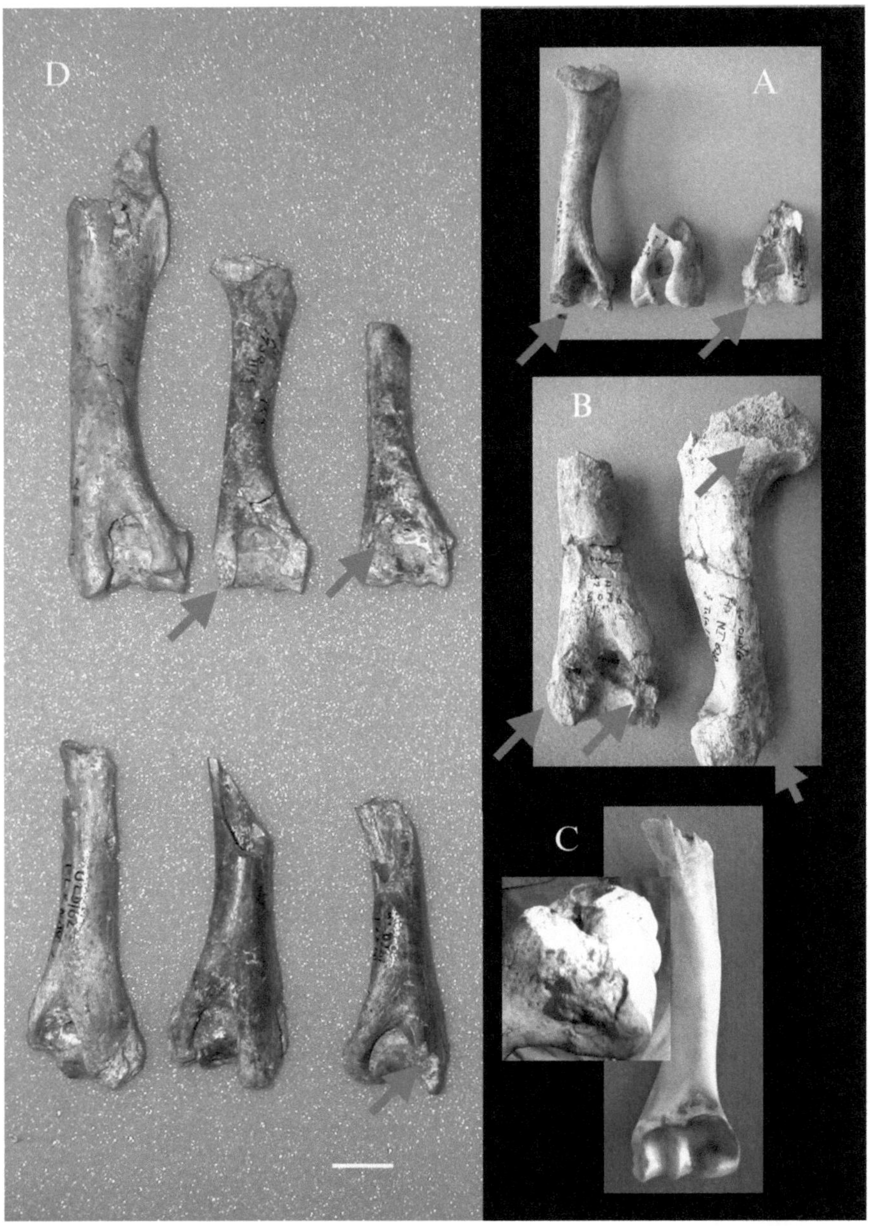

Figure 109. Humeri showing various stages of preservation from FLK North 3. Notice furrowing on some caudal epicondyles. They belong to *Parmularius* (A and B), recent impala (C) and *A. recki* (D). Arrows show presence of furrowing. Furrowing on the medial caudal epicondyle of humeri is typical of felid damage to this kind of bone (scale = 2 cm).

more eclectic prey/scavenged species range.
2. There is a higher survival rate of complete bones, especially from small carcasses, than has been reported for hyena dens.
3. There is absence of digested bone, coprolites and infantiles of hyaenids in both assemblages.
4. The frequencies of tooth marks are much lower than those reported from hyena-ravaged assemblages either in hyena

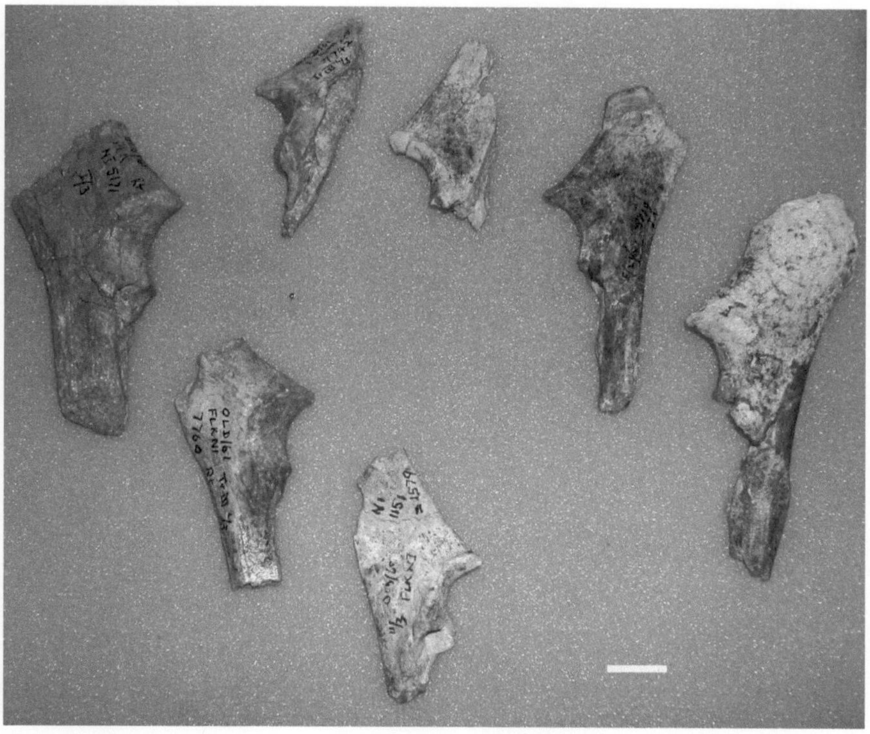

Figure 110. A selection of ulnae from FLK North 3 and 4 from small and larger bovids, showing furrowing on their proximal ends (scale = 1 cm).

Figure 111. Example of puncture on the neck of a scapula of *Antidorcas* from FLK North 3 (scale = 1 cm).

dens (Brain, 1981; Bunn, 1982; Brugal *et al.*, 1997; Fosse *et al.*, 1998; Marra *et al.*, 2004; Villa *et al.*, 2004) or in experiments (Brain, 1981; Bunn, 1982; Capaldo, 1995).

5. The survival of axial elements is also higher than at hyena dens (Brain, 1981; Bunn, 1982; Brugal *et al.*, 1997; Fosse *et al.*, 1998; Marra *et al.*, 2004; Villa *et al.*, 2004).

The interaction between felids and hyenas in the assemblages analyzed has been taphonomically supported. The next question is, was there any interaction between felids and hominids? Hominid presence at the site left a particular trace: stone tools. De la Torre's (2006) analysis of these tools suggested that most artifacts at the site represent traces of

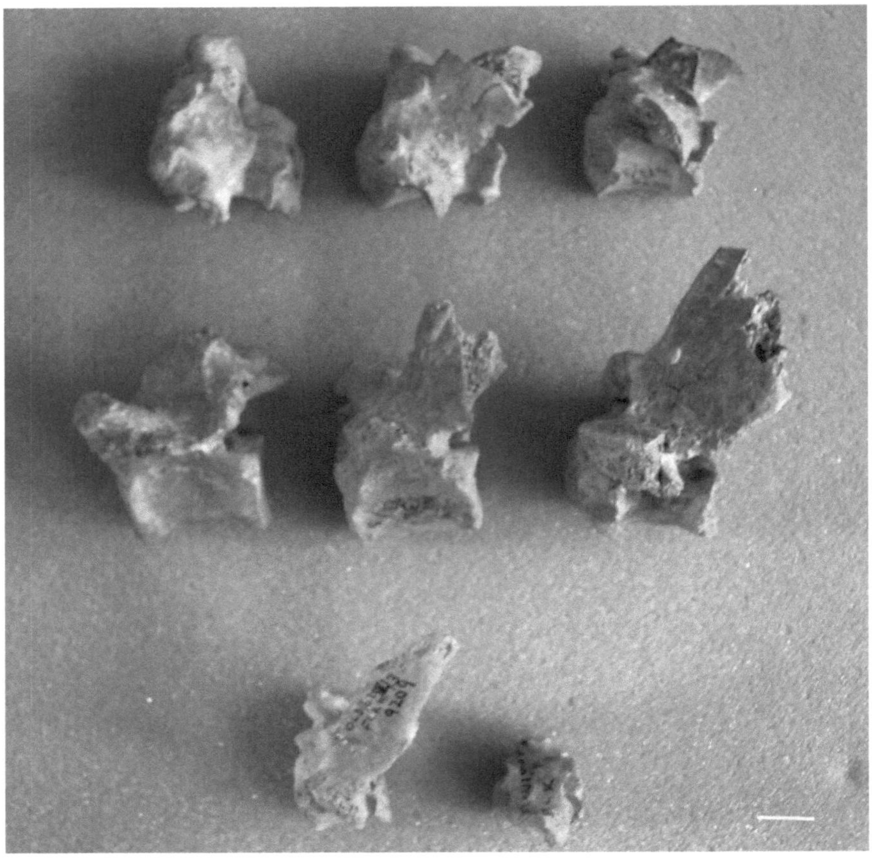

Figure 112. Examples of furrowing on the vertebral apophyses of *Parmularius* and *Antidorcas* at FLK North 3 (scale = 1 cm).

battering activities, which are obviously unrelated to marrow extraction, given the lack of taphonomic evidence on the spatially associated bones (see Tables 25–27 from Chapter 7). FLK North 4 only had four flakes, the remaining tools being core types and angular fragments probably detached through percussion activities. FLK North 3 only had 16 flakes, which would not even have been enough to butcher a single complete carcass, according to our experiments which show that a minimum of 20 flakes is necessary (work in progress). This shows that cutting tools are virtually absent from both archaeological levels. The virtual lack of edges on those artifacts clearly show that hominids were not targeting carcass exploitation in their activities at the site. This indicates that the main stone tool-using activity is unknown and that no interaction between hominids and felids occurred in either archaeological level.

Only three bones in FLK North 3 (and none in FLK North 4) showed traces of hominid butchery; of these, one is a pelvis. Pelves were shown to be a primary target for felids and they remain defleshed after felids abandon their prey (Blumenschine, 1986; Cavallo, 1998). The presence of cut marks on a pelvic specimen would therefore suggest that hominids did not acquire it from felids at the site (see Figure 95). If hominids were scavenging from felids, traces of the scavenging "deed" should appear reflected on the bones. It seems more likely that hominids acquired this carcass independently from felids and that they contributed to the bone assemblage without any interaction with carnivores that might be taphonomically visible. The absolute lack of these traces implies that the depositions of stone tools and bones at FLK North 3 and 4 are not functionally related and correspond to independent depositional episodes.

FLK North 1–2, FLK North 3, and FLK North 4 represent some of the best examples of carnivore (felid)-accumulated bone assemblages in Olduvai Bed I. One of the felids involved was very likely leopard. That is, the theoretical opportunity for scavenging meaty carcasses by hominids, as hypothesized by Cavallo and Blumenschine (1989), probably existed at Olduvai. However, despite the fact that hominids visited the sites and performed stone tool-discarding activities, none of these activities indicates any interaction with felid-transported carcasses. This can be taken as one of the strongest indicators that hominids were not involved in passive scavenging activities at Olduvai.

Further support for this interpretation can be found in the analysis of the underlying level at FLK North 5. The next chapter presents the taphonomic study of this level, and shows that in this time-averaged site (i.e., all of FLK North), felids must have been the principal carcass accumulators for a very long time period. We currently lack any referential framework which can fully replicate this process over such a vast time span. However, the repeated use of the FLK North area by carnivores implies the presence of an ecological magnet – such as a spring – that prompted carnivores, hominids, and herbivores to overlap in that space over a long amount of time. This should be archaeologically and geologically tested in the future.

10. Zooarchaeology and taphonomy of FLK North 5

C.P. EGELAND

Introduction

The assemblage from Level 5 of the FLK North locality is particularly interesting because unlike many Bed I sites (except FLK North North Level 1, see Chapter 12), no systematic taphonomic data have been presented for the large mammal subassemblage since Leakey's (1971) monograph (see Fernández-Jalvo *et al.*, 1998 for a discussion of small mammals). Therefore, and despite the fact that both cut marks and tooth marks have been identified in the large mammal subassemblage (Potts and Shipman, 1981; Shipman, 1983, 1986), the roles of hominids and carnivores in site formation are poorly understood. Although Shipman (1986) identified three cut-marked bovid specimens, this expanded analysis identified only one specimen preserving definitive cut marks and little to no evidence for hominid bone-breaking activities. Therefore, like the other large mammal subassemblages from the FLK North site, Level 5 can securely be attributed almost solely to the activities of carnivores.

The FLK North 5 Faunal Assemblage

SITE DISTURBANCE AND INTEGRITY

A predominance of fragments preserving less than half the original diaphyseal circumference (Type 1) indicate that the assemblage is unbiased by selective retention (Marean *et al.*, 2004) (Figure 113). Figure 114 shows that fragments < 4 cm are slightly less well-represented than expected based on comparisons with experimental assemblages (Blumenschine, 1995; Pickering and Egeland, 2006). Only three specimens (0.2% of total NISP) exhibit the polishing indicative of extensive water transport. Compared to other sites such as DK (see Chapter 15), FLK North 5, along with the other assemblages in the FLK North sequence (see Chapters 7–9), experienced relatively low levels of postdepositional disturbance.

TAXONOMIC AND SKELETAL REPRESENTATION

Table 47 summarizes the relative abundance of major faunal groups at FLK North 5. Bovids and carnivores are particularly well represented in the assemblage. Table 48 provides minimum number of individuals (MNI) estimates. Like the levels below it, the FLK North 5 fauna is dominated by *Antidorcas recki* and *Parmularius altidens*, perhaps suggesting a specialized bone-collecting agent.

Patterns of skeletal element representation also suggest that the bovids experienced a different taphonomic history than the suid subassemblage, which is dominated by teeth

M. Domínguez-Rodrigo, R. Barba, and C.P. Egeland, Deconstructing Olduvai: A Taphonomic Study of the Bed I Sites, 191–200.

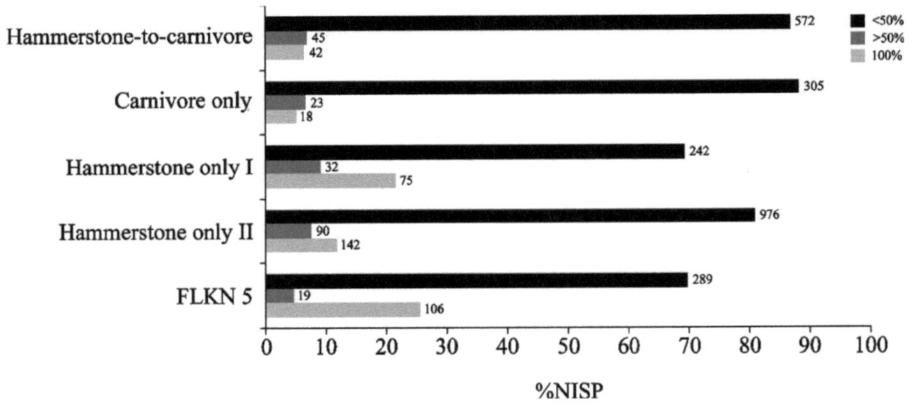

Figure 113. The percentage of Bunn's (1982) limb bone circumference types in the FLK North 5 assemblage compared to several experimental samples. Carnivore only, hammerstone-to-carnivore, and hammerstone only I samples from Marean *et al.* (2004). Hammerstone only II sample from Pickering and Egeland (2006).

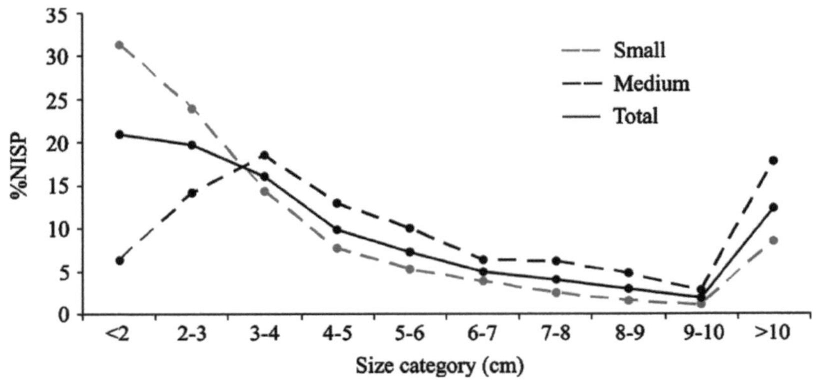

Figure 114. Percentage of specimens in each size range category at FLK North 5. Data are divided into specimens from small carcasses, medium carcasses, and combined.

Table 47. Number of identified specimens (NISP) for major faunal groups at FLK North 5

Faunal group	NISP	%
Bovidae	909	76.2
Suidae	81	6.8
Equidae	1	0.1
Carnivora	202	16.9

(Table 49). Relative to the number of individuals, skulls and limb bones are the most well-represented skeletal elements for both small (Size Class 1 and 2) and medium (Size Class 3a and 3b) carcasses, whereas axial bones are poorly represented (Table 50). According to Marean and Spencer's (1991) limb bone portions, midshafts provide the highest minimum number of element (MNE) estimate for all small-sized limb bones and for all but the metapodials for medium-sized carcasses (Table 51).

Table 48. Minimum number of individuals (MNI) represented at FLK North 5

Taxon	MNI
A. recki	18
P. altidens	5
Beatragus sp.	1
Size 1 Neotragini	1
Size 3b Tragelapini	2
Size 3b Alcelaphini	3
Size 3 Hippotragini	1
Prototocyon recki	1
Kolpochoerus limnetes	2

Table 49. Number of identified specimens (NISP) by skeletal element for bovids and suids at FLK North 5

Element	Bovidae	Suidae
Cranium	40	2
Teeth	165	49
Mandible	22	5
Vertebrae	19	2
Ribs	10	3
Innominate	23	–
Scapula	15	3
Humerus	21	2
Radio-ulna	51	2
Carpals/tarsals	127	–
Metacarpal	42	–
Femur	32	–
Tibia	57	–
Metatarsal	61	–
Patella	5	–
Phalanges	145	13
Sesamoids	54	–
Metapodial	18	–
Limb bone shaft	2	–

BONE SURFACE MODIFICATIONS

As in many of the Bed I assemblages, cortical surface preservation at FLK North 5 is good. Of those specimens scored for surface preservation, only 30.8% were considered to possess poorly preserved cortices, indicating that over two thirds of the assemblage is conducive to secure surface-mark identification. Only a single specimen, a cut-marked humerus from a Size Class 3a bovid, was found to preserve definitive hominid surface modifications.

Table 50. Minimum number of element (MNE) estimates for small (Size Class 1 and 2) and medium (Size Class 3) carcasses at FLK North 5

Element	Small	Medium	Total
Cranium	19	7	26
Mandible	16	7	23
Vertebrae	14	18	32
Innominate	17	4	21
Ribs	5	8	13
Scapula	13	3	16
Humerus	13	14	27
Radius	19	10	29
Ulna	9	9	18
Carpals	33	17	50
Metacarpal	17	13	30
Femur	10	12	22
Patella	6	2	8
Tibia	23	17	40
Tarsals	47	28	75
Metatarsal	20	16	36
Phalanges	119	18	137
Sesamoids	35	20	55

Therefore, Table 52 summarizes the raw incidence of tooth marks in the FLK North 5 assemblage. Tooth marks concentrate on limb bones and forelimbs in particular. The limb bones from small carcasses are tooth-marked at higher frequencies than medium carcasses.

Given the lack of hominid surface modifications, one would expect tooth-mark frequencies on limb bone midshaft fragments (following Blumenschine's [1988, 1995] bone segment system) to mirror experimental "carnivore-only" scenarios where carnivores enjoyed sole access to flesh and marrow. However, as Figure 115 illustrates midshaft tooth-mark frequencies on small and medium carcasses fall outside the range of variation for these experiments and, for medium carcasses at least, match more closely "hammerstone-to-carnivore" experiments where hyenas ravaged defleshed and demarrowed bones. In the absence of hominid surface modifications and a lack of any other evidence for hominid bone-breaking (see later), it is unlikely, as the midshaft tooth-mark frequencies would initially suggest, that carnivores were ravaging

Table 51. Minimum number of element (MNE) estimates for limb bones by bone portion for small (Size Class 1 and 2) and medium (Size Class 3) carcasses at FLK North 5

Element	Small	Medium	Total
Humerus			
PR	2	0	2
PRS	3	2	5
SH	11	14	25
DSS	12	8	20
DS	12	5	17
Radius			
PR	12	5	17
PRS	12	7	19
SH	16	10	26
DSS	8	4	12
DS	3	3	6
Metacarpal			
PR	14	12	26
PRS	12	8	20
SH	16	12	28
DSS	12	7	19
DS	6	7	13
Femur			
PR	5	3	8
PRS	1	2	3
SH	8	11	19
DSS	7	9	16
DS	7	5	12
Tibia			
PR	4	4	8
PRS	7	14	21
SH	19	15	34
DSS	10	12	22
DS	10	8	18
Metatarsal			
PR	17	12	29
PRS	16	13	29
SH	20	12	32
DSS	18	7	25
DS	13	5	18

Abbreviations: PR, proximal; PRS, proximal shaft; SH, midshaft; DSS, distal shaft; DS, distal

Table 52. Tooth-mark frequencies for small (Size Class 1 and 2) and medium (Size Class 3) carcasses at FLK North 5

Element	Small	Medium	Total
Mandible	0/24 (0.0)	0/8 (0.0)	0/32 (0.0)
Vertebrae	3/16 (18.8)	1/27 (3.7)	4/43 (9.3)
Innominate	4/25 (16.0)	3/7 (42.9)	7/32 (21.9)
Ribs	3/23 (13.0)	1/22 (4.5)	4/43 (9.3)
Scapula	2/14 (14.3)	0/12 (0.0)	2/26 (7.7)
Humerus	7/16 (43.8)	2/28 (7.1)	9/44 (20.5)
Radius	8/26 (30.8)	1/21 (4.8)	9/47 (19.1)
Ulna	5/10 (50.0)	2/13 (15.4)	7/23 (30.4)
Carpals	0/33 (0.0)	0/17 (0.0)	0/50 (0.0)
Metacarpal	14/24 (58.3)	3/19 (15.8)	17/43 (39.5)
Femur	4/23 (17.4)	9/30 (30.0)	13/53 (24.5)
Patella	0/6 (0.0)	0/2 (0.0)	0/8 (0.0)
Tibia	11/36 (30.6)	6/57 (10.5)	17/92 (18.5)
Tarsals	2/49 (4.1)	1/28 (3.6)	3/77 (3.9)
Metatarsal	10/38 (26.3)	4/22 (18.2)	14/60 (23.3)
Phalanges	0/127 (0.0)	1/21 (4.8)	1/148 (0.7)
Sesamoids	0/35 (0.0)	0/20 (0.0)	0/55 (0.0)
Limb bone shaft	2/68 (2.9)	9/81 (11.1)	11/149 (7.4)
Total	75/593 (12.6)	43/435 (9.9)	118/1,028 (11.5)

Numerator denotes number of marked specimens; denominator denotes total NISP for each skeletal element; percentage is in parentheses

hominid food refuse. However, if felids are considered a primary agent of bone modification at FLK North 5, which seems likely given the evidence presented for Levels 1–4 (see Chapters 8 and 9), the tooth-mark data make more sense. Felids are known to tooth mark limb bones and midshafts in particular at lower frequencies (Selvaggio, 1994; Domínguez-Rodrigo *et al.*, in press) than the hyenas used in previous experimental studies (Blumenschine and Marean, 1993; Blumenchine, 1995; Capaldo, 1997, 1998b) (see Figure 115).

Tooth-mark frequencies on epiphyseal and near-epiphyseal fragments at FLK North 5 fall outside the 64–100% range observed in hyena-ravaged assemblages (Blumenschine, 1995; Capaldo, 1997, 1998b) (Table 53). They are also different from frequencies generated by leopards and cheetahs, which tooth mark these portions at rates of 15.9% and 16.8%, respectively (Domínguez-Rodrigo *et al.*, in press). The fact that tooth-mark frequencies do not match exactly with any of the published actualistic control samples indicates that FLK North 5 represents a complex palimpsest in which: (1) some felid-modified carcasses were subsequently ravaged by hyenas; (2) some felid-modified carcasses were not ravaged; and (3) some carcasses represent natural deaths that were not processed by carnivores at all.

Figure 116 summarizes tooth pit dimensions on small carcasses in the FLK North 5

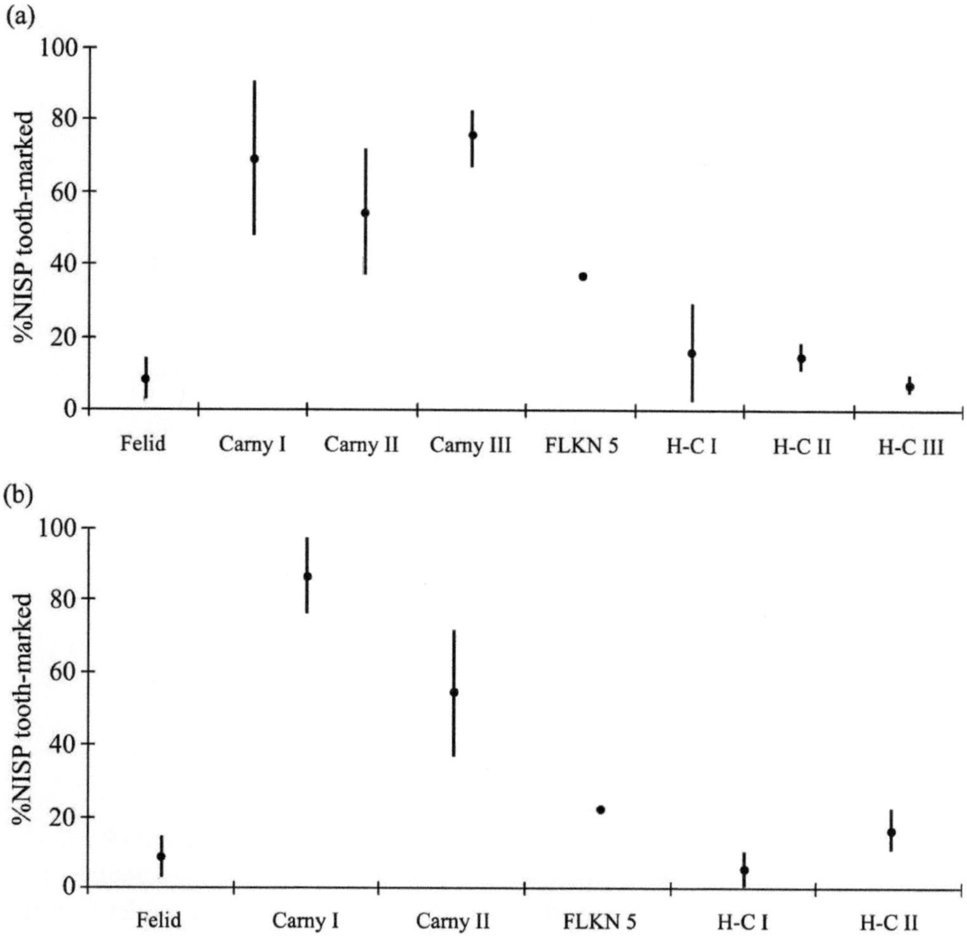

Figure 115. Incidence of tooth-marked midshaft fragments at FLK North 5 for (A) small-sized carcasses and (B) medium-sized carcasses compared to the mean and 95% confidence intervals for actualistic control assemblages. Only fossil specimens with green breakage and good cortical surface preservation are considered. *Abbreviations*: Felid, felid-consumed carcasses; Carny, carnivore-only; H-C, hammerstone-to-carnivore. (Data sources: Blumenschine [1995]; Capaldo [1997, 1998a, 1998b]; Marean *et al.* [2000]; Domínguez-Rodrigo *et al.* [in press].) *Note*: The range of variation from Marean's experiments ("Carny III" and "H-C III") are somewhat smaller because confidence intervals were calculated by bootstrapping a single sample (Marean *et al.*, 2000: Table 3).

Table 53. Percentage of epiphyseal, near-epiphyseal, and midshaft specimens bearing tooth marks at FLK North 5

	EP	NEP	MSH	Total
Small carcasses				
NISP	21	8	52	81
NISP TM	13	2	19	34
%	61.9	25.0	36.5	42.0
Medium carcasses				
NISP	5	18	36	59
NISP TM	1	9	8	18
%	20.0	50.0	22.2	30.5

NISP counts include only those specimens with green breakage and good cortical surface preservation
Abbreviations: EP, epiphyseal; NEP, near-epiphyseal; MSH, midshaft

assemblage. It seems likely that smaller carnivores with less robust dentitions, such as cheetahs and leopards, modified many of the small carcasses at FLK North 5. This is consistent with data presented for the lower levels of FLK North (Chapters 8 and 9) and contrasts with DK, where both small- and medium-sized carcasses were modified by larger carnivores such as lions and hyenas (see Chapter 15). The few tooth pits on medium carcasses (n = 3) at FLK North 5 match better with those created by larger carnivores.

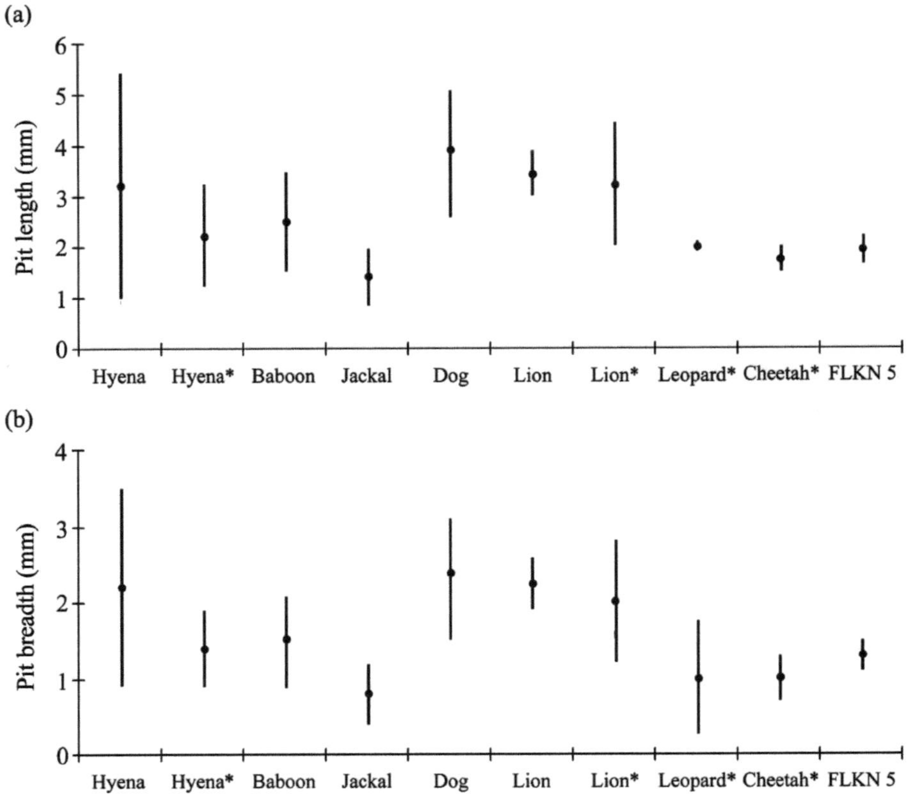

Figure 116. Mean and 95% confidence intervals for tooth pit (A) lengths and (B) breadths on the limb bone diaphyses of small-sized carcasses at FLK North 5 compared to actualistic samples. (Data indicated with an asterisk (*) are from Selvaggio [1994]; all other modern carnivore data are from Domínguez-Rodrigo and Piqueras [2003].)

BONE BREAKAGE

Unlike other levels from the FLK North site, FLK North 5 shows relatively high levels of diagenetic breakage, especially for medium carcasses (Figure 117). Complete bones make up a significant proportion of the limb bone MNE (Table 54), and it appears that they were being broken in direct proportion to their nutritional content. The relative proportions of Bunn's (1982) circumference types are similar to Levels 1 and 2 of the FLK North sequence (see Chapter 8) and differ from experimental assemblages with intensive hyena ravaging (Figure 113). The epiphysis-to-shaft fragment ratios for small (0.38) and medium (0.26) carcasses are higher than carnivore-only assemblages where ravaging is very intense (Blumenschine and Marean, 1993; Capaldo, 1998b). These data all suggest a relatively low level of ravaging in the FLK North 5 assemblage.

Fracture plane analysis can provide information on the identity of the bone-breaking agent. Although fracture planes from small carcasses overlap with both static (carnivore) and dynamic (hammerstone) experiments (Figure 118), fracture planes from medium-sized carcasses are more consistent with carnivore breakage (Figure 119).

Although only four notches were complete enough to calculate notch breadth:notch depth and scar breadth:notch depth ratios, all fall comfortably within the carnivore range (not pictured). Micronotches, which are characteristic of carnivore-broken assemblages,

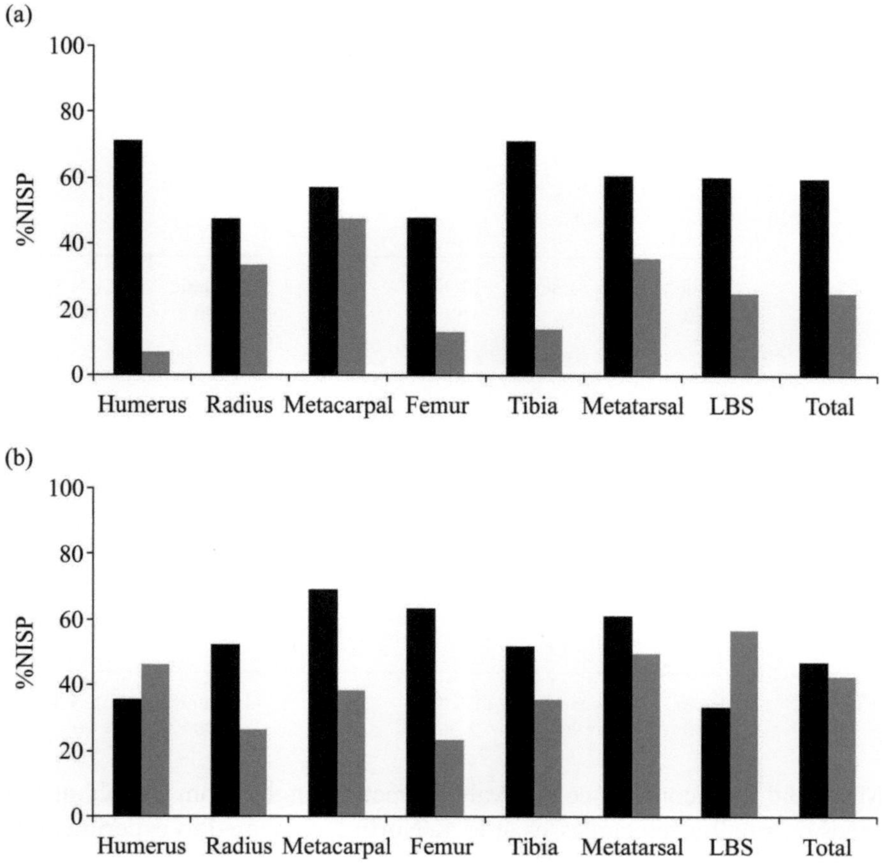

Figure 117. Incidence of green and diagenetic breakage on limb bones from (A) small and (B) medium carcasses at FLK North 5. Percentages may not add to 100% as specimens with recent breakage are not considered. *Abbreviations*: LBS, unidentified limb bone shaft.

Table 54. Frequencies of complete limb bones for small (Size Class 1 and 2) and medium (Size Class 3) carcasses at FLK North 5

Element	Small	Medium	Total
Humerus	2/13 (15.4)	0/14 (0.0)	2/27 (7.4)
Radius	2/19 (10.5)	1/10 (10.0)	3/29 (10.3)
Metacarpal	3/17 (17.6)	6/13 (46.2)	9/30 (30.0)
Femur	0/10 (0.0)	0/12 (0.0)	0/22 (0.0)
Tibia	1/23 (4.3)	1/17 (5.9)	2/40 (5.0)
Metatarsal	10/20 (50.0)	4/16 (25.0)	14/36 (38.9)
Total	18/102 (17.6)	12/82 (14.6)	30/184 (16.3)

Numerator denotes number of complete bones; denominator denotes total MNE for each limb bone; percentage is in parentheses

predominate in FLK North 5 and Incomplete Type C notches, also common in carnivore-broken assemblages, are abundant among medium-sized carcasses (Figure 120).

Discussion and Conclusions

The bone assemblage from FLK North 5 is almost exclusively the result of carnivore activities. The lack of cut marks fits well with the fact that only ten flakes (presumably the main butchering tool of the Oldowan toolkit) occur in the 132-piece lithic assemblage (de la Torre, 2006). Although anvils and lithic pieces with percussion damage are present in the FLK North 5 assemblage (Leakey, 1971; de la Torre, 2006), the absence of percussion marks on bones and any evidence for hominid bone breakage suggests that these tools were used for a purpose other than carcass butchery.

A vast majority of the assemblage derives from just two bovid size classes from two

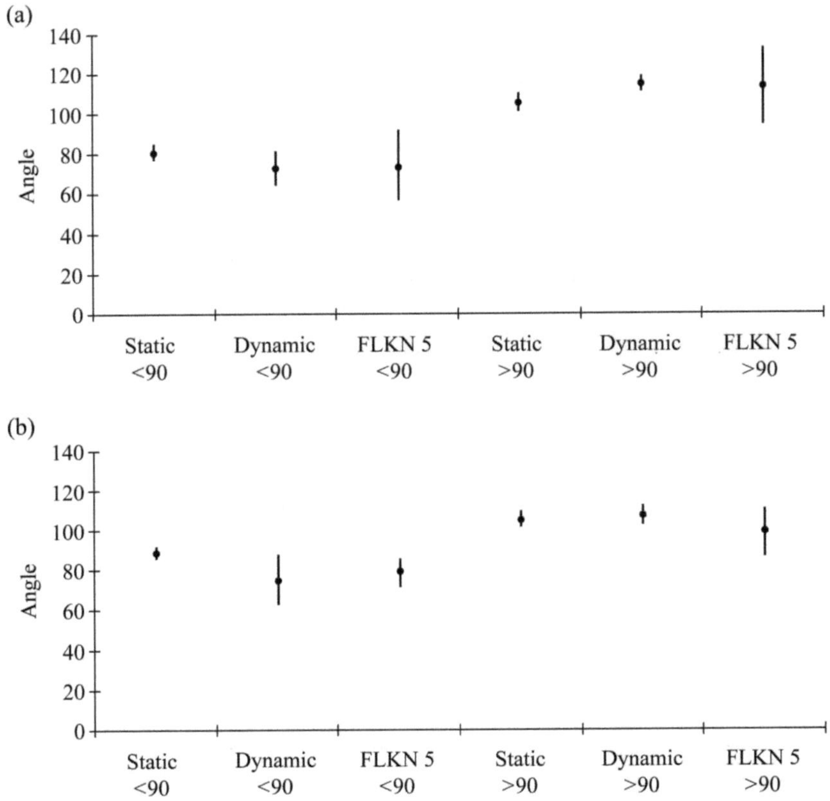

Figure 118. Mean and 95% confidence intervals of fracture angles from (A) oblique and (B) longitudinal planes on small-sized carcasses at FLK North 5 compared to experimental samples.

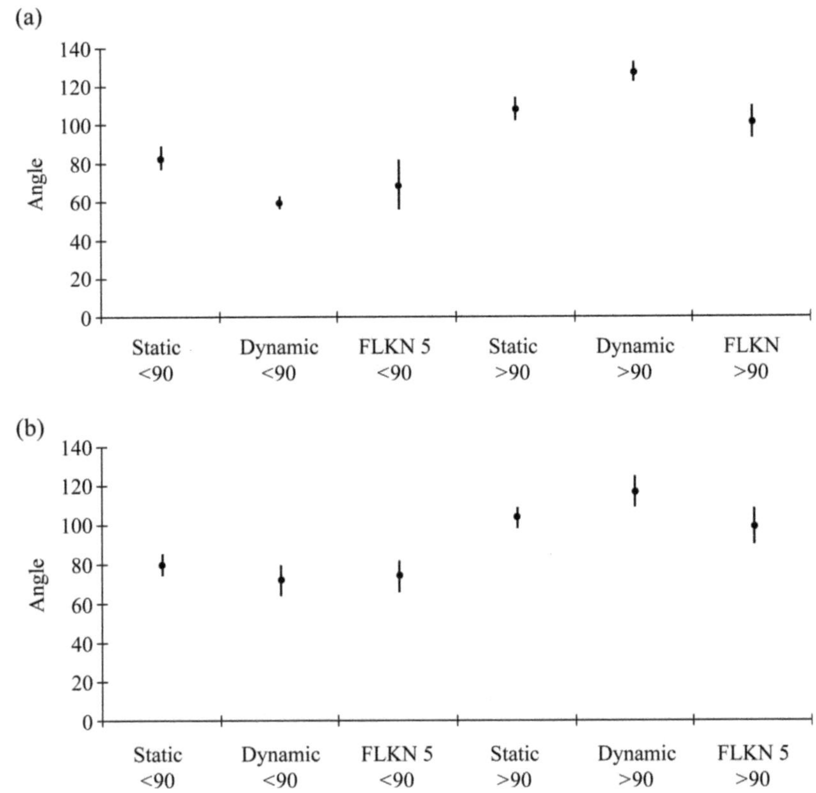

Figure 119. Mean and 95% confidence intervals of fracture angles from (A) oblique and (B) longitudinal planes on medium-sized carcasses at FLK North 5 compared to experimental samples.

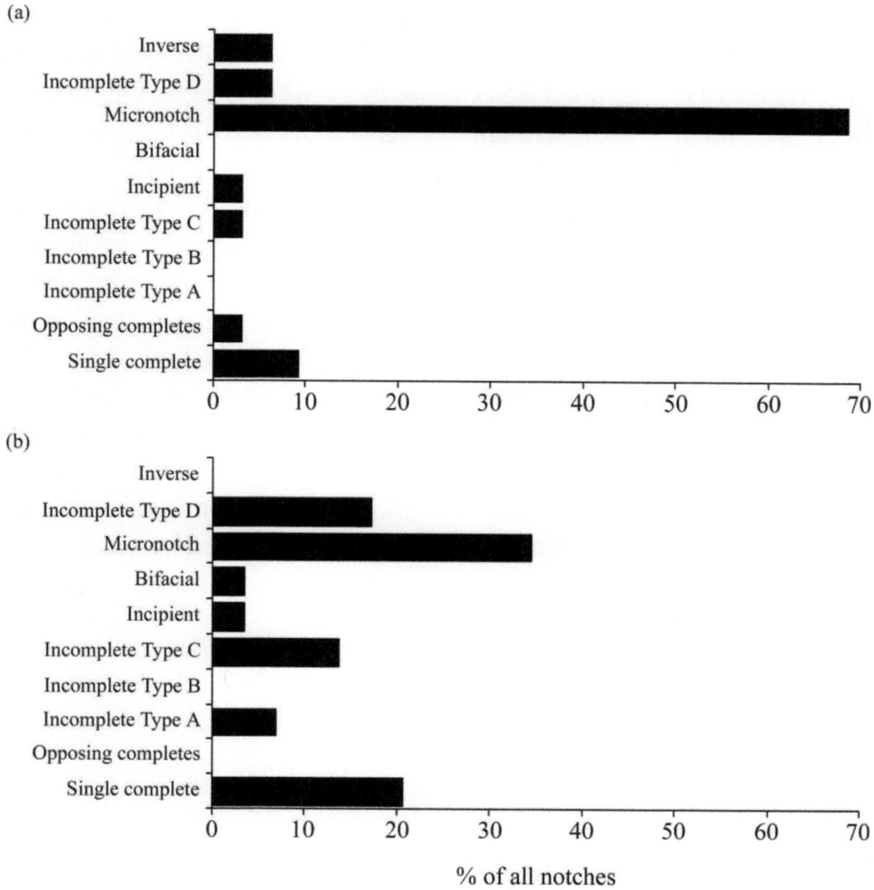

Figure 120. Incidence of notch types (modified from Capaldo and Blumenschine, 1994) from (A) small and (B) medium carcasses in the FLK North 5 assemblage.

species, *A. recki* (Size Class 1) and *P. altidens* (Size Class 3a). The low diversity of size classes and species strongly suggests some sort of specialized bone collector. A felid is the most likely agent here given the low overall incidence of tooth marks in the FLK North 5 assemblage. Tooth pit dimensions on small carcasses point to a medium-sized felid, and, given their propensity for specializing on two or three different prey species (Scott, 1987; Cavallo, 1998), leopards are a likely candidate. The modification of medium carcasses may be attributable to the larger bodied felid *Dinofelis*, which appears similar in its strategy of carcass acquisition and processing to leopards (Lewis, 1997).

As noted earlier, a number of factors suggest that the overall intensity of carnivore ravaging at FLK North 5 is low to moderate. The fact that limb bones were broken in proportion to their nutritional content indicates that carnivores were only processing higher yield elements for within-bone nutrients. This suggests further that when carcasses where discovered and processed, group size and/or levels of competition were low. Moreover, the large number of accumulated carcasses coupled with the fact that almost 1.5 ft of Level 5 deposit overlie the similarly thick deposits of Levels 1–4, all of which display high MNIs and felid taphonomic signatures, demonstrates that this area was favorable to felid carcass transport over a long period of time. It is therefore likely that trees were abundant in the vicinity, although within the broader context of a relatively open habitat, as FLK North 5 plots with Levels 1–2 in Domínguez-Rodrigo and Organista's ravaging stage 2 (see Chapter 11). This is in contrast to DK (Chapter 15) or levels from the FLK North North sequence (Chapters 12–14), which seem to indicate mixed and closed environments, respectively. The FLK North area would have provided hominids ample opportunities to passively scavenge from felid kills. The fact that little to no evidence for hominid carcass exploitation has been documented in these assemblages, including Level 5, indicates that this was not a commonly employed carcass acquisition strategy.

In order to understand the trophic dynamics that created the accumulations of carcasses described in this and previous chapters, it is necessary to know what the background bone scatters were like at Olduvai (some of which were addressed in Chapter 7), and how these scatters compare with the modern savanna ecosystems used as analogs for understanding Plio-Pleistocene savanna environments. Chapter 11 will provide new data on background bone scatters at Olduvai during Bed I times, and will show that studies of trophic dynamics and competition in modern savannas can be applied to interpret the paleoecological situation in the Bed I sites, and therefore to understand the reasons why bones appeared more concentrated in some spots than in others.

11. Natural background bone assemblages and their ravaging stages in Olduvai Bed I

M. DOMÍNGUEZ-RODRIGO AND E. ORGANISTA

Introduction

One pillar of Potts' (1988) argument that several Olduvai Bed I sites were hominid-made was that the dense concentrations of bones seemed to have no equivalent in natural processes of bone dispersal and concentration in modern savannas (see Chapter 1). The ongoing landscape archaeology project at Olduvai has obtained a wealth of information about bone scatters on the landscape during that time. Cushing's (2002) study of these scatters reveals that bone density on the landscape during the formation of Lowermost Bed II varied according to geological facet (including habitat type), and that natural bone clusters on the landscape were sometimes fairly dense. Of all the trenches analyzed by Cushing, those belonging to Olduvai Bed I Middle Bed – the same stratigraphic interval in which archaeological sites such as FLK *Zinj* are situated – show the highest densities of bone scatters. This suggests that there were *loci* on the landscape in which natural bone accumulations were substantially high and in which no hominid participation was required (see Chapter 7 for an extensive explanation of the FLK North 6 case).

Bunn (1986: 686) argued that some bone accumulations with no stone tools or associated with very few artifacts were natural background scatters "possibly derived predominantly from carnivore activity or other non-hominid natural background processes of accumulation," and distinguished from classical "living floors" or "central-foraging places" by exhibiting different patterns of skeletal part representation. Bunn included sites like FLK North North 2 and FLK 13 among such assemblages. Nevertheless, there are a few more. It could be argued that most archaeological levels overlying the FLK *Zinj* level (i.e., level 22), for instance, are natural background accumulations, as Leakey (1971) argued. Here we present an analysis of FLK 10, FLK 11-12, FLK13, and FLK 15. We will show that skeletal part profiles are not very different from those documented at classical sites and that dense accumulations of bones were a natural part of the Olduvai ecosystem during the formation of Bed I. Some carnivores (probably felids) were extremely active in accumulating carcasses, while others (hyaenids) also seemed very active in modifying them.

FLK 10 was found within a consolidated reworked tuff under Tuff IF. FLK 11 was found within another yellowish tuff interbedded with a greenish sandy clay deposit. FLK 12 was underlying the previous level within the yellow tuff. Given its stratigraphic proximity to FLK 11 and the paucity of materials, both levels 11 and 12 were analyzed together with FLK 10 in the present study. FLK 13 was a 2-ft (about 61 cm) deposit of interbedded clays and tuffs. FLK 15 consisted of a clayey tuff. Leakey (1971) provides no information regarding the vertical distribution of materials in this site.

M. Domínguez-Rodrigo, R. Barba, and C.P. Egeland, Deconstructing Olduvai: A Taphonomic Study of the Bed I Sites, 201–215.

Samples, Analysis, and Results

Four major bovid tribes (antilopini, alcelaphini, reduncini, and tragelaphini) are represented in the four archaeological levels studied (Table 55). The number of individuals varies in each level. The FLK 10-11-12 bone assemblage is very small (most of the elements are from five individuals), and therefore, the number of individuals represented is also fewer than at the other sites (12 at FLK 13 and 8 at FLK 15, although most postcranial elements are from six individuals). The number of bone specimens also varies among sites (Table 56). At FLK 10-11-12 a total of 64 specimens have been documented, whereas at FLK 13 and 15 there are 128 and 199 specimens, respectively. Despite variation in sample size, the three sites seem to show similar differential preservation of skeletal elements, and especially long limb bone portions (Figure 121). Proximal humeri and distal radii seem to be underrepresented compared to distal humeri and proximal radii, respectively. Femora are the least represented elements. Distal ends of tibiae are between two and three times more represented than proximal ends. Metapodials are well-represented with small differences between ends. This distribution of bone portions is similar to that documented by Brain (1981) in carnivore-ravaged assemblages and shows identical representation profiles to those bone accumulations which were traditionally considered "sites," like FLK *Zinj* or FLK North 1–2 (see Chapters 6, 8). This suggests carnivore ravaging in the assemblage.

Table 55. Minimum number of individuals (MNI) in each FLK level analyzed

	FLK 10–11–12	FLK 13	FLK 15
Antilopini	2	1	1
Antidorcas recki	1	3	1
Parmularius altidens	1	1	1
Connochaetes sp.	0	0	1
Kobus sigmoidalis	1	1	2
Tragelaphus strepsiceros	0	3	2
Hippotragini	0	1	0
Bovini	1	0	0
Equus sp.	0	0	1
Giraffa sp.	0	1	1
Hippopotamus gorgops	0	1	0
Ceratotherium simum	0	1	0
Suidae size 3A	1	2	2
Canis sp.	0	1	0
Crocuta cf. *crocuta*	1	0	0
Carnivore indet	0	0	1

Representation of skeletal elements is shown in Table 57 and Figure 122. Distributions of elements between the "natural accumulations" and "sites" can be used to observe the degree of variability in skeletal element representation, using frequencies calculated by dividing the MNE for each element by the total MNE that would be expected for that element, given the total MNI for the assemblage. Bunn (1986) remarked that one of the differences between natural and anthropogenic sites was that natural sites, as representative of death places, had a higher presence of skull elements and proportionally fewer mandibles, which are more easily transported. However, that does not seem to be the case for all the sites that Bunn presented as anthropogenic. FLK *Zinj* has an intermediate frequency of skulls, similar to that of mandibles; FLK North 1–2, argued by Bunn to be another hominid-made site, has, together with FLK 13, the highest number of skulls and a much lower frequency of mandibles.

By contrast, "natural" sites like FLK 10-11-12 and FLK 15 have a much lower frequency of skulls than is found at FLK *Zinj*. Skull frequencies in both types of sites are thus highly variable, whereas mandibles are represented similarly within a 20% range. All sites are almost identical when it comes to axial skeleton representation. Bunn (1986) also mentioned that death sites would have a higher presence of axial bones (namely vertebrae) than transported assemblages. However, when the actual number of axial bones is compared to the

Table 56. Number of identified specimens (NISP) from each FLK level analyzed

	FLK 10–11–12		FLK 13		FLK 15	
	Size 1/2	Size 3	Size 1/2	Size 3	Size 1/2	Size 3
Horn	0	0	3	4	0	2
Skull	0	1	1	6	0	1
Mandible	0	3	1	5	2	4
Teeth	0	6	1	17	3	24
Vertebra (atlas, axis, cervical)	0	0	2	2	0	0
Vertebra (thoracic)	0	3	0	6	0	1
Vertebra (lumbar, sacral)	1	2	0	2	0	3
Vertebra indet.	0	0	0	8	0	0
Ribs	3	4	0	4	1	5
Scapula	2	7	0	5	2	6
Pelvis	0	1	0	4	0	2
Humerus complete	0	1	0	0	0	0
Humerus prox. end	0	0	0	1	0	1
Humerus shaft	0	0	0	1	1	14
Humerus dist. end	0	1	0	5	0	4
Radius complete	0	1	0	1	0	0
Radius prox. end	1	2	0	5	2	10
Radius shaft	0	0	0	3	1	7
Radius dist. end	0	0	0	0	2	0
Metacarpal complete	0	1	0	1	0	2
Metacarpal prox. end	1	1	0	3	1	5
Metacarpal shaft	1	0	1	4	1	7
Metacarpal dist. end	0	0	0	1	2	3
Femur complete	0	0	0	0	0	0
Femur prox. end	1	0	0	0	0	0
Femur shaft	0	1	2	3	4	6
Femur dist. end	0	0	0	2	1	0
Tibia complete	0	0	0	0	0	1
Tibia prox. end	0	0	1	2	0	1
Tibia shaft	1	1	1	11	2	29
Tibia dist. end	0	0	3	2	2	4
Metatarsal complete	0	1	0	0	0	0
Metatarsal prox. end	1	0	1	6	2	4
Metatarsal shaft	0	1	1	3	2	4
Metatarsal dist. end	0	0	3	3	0	1
Carpals	0	4	0	2	2	2
Tarsals	0	3	0	1	0	3
Phalanges	0	4	0	1	3	3
ULB*	0	0	0	0	0	2
ILB*	1	1	0	0	0	1
LLB*	0	0	0	0	3	5
Indeterminate	0	2	0	0	0	5
Total	13	52	21	124	39	172

*ULB, upper limb bones; ILB, intermediate limb bones; and LLB, lower limb bones

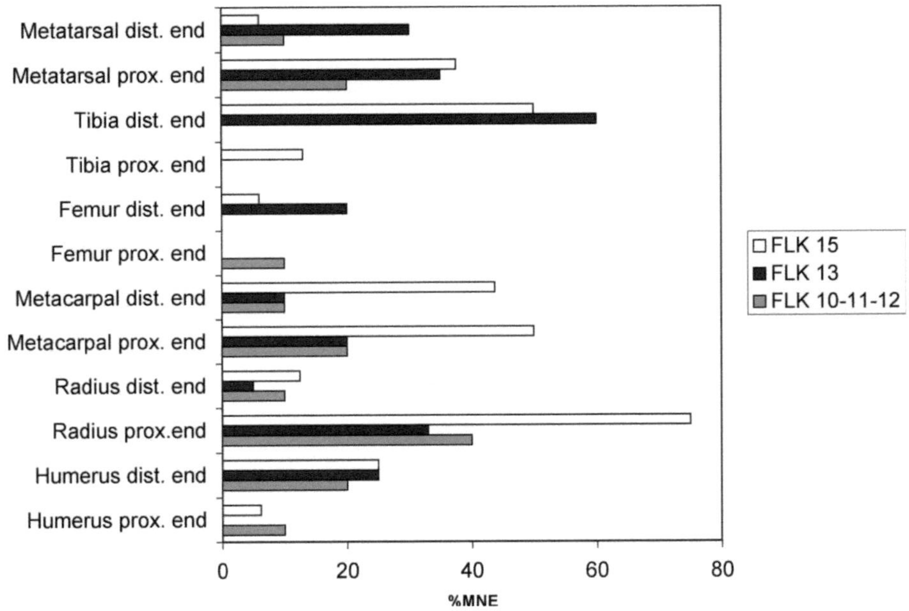

Figure 121. Representation of long limb bone portions (as %MNE) at FLK 10-11-12, 13, and 15, according to the number expected based on MNI.

Table 57. Minimum number of elements (MNE) in each FLK level analyzed

	FLK 10–11–12		FLK 13		FLK 15	
	Size 1/2	Size 3	Size 1/2	Size 3	Size 1/2	Size 3
Horn	0	0	3	4	0	2
Skull	0	1	1	11	0	1
Mandible	0	6	2	11	2	3
Vertebra (atlas, axis, cervical)	0	0	2	3	0	0
Vertebra (thoracic)	0	3	0	10	0	1
Vertebra (lumbar, sacral)	1	0	0	1	0	1
Ribs	3	4	0	4	2	2
Scapula	2	6	0	4	2	3
Pelvis	0	1	0	4	0	1
Humerus	0	2	0	7	1	6
Radius	2	3	0	7	2	5
Metacarpal	2	2	1	5	2	6
Femur	1	0	2	2	2	4
Tibia	1	1	5	7	3	6
Metatarsal	0	2	3	6	3	4
Carpals	0	4	0	2	1	2
Tarsals	0	3	0	2	0	3
Phalanges	0	2	0	0	2	3
ULB*	0	0	0	0	0	0
ILB*	1	1	0	0	0	0
LLB*	0	0	0	0	3	5
Indeterminate	0	0	0	0	3	4
Total	13	41	19	90	28	62

*ULB, upper limb bones; ILB, intermediate limb bones; and LLB, lower limb bones

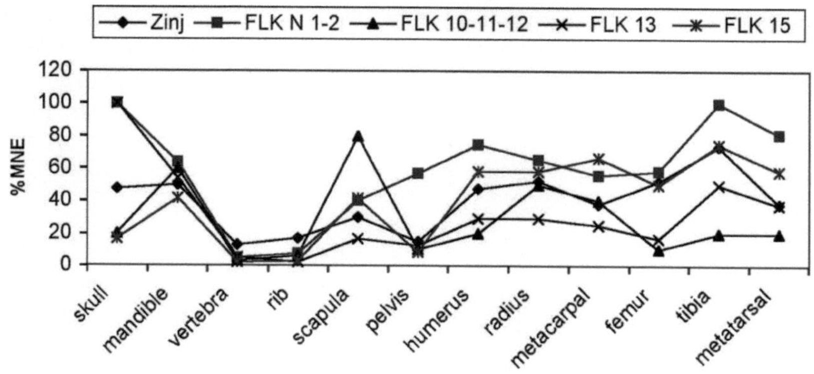

Figure 122. Representation of frequencies of element distribution (as %MNE) in the FLK levels, according to estimated MNI.

number of carcasses represented (Figure 122), we see that axial skeleton representation is highly variable between sites (FLK *Zinj* and FLK North 1–2) and that "natural" assemblages show lower frequencies of vertebrae and ribs. Scapulae and pelves also vary between archaeological sites, although pelvic representation is very similar in natural sites. Frequencies of limb bones also vary among sites, although almost all of them share a similar profile in which femora are the least represented elements, and humeri and tibiae are most represented.

Therefore, skeletal part profiles are similar and equally variable when comparing "archaeological" and "natural" sites, which prevents any differentiation using skeletal parts alone as diagnostic criteria. This also suggests that natural processes, operating in the same ecological scenarios as hominids, were capable of creating bone accumulations with the same skeletal part profiles as have been traditionally interpreted as belonging to anthropogenic sites. Meaty bones in these natural sites are also as abundant as they are in hominid-made assemblages. Therefore other criteria, namely bone surface modifications, should be used to differentiate between hominid- and carnivore-made accumulations.

Turning to bone surface modifications, tooth marks are very abundant in the analyzed assemblages (Table 58). Almost all limbs present in the site are tooth-marked. This, together with the high degree of green fractures on shafts, suggests hyena ravaging of bones in most levels (Figure 123). Green fractures and associated tooth-marking occur both on ends and shafts of almost all limb bones. Therefore, the only taphonomic interpretation that can be made is that hyena ravaging took place on those carcasses after their deposition in (or perhaps transport to) those spots. How were these carcasses accumulated? The possibility of hyena denning is once again excluded for the following reasons (Brain, 1981; Bunn, 1982; Brugal *et al.*, 1997; Fosse *et al.*, 1998; Kerbis, 1990; Marra *et al.*, 2004; Villa *et al.*, 2004):

1. There is no significant number of carnivore remains, especially from subadults, associated with the herbivores' bones.
2. There are neither coprolites nor digested bone, which are abundant in hyena dens (see Egeland *et al.*, in preparation).
3. There is virtual lack of gnawed bone, defined by specimens bearing >10 tooth marks, which makes up an important part of the tooth-marked specimens in hyena dens.

Table 58. Frequencies of tooth-marked specimens in each level. Numerator is total number of tooth-marked specimens, denominator is total number of specimens of each type. Percentages are in parentheses

	FLK 10–11–12		FLK 13		FLK 15	
	Size 1/2	Size 3	Size 1/2	Size 3	Size 1/2	Size 3
Horn	–	–	–	–	–	–
Skull	–	–	–	–	–	–
Mandible	–	–	–	–	–	–
Teeth	–	–	–	–	–	–
Vertebra (atlas, axis, cervical)	–	–	1/2 (50)	2/2 (100)	–	–
Vertebra (thoracic)	–	–	–	–	–	1/1 (100)
Vertebra (lumbar, sacral)	–	–	–	–	–	–
Ribs	–	–	–	–	–	1/5 (20)
Scapula	–	2/6 (33.3)	–	2/5 (40)	2/2 (100)	1/6 (16.6)
Pelvis	–	–	–	1/4 (25)	–	1/2 (50)
Humerus complete	–	–	–	–	–	–
Humerus prox. end	–	1/1 (100)	–	–	–	–
Humerus shaft	–	–	–	–	1/1 (100)	1/14 (7.14)
Humerus dist. end	–	–	–	4/5 (80)	–	1/4 (25)
Radius complete	–	–	–	–	–	–
Radius prox. end	–	2/2 (100)	–	–	1/2 (50)	5/10 (50)
Radius shaft	–	–	–	–	–	3/7 (42.8)
Radius dist. end	–	–	–	–	–	–
Metacarpal complete	–	–	–	–	–	–
Metacarpal prox. end	–	–	–	–	–	1/5 (20)
Metacarpal shaft	–	–	–	–	–	–
Metacarpal dist. end	–	–	–	–	–	–
Femur complete	–	–	–	–	–	–
Femur prox. end	–	–	–	–	–	–
Femur shaft	–	–	–	–	2/4 (50)	1/6 (16.6)
Femur dist. end	–	–	–	–	–	–
Tibia complete	–	–	–	–	–	1/1 (100)
Tibia prox. end	–	–	–	–	–	–
Tibia shaft	–	–	–	2/11 (18)	1/2 (50)	7/29 (24)
Tibia dist. end	–	–	1/3 (33)	1/2 (50)	–	–
Metatarsal complete	–	–	–	–	–	–
Metatarsal prox. end	–	1/1 (100)	–	–	–	–
Metatarsal shaft	–	–	–	–	–	–
Metatarsal dist. end	–	–	–	2/3 (66.6)	–	–
Carpals	–	–	–	–	–	–
Tarsals	–	–	–	–	–	1/3 (33)
Phalanges	–	–	–	–	–	–
ULB*	–	–	–	–	–	–
ILB*	–	–	–	–	–	–
LLB*	–	–	–	–	1/5 (20)	–
Indeterminate	–	–	–	–	–	–

*ULB, upper limb bones; ILB, intermediate limb bones; and LLB, lower limb bones

The tooth-marking observed in these bone assemblages (Table 60) is more similar to that documented in carnivore ravaging experiments where tooth-marked specimens frequently bear no more than one tooth mark resulting from fracturing the bone; this contrasts with the constant gnawing that hyenas do in their dens. Given the small area excavated (Leakey, 1971) and the mixing of four tribes with different ecological adaptive patterns, the carcass

Figure 123. Various stages of completeness of humeral shafts. Note the two humeri on the right showing furrowing on the caudal side of the epicondyles (see also Figure 125) (scale = 1 cm).

accumulations in some of the sites (FLK 13 and 15) should probably be the result of transport, as suggested by Potts (1988) for other Bed I sites. We have no data to make a case for complete carcass transport. Two observations may support the candidacy of felids, but neither of them is conclusive: the first is the presence of articulated elements, like the complete limb found in FLK 11-12 (Figure 124). It belongs to a *Parmularius*, whose mandible is also adhered to it by the carbonated sediment matrix. This suggests that the limb was not deposited at the site individually, but that it was very likely attached to the carcass to which the mandible belongs. On the proximal articular end of the humerus, some moderate furrowing (like that described in Chapters 8 and 9 for felids) can be observed. Complete articulated limbs occasionally form part of the assemblage in hyena dens, but they usually are not transported with their corresponding heads. The second observation lies in the location of furrowing on caudal distal ends of humeri (Figure 125), where once again the condyles remain intact in their lateral-cranial aspects; these condyles should be the first ones modified by hyenas. The few instances of furrowing on these caudal epicondylar areas could be the result of undocumented or marginal modification by hyenas, or they could be the result of a felid stage of bone modification before hyena intervention. In Figure 125, the humerus on the right is a specimen from FLK 13 (there are more in FLK 13 and 15) bearing furrowing just on the caudal epicondylar area, which is the same pattern as is found in felid-modified carcasses. The humerus from the left also bears a similar modification, but given that the latero-cranial aspect of the condyle has been ravaged, it cannot be ruled out that it was made by hyenas.

The samples from these archaeological levels are too small to gain any further resolution. However, they seem to be modified in ways similar to those fauna from other assemblages which have been traditionally interpreted as archaeological sites.

Figure 124. Example of a complete articulated limb from a *Parmularius* from FLK 11-12. The mandible is also attached. Note the intense furrowing on the proximal end of the humerus (scale = 1 cm).

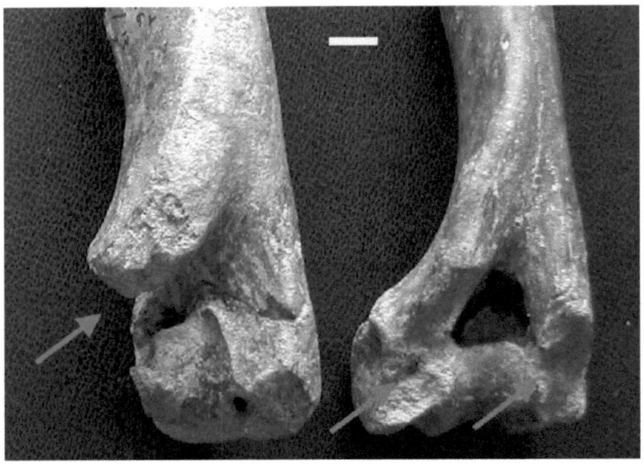

Figure 125. Ravaging on the caudal side of the distal epicondyles of humeri (detail from Figure 3) (scale = 1 cm).

Measuring Carnivore Ravaging

Many of the differences that account for variation in bone representation and modification, both in sites interpreted as hominid refuse and in those interpreted as background accumulations, lie in the degree of ravaging. Archaeologists have experimented with carnivores to a very limited degree, and they tend to think that when ravaging occurs, the results are the same irrespective of context. However, in Chapter 8, it was shown that the palimpsest at FLK North 1–2 was the product of multiple depositional events taking place over an extended time period during which trophic dynamics fluctuated, resulting in a high degree of variability in bone modification. In one depositional event, depending on the ecological conditions (i.e., the degree of competition), hyenas may either: intensively ravage every single bone; only scavenge for either marrow or epiphyseal grease; or ignore defleshed carcasses completely. The senior author of this work witnessed hyenas in Maasai Mara overlooking the rich fat resources in defleshed carcasses during times of abundant biomass and mass killings by lions (Figure 126). Therefore, hyena damage to bones may vary with changes in ecological conditions. This may be reflected in a bone accumulation that results from multiple depositional events along extended time spans.

In order to evaluate the degree of ravaging intensity in any given assemblage (excluding other processes that affect bone preservation), we propose using the following indices (Figure 127):

Figure 126. Hyena in Maasai Mara neglecting a defleshed carcass due to the abundance of food in the area.

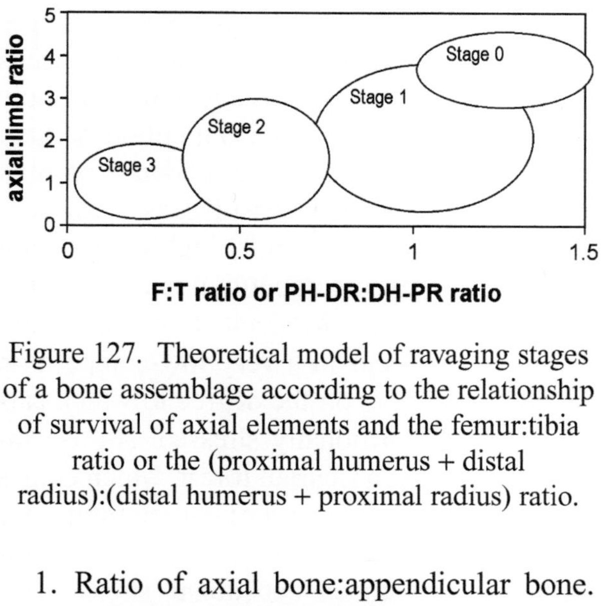

Figure 127. Theoretical model of ravaging stages of a bone assemblage according to the relationship of survival of axial elements and the femur:tibia ratio or the (proximal humerus + distal radius):(distal humerus + proximal radius) ratio.

1. Ratio of axial bone:appendicular bone. In the absence of ravaging, a carcass that has been transported complete, or that has died in a setting devoid of competition among carnivores, would be expected to show a higher frequency of axial bones than appendicular ones, given their higher numbers in a complete skeleton. The ratio would be obtained by dividing axial bones (ribs and vertebrae, excluding sacral and caudal vertebrae) by long limb bones. The results would range from 4.25 as the expected ratio for a complete skeleton, to 0 for a completely ravaged skeleton.
2. Ratio of femur:tibia. Despite previous claims by Marean and Kim (1998) that shafts always reflect the original number of bones present at a site, this is only the case when hominids broke open bones before carnivore intervention; these bone fragments would then lack resources that would have been attractive to scavenging carnivores. In accumulations where bones were abandoned whole and postdepositional ravaging took place, hyenas may proceed to chew a bone from one end to the other and consume it. The less dense bones (e.g., femora) would thus be more prone to disappear than denser elements (e.g., tibiae). A femur:tibia ratio of 1 would be expected in ravaging-free assemblages

and a ratio of 0 would indicate maximum ravaging, with intermediate values reflecting different degrees of ravaging. This ratio is only relevant when carnivores have primary access to complete bones and not to bones already broken by humans. In the latter case, a bone portion approach (see below) would be most informative.

3. The ratio of (proximal humerus + distal radius):(distal humerus + proximal radius). Following the argument above, carnivores, including hyenas and dogs, will preferentially delete the least dense portions; in the forelimbs these are the proximal humerus and distal radius (Brain, 1981). Therefore in an undisturbed carcass, the expected ratio would be 1. The smaller the number of the least dense portions, the higher the degree of ravaging. This ratio should be applied to contexts in which it is taphonomically discernible that hominids broke long limb bones and that only bone portions instead of whole bones were available for ravaging.

A theoretical model is illustrated in Figure 127. The decrease in axial elements would be expected to be related to the decrease in the less dense elements or element portions. However, this relationship is not modeled as a perfect straight line. It is known that hyenas delete axial elements more rapidly than limb bones (Capaldo, 1995). It is also true that axial bones may disappear during prolonged subaerial exposure more easily than appendicular bones because of their cancellous bone structure. Furthermore, quantifying some axial elements like ribs is difficult and estimates are usually lower than the actual numbers deposited. For all these reasons, the stages are modeled as ellipsoid areas in which the axial:appendicular ratio is allowed a high degree of variation. This model enables us to establish stages of bone destruction and deletion, and thereby of ravaging.

Stage 0 represents no ravaging at all. It would reflect ecological settings where competition is nonexistent or negligible. Stage 1 represents some minor deletion of axial bones and some soft bone portions. Prolonged subaerial exposure may also reduce the number of axial elements. Stage 2 represents moderate ravaging where at least half of the bones of each part (axial and soft portions/elements) have reduced their numbers by half. Finally, Stage 3 shows a range of deletion of axial elements and soft appendicular parts that is due to intense ravaging, in which both axial and limb elements have reduced their frequencies by one third or less. It is the first part of a process of intense ravaging, which at this stage can be considered moderate to high, and which finds its maximum expression in Stage 4, the complete deletion of the carcass.

The relationship between ravaging stages and ecological trophic dynamics still needs further study. In theory, Stage 1 would imply low ravaging and a low-competition setting. Stage 2 would imply some moderate to high competition among carnivores, whereas Stages 3 and 4 would imply high-competition settings. However, competition is not the only conditioning factor. Moderate to intensive ravaging might occur in a very low-competition setting depending on the degree to which carnivores are nutritionally stressed. An assemblage formed in a riparian forest, which enjoys one of the lowest degrees of competition, may still undergo severe ravaging if hyenas, for instance, are under nutritional stress. This seems however to be the exception rather than the rule; normally, these stages reflect degrees of competition. As an exception, hyena dens could be considered low-competition settings, and yet ravaging is often moderate to intense, but is less intense than the ravaging of a single carcass found by hyenas in an open habitat (see Figure 128).

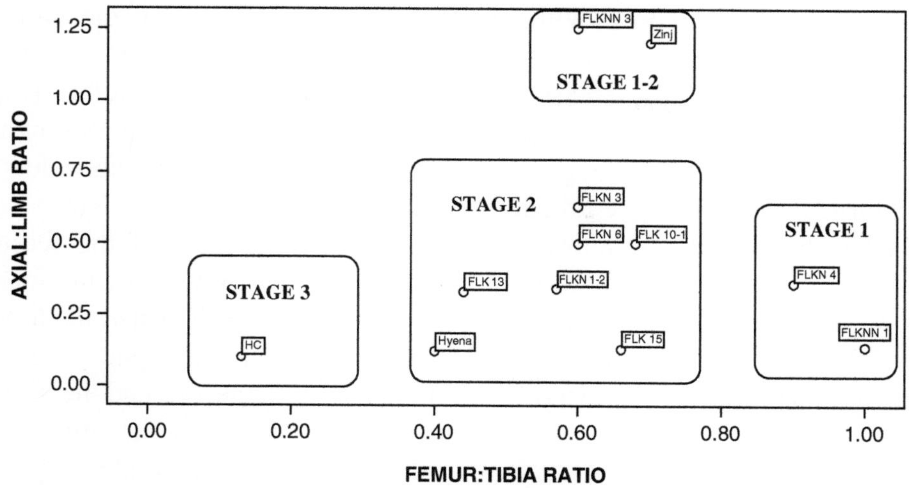

Figure 128. Ravaging stages of some of the bone assemblages from the Olduvai Bed I sites, using the femur:tibia ratio. (Data from human (hammerstone)-to-carnivore experiments [Capaldo, 1995] (HC) and from the Maasai Mara hyena den (Hyena) have also been included for comparison.)

Recently, Faith and Behrensmeyer (2006) presented data from two different ecological situations at Amboseli National Park in Kenya: the first was in the 1970s, when the park was dominated by lions and had few hyenas; later, in the 2000s, the situation was reversed and the hyena population became more significant. They argued that the degree of ravaging is not strongly reflected in element representation and is not correlated to bone density; they also claimed that distal ends of humeri are the least consumable of all the limb ends, even when carnivore impact is high. This statement hinges on one detail: when one looks at the number of long limb bone specimens retrieved in the 1970s and 2000s, the most striking feature is that metapodials are systematically underrepresented. Being the densest long limb elements, one would expect them in greater numbers than less dense bones, following the carnivore-ravaging patterns experimentally documented and which seem to be sensitive to bone density (Brain, 1981; Marean *et al.*, 1992; Capaldo, 1995, 1998b). Faith and Behrensmeyer's (2006) study assumes that the bone samples they recovered in their transects accurately reflect the complete result of carnivore competition and ravaging in the landscape. However, that is probably not the case. Kerbis' (1990) data from hyena dens often included high numbers of metapodials regarding specific taxa. It is very likely that the "disappearance" of metapodials in the landscape that Faith and Behrensmeyer (2006) sampled has more to do with selective transport of those elements by carnivores to dens or, more likely, somewhere else.

As support for this idea, we carried out a study of carcass dispersal in the modern floodplain of Lake Eyasi (Tanzania; work in progress). As a result of a severe drought beginning in February 2005, several cattle died over the course of the following year and their carcasses could be seen along several kilometers of the North-East Bay. Some carcasses that died early in 2006 were monitored during February and then July to document bone disarticulation and dispersal. In July, carcasses were disarticulated but most of the bones were preserved in the clusters where they were originally documented. Carnivore modification of these bones was minimal, since most of the axial skeleton was present and most limb bones remained complete.

Tooth marking was also very low. With this minimal carnivore disturbance the most striking feature was the under-representation of metapodials compared to the rest of the skeleton. Carnivores (i.e., hyenas and dogs) may have selected these bones and moved them far away from the North-East Bay, which was surveyed without finding more bone. Metapodials were, approximately, 40% less represented than upper limb bones (Prendergast and Domínguez-Rodrigo, in preparation).

Further documentation is necessary to know if a lack of metapodials is a common feature in open savanna landscapes, but this certainly appears to be the case in both Lake Eyasi and Amboseli. The bias that this introduces in bone quantification causes skeletal part profiles, when using long limb %MAU/density, to produce negative relationships. This could be solved by only considering non-metapodial epiphyses in the long limb sample. Then %MAU/density would produce a more positive correlation between bone representation and density.

The application of the ratios described above to the Olduvai Bed I sites in this study yields interesting results (Figure 128). Comparing a femur:tibia ratio to an axial:limb ratio, some assemblages show low to moderate ravaging: FLK *Zinj*, FLK North North 1–3 and FLK North 4. Even if the femur:tibia ratio indicates very low ravaging for FLK North North 1 and FLK North 4, the low frequencies of axials suggest a higher degree. FLK *Zinj* falls into in Stage 1–2 and the high presence of axials suggests low ravaging. Most of the other sites show moderate to higher ravaging, given their low axial:limb ratios. FLK 13 shows the most intense degree of ravaging, with FLK North 1–2 and FLK North 6 (the bovid sample) being very close to it. If we replace the femur:tibia ratio with the (proximal humerus + distal radius):(distal humerus + proximal radius) ratio, sites remain classified in similar stages, with the exception of FLK North 13–15 and FLK North 3, which appear in Stage 3 (Figure 129). This relative consistency in the stages of each site suggests that changes in the femur:tibia ratio correlate with

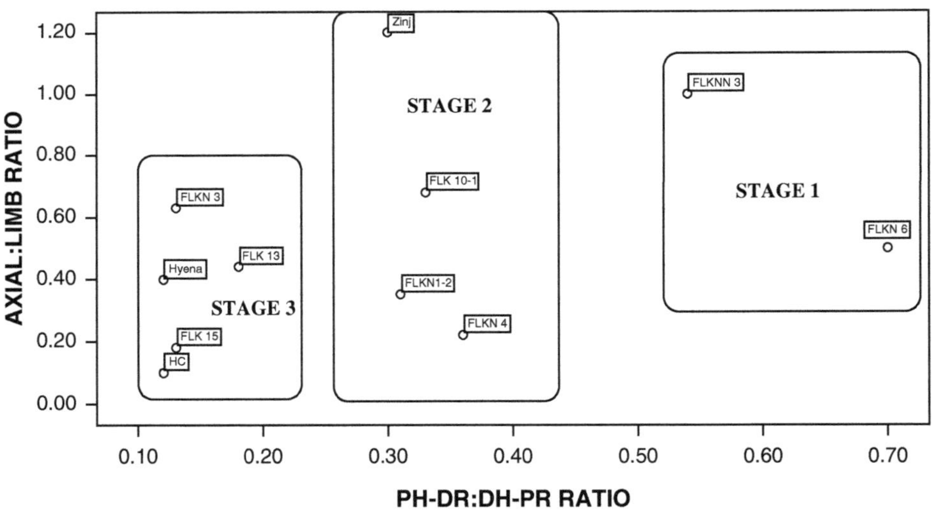

Figure 129. Ravaging stages of some of the bone assemblages from the Olduvai Bed I sites, using the (proximal humerus + distal radius):(distal humerus + proximal radius) ratio. (Data from human (hammerstone)-to-carnivore experiments [Capaldo, 1995] (HC) and from the Maasai Mara hyena den (Hyena) have also been included for comparison.)

changes in the ratio of epiphyses in the humerus and radius. In the Olduvai Bed I sites plotted, this relationship is significantly positive, showing a Pearson's correlation coefficient of 0.74 (two-tailed sig. = 0.009).

This theoretical exercise allows us to argue that differences in element representation and element abundance are not related to different degrees of ravaging intensity, given the overall proximity in plotted location between the "natural assemblages" and the "hominid-made" ones (according to traditional interpretations). Both types of sites cluster in Stages 1 and 2, indicating low to moderate ravaging. Sites occurring in one particular stage when using the femur:tibia ratio tend to be displaced by one stage when using the ratio of epiphyses in the humerus and radius. This raises the question of which ratio more accurately reflects ravaging. In general, the persistence of shafts has helped to determine the reconstruction of long limb bone MNE. This means that the chances of survival after hyena deletion for a shaft fragment (high density portion) from any given bone are higher than for a fragment from a cancellous portion. For this reason, ravaging might be more accurately reflected in limb bone end deletion and differential preservation. If so, most Olduvai sites would be characterized by moderate to high ravaging. However, if we consider that most of these sites have a portion of long limb elements that have been preserved complete (between one third and one fifth depending on the site) the overall estimation is that ravaging is moderate. Therefore, the ecological proxy for such a type of ravaging would neither be a dense riparian forest nor an open environment. The possibility of an ecotone should be raised (see below). FLK North 6, FLK North North 3, and FLK *Zinj* appear as the sites showing the lowest degree of ravaging. It could be argued that this might reflect a more closed environment and, therefore, less competitive habitat.

It is tempting to relate ravaging stages to competition, but as mentioned above, the match is not always perfect. Low-competition settings like hyena dens can show high ravaging stages, whereas high-competition settings (such as open plains) may show moderate or low ravaging stages, according to trophic dynamics and the nutritional stress of carnivores. A good example lies in the comparison between the landscape taphonomic studies carried out in the Serengeti (Tanzania; Blumenschine, 1989) and in Galana and Kulalu (Kenya; Domínguez-Rodrigo, 1996). If we plot data from each habitat type in the Serengeti, it becomes clear that the higher competitive settings show more intense ravaging stages (Figure 130). Something similar, although not as clear-cut, appears in Galana and Kulalu. The lowest-competition habitat in this ecosystem (Lali Hills) shows the lowest stage of ravaging. However, the second lowest (riparian woodland) shows similar ravaging stages to those documented in the open grassland, implying that carnivore deletion of bones might have been similar in both settings irrespective of the degree of competition. This is explained by the low degree of competition in open grasslands in Galana and Kulalu, compared with similar habitats in the highly competitive ecosystem of the Serengeti. Therefore, there seems to be some degree of relatedness between competition degree and ravaging stage. Low-competition degrees should generate low ravaging stages, irrespective of whether this happens in open or closed environments.

Conclusions

The study of "natural background" assemblages shows that no significant differences exist between these natural sites and "traditional" hominid-made sites, either in terms of skeletal representation or bone attrition. Element presence according to the number of carcasses represented by more than one bone, and the distribution of tooth-mark percentages

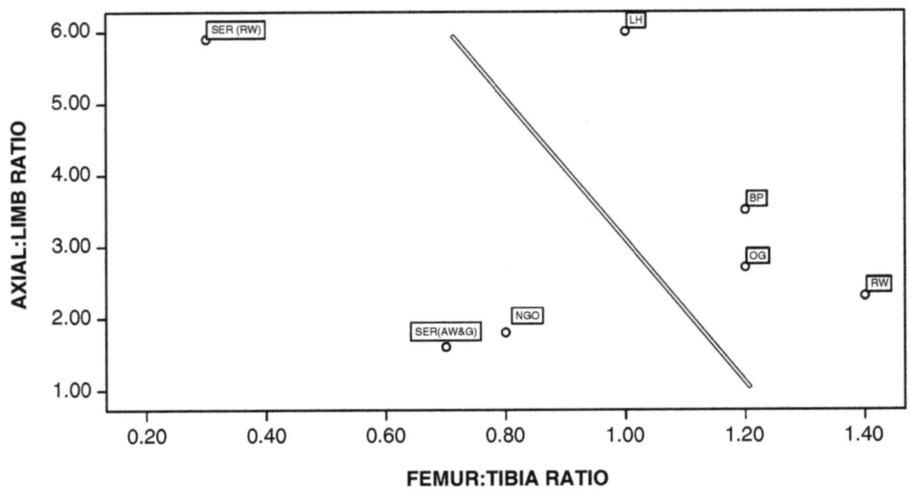

Figure 130. Ravaging stages of bone assemblages located in different habitats of modern savanna ecosystems. (Data for Serengeti and Ngorongoro are from Blumenschine [1989].) SER, Serengeti; NGO, Ngorongoro; AW, Acacia Woodland; RW, Riparian Woodland; GP, Grass Plain. (Data for Galana and Kulalu are from Domínguez-Rodrigo [1996]: RW, Riparian Woodland; OG, Open Grassland; BP, Bushy Plain; LH, Lali Hills. The line separates the habitats with low competition (*to the right*) from those with higher competition (*to the left*).)

are similar to those reported for other sites such as FLK North 1–2. The difference between "natural" and "archaeological" bone assemblages lies in the density of remains. Natural assemblages have lower numbers of specimens (NISP), MNE and MNI. This suggests that these accumulations represent natural attritional carcass deposition in the landscape, whereas other assemblages with significantly higher numbers of elements and individuals are very likely the result of repeated carcass transport to the same location by biotic agents (Potts, 1988). These agents seem to have been felids at sites like FLK North 1–2 (Chapter 8) and FLK North North 3, and these could have also been responsible for some of the attritional accumulations documented in natural background assemblages, as suggested above.

A well-supported example of this would be the bones from a carcass accumulated at FLK 7, which was not included in the present study. The level consists of a couple of background bones belonging to a small antilopini, as well as a humerus, a radius, a ulna, a femur, and a tibia from one *Parmularius*. The tibia shows perfect preservation of the proximal end, although substantial furrowing can be seen on the crest. The humerus also shows extensive furrowing of the caudal epicondylar area and an intact condyle, especially from the craniolateral aspect, which is the first one that usually appears modified by hyenas. The analysis of these small bone accumulations shows that the same taphonomic dynamics that created large assemblages (devoid of hominid intervention) were also at play in making smaller ones. These can be taphonomically linked to one another, assemblage size being the only difference.

Another lesson to be learned from studying small background assemblages is that it is common to find palimpsests mixing independent episodes of discard by carnivores (discarding bones) and hominids (discarding stones). In FLK 13 and FLK 15 there are ten and nine artifacts, respectively. Only one tibial shaft bore any trace of hominid bone modification (a cut mark) at FLK 13. No other specimen showed any indication that there might have been defleshing or breakage at the site by hominids. This cautions against assuming functional relationships between stones and bones in the same archaeological level; however, it also cautions against assuming that all "natural background assemblages" are free from hominid presence, though this presence may be unrelated to most of the fauna present.

In other larger sites, the few artifacts preserved with the faunal remains are mainly unmodified blocks and cobbles, flaked pieces, hammerstones, anvils and, to a lesser degree, flakes. In FLK 15, only nine artifacts were found; 11 in FLK 10 and 10 in FLK 13. In these three sites, only one flake was found. Assemblages are composed of large blocks, anvils, hammerstones, choppers, cores, and occasionally polyhedrons. A good example is FLK 13, where all these types are present. Pitting on anvils provides evidence of battering activities. However, the pitting is not restricted to anvils. Percussive pitting has been spotted on the damaged edges of one chopper, on one edge of a polyhedron, and on the edges of two blocks. This was also observed by Leakey (1971:59), who noted that the chopper showed "several small crushed indentations on the working edge and on two short ridges on the butt end."

In sum, the flaked artifacts found in association with fauna at FLK 10, FLK 13, and FLK 15 are indicative only of battering activities. These activities were not functionally related to carcass exploitation since no percussion mark, percussion notch, or percussion breakage plane was observed. Rather, a significant number of tooth marks and notches on the shaft fragments of long limb bones suggest that the agent of breakage was a hyaenid. These natural background accumulations and their lack of functional link to the associated artifacts are an eye-opener for the larger accumulations of fauna and stone tools, where (as shown in previous chapters) no functional link between stones and bones could be found. Carnivores and hominids were overlapping in the use of alluvial spaces at Olduvai Bed I, each leaving traces of their presence. One was consuming carcasses, the other using stone tools for battering activities whose functional meaning remains unknown, although the exploitation of plants remains a probable hypothesis.

Therefore, hyenas seem to have been present in the modification of faunal assemblages at sites during Bed I times. However, in the FLK *Zinj* and FLK North archaeological levels presented in previous chapters, their impact seems to have been moderate to low, allowing many complete bones to enter the lithosphere. The chapter 12 shows a location (the FLK North North site) in the paleo-lake margin ecosystem where hyenas were more active. It also provides further evidence of the characteristics of natural bone clusters, where felids and time-averaging can account for the faunal accumulations documented at FLK North North 1.

12. FLK North North 1: "living floor" or natural accumulation?

R. BARBA AND M. DOMÍNGUEZ-RODRIGO

Introduction

FLK North North (henceforth called "FLK NN"), the northernmost site in the FLK gullies at Olduvai, was discovered in 1960 by J. H. E. Leakey upon finding a mandibular fragment of a saber-toothed felid. The site was excavated in trenches. Two clay levels containing fossil remains were identified in the Trial Trench. A tuff containing birds, fish, rodents, and a few other mammal remains lay below these clay levels. The excavation was expanded after unearthing a hominid clavicle and several skull remains. A second trench was opened (Trench II), followed by two more to the west (Trenches III and IV). The total surface excavated was 18 m × 13 m. Fieldwork continued during 1962 in Trench IV and a further 32 m^2 was excavated, reaching the southernmost limit of the Bed I deposits. The last trench (Trench V) was opened parallel to the previous trench (Leakey, 1971).

Three archaeological levels were identified (Leakey, 1971):

1. Level 1 (FLK NN1). This level is in the upper part of the sequence and is the focus of this chapter. The stone tools and bones discovered in a vertically discrete horizon were interpreted as a living floor (Leakey, 1971: 258). The clay sediments suggest that this level was formed in the lake-margin zone (Hay, 1976). It is stratigraphically correlated with FLK *Zinj*.
2. Level 2 (FLK NN2). This level is a fine-grained tuff with clay patches in the central and bottom parts of the stratum. The clay sediments contained only bones: mainly macromammals, with rodents and birds also occurring in the tuff (Leakey, 1971). The lack of artifacts was interpreted as a lack of hominid occupation, and thus this level was considered a paleontological locality (Leakey, 1971; Potts, 1988; see Chapter 13).
3. Level 3 (FLK NN3). This level overlies Tuff IF and is mainly composed of clay. Several remains of *Homo habilis* were found, with which the stone tools and bones were associated. It was classified as a living floor (see Chapter 14).
4. Level 4 (FLK NN4). This level, underlying Tuff IF, corresponds to the upper levels of the DK and MK sites. This level yielded fewer materials than the overlying levels. Most of the bones belonged to birds (flamingoes and other aquatic birds), crocodile and very rarely, bovids and equids. For this reason, it was not included in the present study.

The clay matrix of FLK NN1 and the presence of fish and aquatic bird remains suggest that this site was formed in a lake-margin zone, although it is not known how far from the actual lake

M. Domínguez-Rodrigo, R. Barba, and C.P. Egeland, Deconstructing Olduvai: A Taphonomic Study of the Bed I Sites, 217–228.

margin. The westernmost part of this level was sterile; most artifacts and bones were found on the eastern part of it (Leakey, 1971). The spatial association of lithic tools and bones prompted Leakey to interpret this level as a living floor. However, despite its stratigraphic correlation with FLK *Zinj*, the abundance of manuports and the virtual lack of débitage made Leakey think that this level was only comparable to the most marginal areas of FLK *Zinj*, since in the latter 92.1% of lithic materials are débitage (Leakey, 1971). This difference between the sites was pointed out by Binford (1981: 273), who suggested that Leakey's interpretation of FLK NN1 as a living floor might be wrong.

FLK NN1 and NN3 were considered to be the result of independent depositional events separated by the natural formation of FLK NN2. The latter site lacks the abundant fish and reptile remains found in the underlying level. FLK NN2 shows a high frequency of complete elements (Leakey reports that 38% of MNE are complete), including abundant suid remains (52.2% of NISP; predominantly juveniles), 3 partially articulated skeletons, and an absence of stone tools. The site was classified as a "site with diffused material" (Leakey, 1971: 248). For a more comprehensive analysis see Chapter 13.

Whereas FLK NN3 was reanalyzed by Potts (1988), the only information available for FLK NN1 was presented by Leakey in her 1971 volume. In this chapter, we present a taphonomic study of this level. Our data include taxonomic representation, quantification of the number of specimens and elements, the number of individuals represented, and representation of limb portions which may be informative about the action of hyenas (Marean *et al.*, 1992; Capaldo, 1995; Pickering *et al.*, 2003; see details in Chapter 3). In this level, no complete notches have been observed. Therefore, no measurements could be taken. In addition, only 3 breakage planes larger than 4 cm could be measured. Given this small simple size they have been excluded from the analysis. Finally, a study of bone surface modifications is included. The absence of hominid-imparted marks (cut marks and percussion marks) on the bones from this assemblage, together with other taphonomic indicators discussed below, suggests that hominids did not participate in either the formation or the modification of the bone assemblage. Also, the virtual lack of flakes advises against any functional relationship between lithics and bones. This level can no longer be interpreted as a living floor.

FLK NN corresponds to a fairly different environment from that of the previously discussed FLK North site (Chapters 7–10). Several studies suggest that the site was formed in a more densely wooded environment, verging on some form of forest (Kappelman, 1984; Plummer and Bishop, 1994; Fernández-Jalvo *et al.*, 1998; Andrews and Humphrey, 1999). The presence of fish remains and bones from aquatic birds does not necessarily mean that the site was very close to the lake margin. Eagle owls are known to have been present during Bed I times, and they habitually prey on large aquatic birds and fish if a lake or river is within their hunting range (Andrews, personal communication). Therefore, these remains might have been transported by these predators.

Results

According to data presented by Leakey (1971), the taxonomic representation of FLK NN1 includes bovids (NISP = 85), carnivores (NISP = 4), suids (NISP = 54), equids (NISP = 8), rhinos (NISP = 2), and a high proportion of turtles (NISP = 98) (Table 59). In the current restudy a higher number of bovids has been identified. Most of them are medium-sized (*Connochaetes* sp., *Tragelaphus strepsiceros maryanus*, Tragelaphini size 3, Hippotragini, *Kobus sigmoidalis*, and Antilopini). We have also identified a smaller frequency of suids and equids than was previously reported

Table 59. Frequencies of specimens identified for each major faunal group, comparing Leakey's (1971) list and the present study

	Leakey (1971)		RB & MDR	
	NISP	%	NISP	%
Bovidae	85	33.9	123	47.9
Carnivora	4	1.6	7	2.7
Suidae	54	21.5	27	10.5
Equidae	8	3.2	1	0.4
Rhinocerotidae	2	0.8	0	0
Hippopotamidae	0	0	1	0.4
Chelonia	98	39	98	38.1
Total	251	100	257	100

Table 60. Minimum number of individuals (MNI) identified at FLK North North 1

	MNI
Connochaetes sp.	1
T. strepsiceros maryanus	2
Tragelaphini size 3	1
Hippotragini	1
K. sigmoidalis	2
Antilopini	1
Carnivora indet.	1
Megantereon sp.	1
Suid	2
Hippopotamidae	1
Equid	1
Total	14

(Table 59); only one vertebra of an equid was found. The lower frequency of suids could be accounted for by the fact that more than half of the suid remains were labeled simply as "FLK NN," without any indication of level; these were excluded from the present study. No rhino remains were found; however, a metacarpal from a hippopotamus was found. M. Leakey (1971) also mentioned a horn frontlet of *T. strepsiceros maryanus*, which we did not find. The only bones from this species identified are a series of articulated vertebrae.

In our restudy of FLK NN1, bovids represent 47.9% of the assemblage, and suids 10.5% (Table 59). If the 98 turtle remains are excluded, bovids would represent 77.4% of the macromammal assemblage (NISP = 123 out of a total of 159) and suids would constitute 17% (NISP = 27). Carnivores are represented only by seven specimens. A total of 14 individuals have been identified (Table 60): eight bovids, which are all are from medium-sized taxa with the exception of one Antilopini, two carnivores (a small one and a *Megantereon* cf. *eurynodon*), and two suids (one adult and one juvenile) as well as one equid and one hippopotamus. The most represented bovid is *K. sigmoidalis* (two individuals), although all bovid tribes except Reduncini are represented by at least one individual and sometimes, as in the case of the only Hippotragini identified, are represented by only one element. A study of bone size representation shows that all sizes are present (Figure 131). The presence of small specimens (< 4 cm), very abundant in small carcasses, suggests minimal postdepositional disturbance. The high frequency of specimens >10 cm can be explained by the abundance of complete elements.

The analysis of skeletal part profiles shows that suids, as in other Bed I sites, are mainly represented by teeth (66.7%) whereas bovids are represented by most skeletal elements (Figure 132). The abundance of vertebrae (26.8%) is particularly remarkable. Some vertebrae appear complete and articulated, protected by carbonate (Figure 133). Most vertebrae appear without apophyses and others show clear evidence of having been modified by carnivores (i.e., tooth marks). Preservation stages differ among the bovids represented, indicating that there were probably multiple depositional events in which postdepositional processes operated distinctly. The difference in representation between suids and bovids may be the result of the poorer preservation of suid bones because of a higher presence of subadult individuals, although subadult bovids in other sites with the same differential representation between both faunal groups are far better represented than suids.

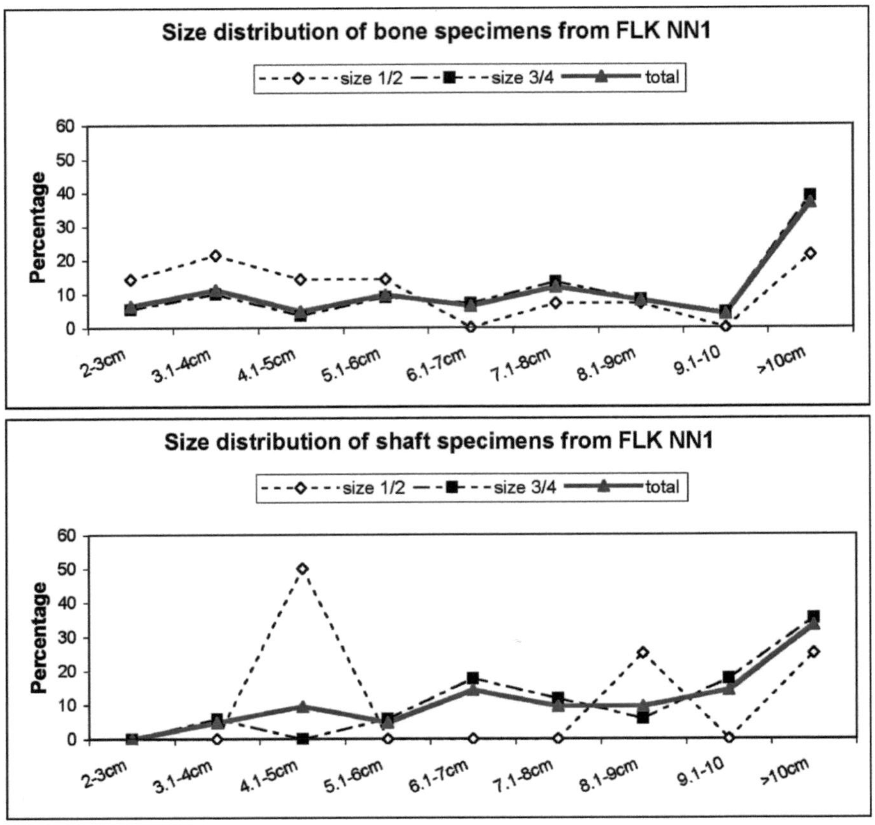

Figure 131. Size distribution of all bone specimens from all carcass sizes at FLK NN1, and size distribution of long limb bone shaft specimens alone.

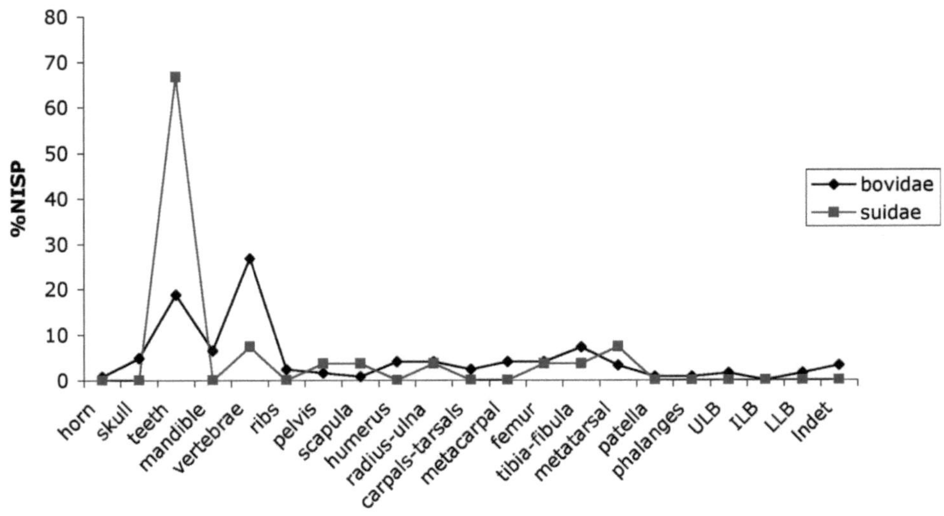

Figure 132. Frequencies of all skeletal elements from suids and bovids at FLK NN1. ULB = upper limb bone, ILB = intermediate limb bone, and LLB = lower limb bone.

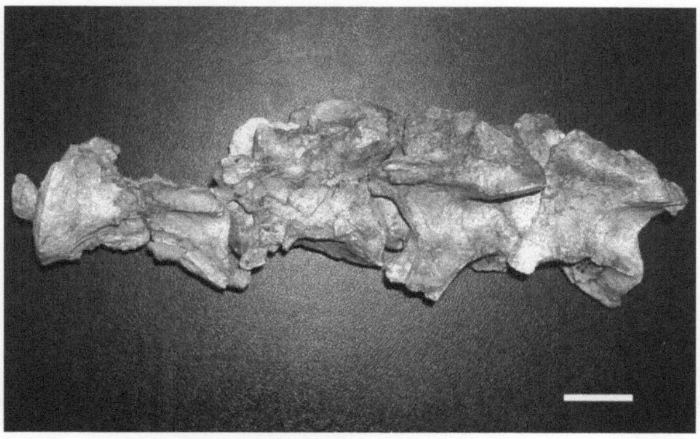

Figure 133. Articulated vertebrae from *T. strepsiceros maryanus*, (scale = 5 cm).

When we consider the number of elements represented, the abundance of vertebrae does not seem to indicate any strong bias against the preservation of the axial skeleton; however, more axial elements would in fact be expected according to the estimated MNI (Tables 61 and 62, Figure 134). This lack of axial elements compared with the expected number suggests carnivore ravaging; the absence of apophyses and the presence of crenulated edges in the axial material support this assertion.

Further support for carnivore modification comes from the analysis of appendicular element and bone portion representation. Table 63 shows the distribution of limb bones (n= 32, 26.8% of the assemblage) by bone portion and carcass size. In addition to the virtual lack of elements from small carcasses (limb NISP = 3), epiphyses in all carcass sizes are more poorly represented than diaphyses. The least dense portions from upper limb bones (proximal epiphyses of humeri and distal epiphyses of femora) are absent. This is typical of carnivore bone deletion (Brain, 1981). Due to the bias created by carnivores, the estimates of limb elements based on epiphyses are significantly lower than estimates obtained when shaft specimens are included. By using both epiphyses and shafts we have obtained a MNE of 22 (excluding carpo-tarsals and phalanges); 19 of these belong to medium-sized carcasses and 3 to small carcasses (Table 62). With respect to

Table 61. Number of identified specimens (NISP) for each skeletal element at FLK North North 1

	Bovidae	Suidae	Carnivora	Hippo	Equidae
Horn	1				
Skull	6		1		
Teeth	23	18			
Mandible	8		1		
Vertebrae	33	2			1
Ribs	3				
Pelvis	2	1			
Scapula	1	1			
Humerus	5		1		
Radius-ulna	5	1	1		
Carpals-tarsals	3		2		
Metacarpal	5			1	
Femur	5	1			
Tibia–fibula	9	1	1		
Metatarsal	4	2			
Patella	1				
Phalanges	1				
ULB*	2				
ILB*	0				
LLB*	2				
Indeterminate	4				
Total	123	27	7	1	1

*ULB, upper limb bones, ILB, intermediate limb bones, and LLB, lower limb bones

Table 62. Minimum number of elements (MNE) for each skeletal element and bone portion (for appendicular bones; epi = epiphysis, msh = midshaft) in small, middle-sized and large carcasses of FLK NN1. MNE estimates are given using epiphyses alone and using epiphyses and shafts combined

	Size 1–2		Size 3–4		Total
	epi	msh	epi	msh	
Humerus	0	0	4	4	4
Radius	0	1	2	3	4
Metacarpal	1	1	3	4	5
Femur	0	0	1	3	3
Tibia	0	1	2	2	3
Metatarsal	0	0	2	3	3
Total	1	3	14	19	22

MNE Axial elements, bovidae only

	Size 1–2	Size 3–4	Total
Scapula	0	1	1
Ribs	1	1	2
Pelvis	0	2	2
Vertebrae	2	18	20

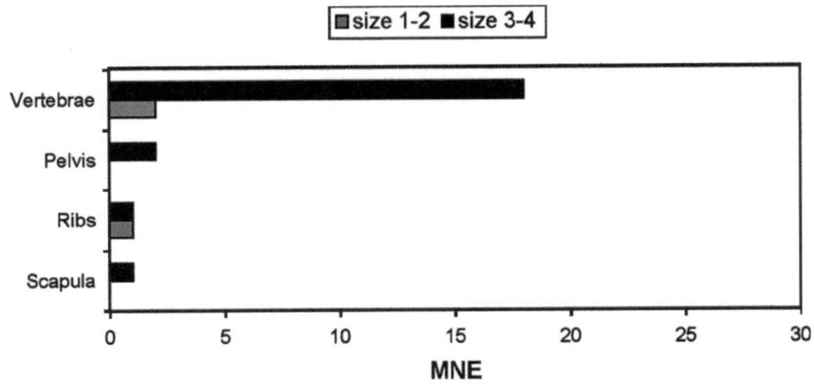

Figure 134. Minimum number of axial elements (including scapulae and pelves for taphonomic purposes) in small- and medium-sized carcasses at FLK NN1.

small carcasses, the difference in MNE estimates is even more marked; when shafts are included the MNE is tripled. In medium-sized carcasses this difference is less sharp given the presence of complete elements, suggesting a less intense ravaging. As was suggested for the axial skeleton, bone portion deletion in limbs is also moderate. The MNE:MNI ratio in medium-sized carcasses shows that the most affected elements are the upper limb bones (Figure 135). In sum, we have a lack of taphonomic evidence for hominid presence, and evidence for moderate modification of the assemblage by carnivores.

The study of bone breakage patterns shows that fracturing is rare. In those long limb bones that were found broken, green fractures (n = 17) are more abundant than diagenetic fractures (n = 9) (Table 64; Figure 136). Specimens from small carcasses showing any type of fracture are so few (n = 3) that no bone breakage pattern can be discerned, but in medium-sized carcasses green fractures are predominant in all limb elements but radii (Figure 136). The physical attributes used to distinguish among bone-breaking agents are, as described in Chapter 3, bone breakage planes and notches. However in this assemblage, from the 17 specimens showing green fractures, only 3 had breakage planes that could be quantified. This, together with the lack of complete notches, did not allow us to infer the agent of bone breakage.

As noted above, the high frequency of specimens <10 cm indicates that there are many complete elements in the assemblage. This is underscored by the relative frequencies of long limb circumference types (following Bunn, 1982; see Chapter 3), which are distinct from any of the three experimental scenarios described by Bunn (Figure 137). Type 3 (100% of shaft section) is the most abundant at FLK NN1, given the abundance of complete elements and the presence of cylinders created by carnivores during end deletion or bone breakage. The ratio of the sum of Type 2 and Type 3 shafts to Type 1 shafts is 2.3, that is, greater than the ratio that would be expected in experimental assemblages fragmented by carnivores (Bunn, 1982, 1983a; Marean *et al.*, 2004). Bone breakage by both carnivores and hominids generates assemblages where Type 1 is predominant. Carnivore-broken assemblages show a higher proportion of Types 2 and 3 than do hominid-broken assemblages (Bunn, 1982). The data from FLK NN1 indicate very marginal intervention by bone-crunching carnivores. The lack of percussion marks (particularly given the well-preserved cortical surfaces) also suggests that hominids did not intervene in assemblage formation. Furthermore, Leakey (1971: 47) reported that

Table 63. NISP of each bovid long limb bone portion according to element type at FLK North North 1

	NISP	
	Size 1/2	Size 3/4
Humerus complete		
Humerus prox.		
Humerus prox. + shaft		
Humerus shaft		1
Humerus dist.		2
Humerus dist. + shaft		2
Radius complete		
Radius prox.		
Radius prox. + shaft		2
Radius shaft	1	1
Radius dist.		
Radius dist. + shaft		
Metacarpal complete		1
Metacarpal prox.		1
Metacarpal prox. + shaft	1	1
Metacarpal shaft		1
Metacarpal dist.		
Metacarpal dist. + shaft		
Femur complete		
Femur prox.		1
Femur prox. + shaft		
Femur shaft		4
Femur dist.		
Femur dist. + shaft		
Tibia complete	1	1
Tibia prox.		1
Tibia prox. + shaft		
Tibia shaft		6
Tibia dist.		
Tibia dist. + shaft		
Metatarsal complete		
Metatarsal prox.		
Metatarsal prox. + shaft		2
Metatarsal shaft		2
Metatarsal prox. + shaft		
Metatarsal dist. + shaft		
Total	3	29

Table 64. Number of green and diagenetic breakages by skeletal element

	Size 1–2		Size 3–4	
	Green	Diagenetic	Green	Diagenetic
Vertebra	0	0	0	0
Ribs	0	0	0	1
Pelvis	0	0	0	0
Scapula	0	0	0	0
Humerus	1	0	3	1
Radius-ulna	1	0	0	2
Metacarpal	0	1	2	1
Femur	0	0	3	2
Tibia–fibula	0	0	4	0
Metatarsal	0	0	3	1
Phalanges	0	0	0	0
Other	0	0	0	0
ULB*	0	0	0	0
ILB*	0	0	0	0
LLB*	0	0	0	0
Indeterminate	0	0	0	0
Total	2	1	15	8

*ULB, upper limb bones, ILB, intermediate limb bones, and LLB, lower limb bones

half of the FLK NN1 lithic assemblage is composed of manuports, as well as of natural basalt blocks, and cobbles without any evidence of use. Only two broken flakes were reported. This tool kit is obviously missing the elements necessary for hominid processing of the carcasses at the site.

The presence of complete elements is important. The graphic representation of complete elements:NISP and complete elements:MNE is shown in Figures 138 and 139. About 21% of long limb elements from

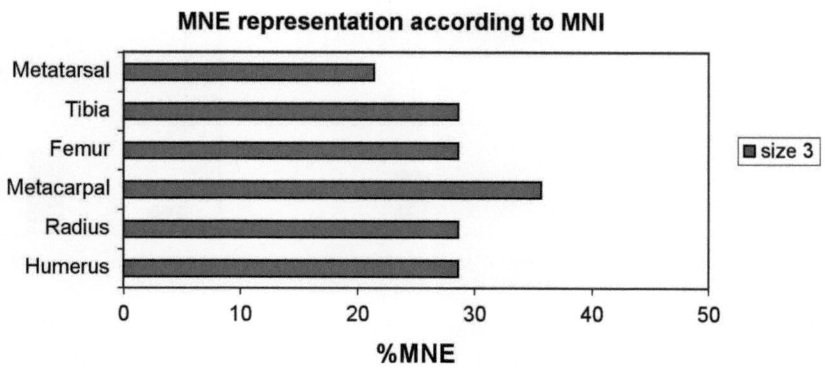

Figure 135. Minimum number of appendicular elements according to MNI at FLK NN1.

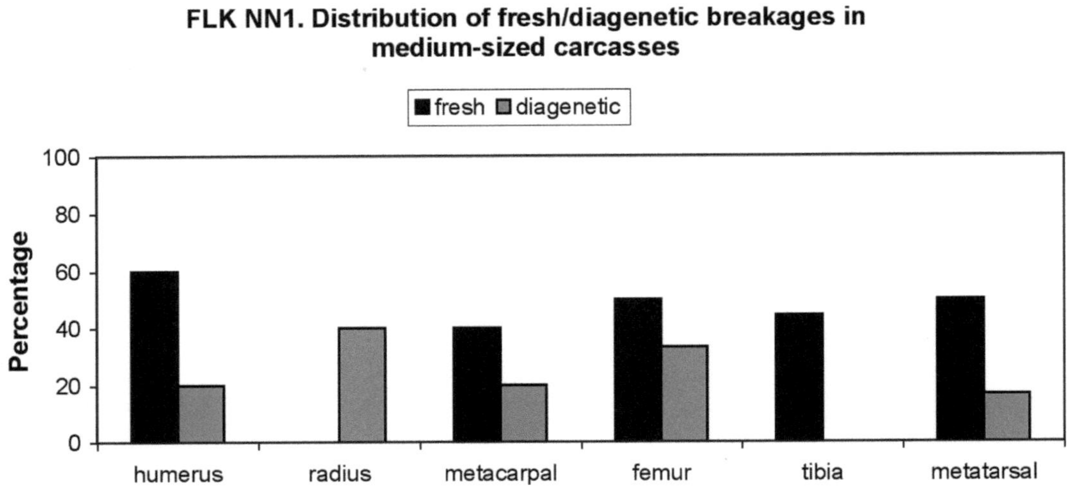

Figure 136. Distribution of frequencies of green and diagenetic fractures on long limb bones from medium-sized carcasses at FLK NN1.

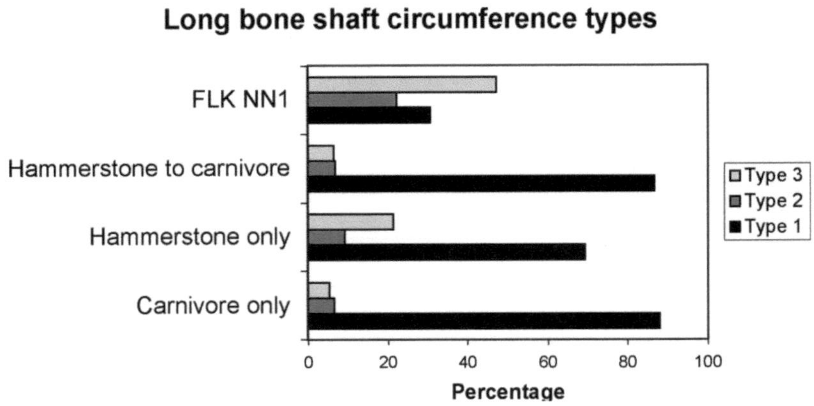

Figure 137. Frequencies of long limb bone shaft circumference types (Bunn, 1982) in experiments and in the FLK NN1 bone assemblage. Data for experimental assemblages are from Marean and Spencer (1991) and Marean *et al.* (2004).

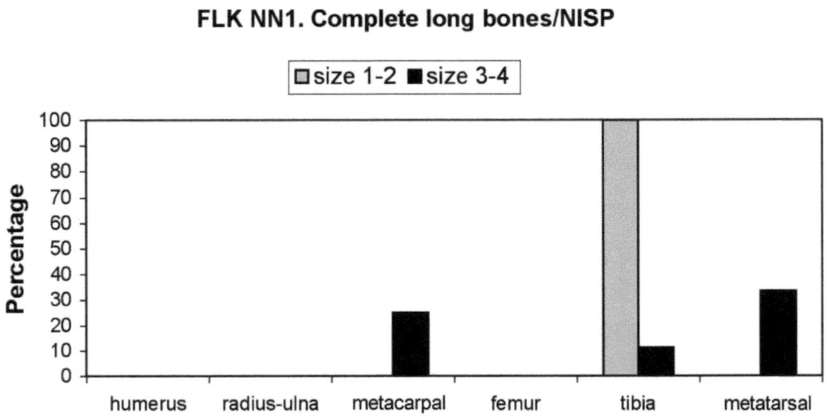

Figure 138. Relationship between the number of complete elements and the number of identified specimens (NISP) in small and medium-sized carcasses at FLK NN1.

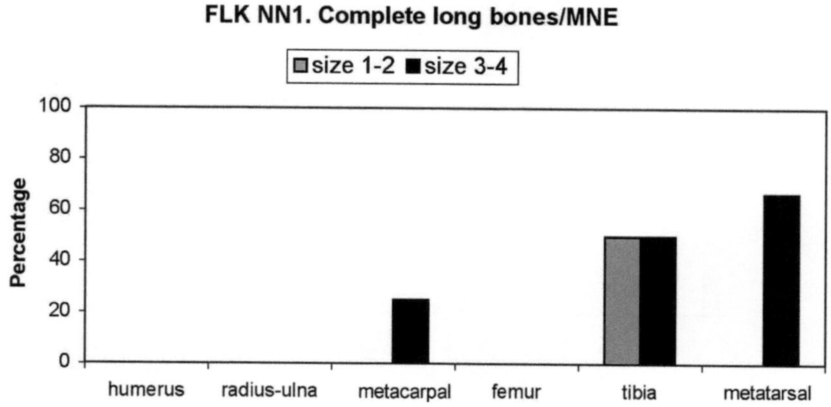

Figure 139. Relationship between the number of complete elements and the number of identified elements (MNE) in small and medium-sized carcasses at FLK NN1.

medium-sized carcasses and 20% of long limb elements from small carcasses are complete, including 3 lower limb bones (42.9% of all metapodials preserved are complete). This excludes hyaenids as the main accumulating agents, since they produce more intensive bone breaking (Brain, 1981; Blumenschine, 1988). Rather, these data support the hypothesis that felids might have been responsible for part of the bone assemblage The exclusion of hyaenids is based not only on the degree of completeness of the assemblage, but also on the relative lack of tooth marks (see below). The diaphysis ($n = 16$):epiphysis ($n = 13$) ratio is 1.2, showing a small bias in the preservation of long limb bone portions. This frequency is similar to that reported for felids (Domínguez-Rodrigo *et al.*, in press). The remains of a *Megantereon* found in this level suggest that this felid might have been a bone-modifying agent at the site.

The study of bone surface modifications does not reveal a single hominid-imparted mark (cut mark, percussion mark) in this assemblage. Tooth marks were documented on 24 specimens (19.5% of bovid remains): 2 specimens from small carcasses (1 rib and 1 astragalus) and 22 on specimens from medium-sized carcasses (Table 65). The distribution of tooth marks on each long limb bone is shown in Figures 140 and 141. Most tooth marks occur on epiphyses from upper limb bones and, to a lesser degree, on intermediate limb bones in the form of furrowing. The type of furrowing observed, in which only parts of

Table 65. Frequencies of tooth-marked specimens in small and large carcasses at FLK North North 1. Numerator is total number of tooth-marked specimens, denominator is total number of specimens for each skeletal element, and percentages are in parentheses

	Size 1/2	Size 3/4	Total
Skull	0/0(0)	0/6(0)	0/6(0)
Horn	0/0(0)	0/1(0)	0/1(0)
Teeth	0/0(0)	0/23(0)	0/23(0)
Vertebrae	0/2(0)	5/33(15.2)	5/35(14.3)
Ribs	1/1(100)	0/2(0)	1/3(33.3)
Pelvis	0/0(0)	1/2(50)	1/2(50)
Scapula	0/0(0)	1/1(100)	1/1(100)
Humerus	0/0(0)	4/5(80)	4/5(80)
Radius-ulna	0/1(0)	1/4(25)	1/5(20)
Metacarpal	0/1(0)	1/4(25)	1/5(20)
Femur	0/0(0)	1/5(20)	1/5(20)
Tibia–fibula	0/1(0)	3/8(37.5)	3/9(33.3)
Metatarsal	0/0(0)	3/4(75)	3/4(75)
Carpo-tarsals	1/2(50)	0/1(0)	1/3(33.3)
ULB*	0/1(0)	0/1(0)	0/2(0)
ILB*	0/0(0)	0/0(0)	0/0(0)
LLB*	0/0(0)	1/2(50)	1/2(50)
Indeterminate	0/0(0)	1/2(50)	1/2(50)
Total	2/9(22.2)	22/104(21.2)	24/113(21.2)

*upper limb bones (ULB), intermediate limb bones (ILB) and lower limb bones (LLB)

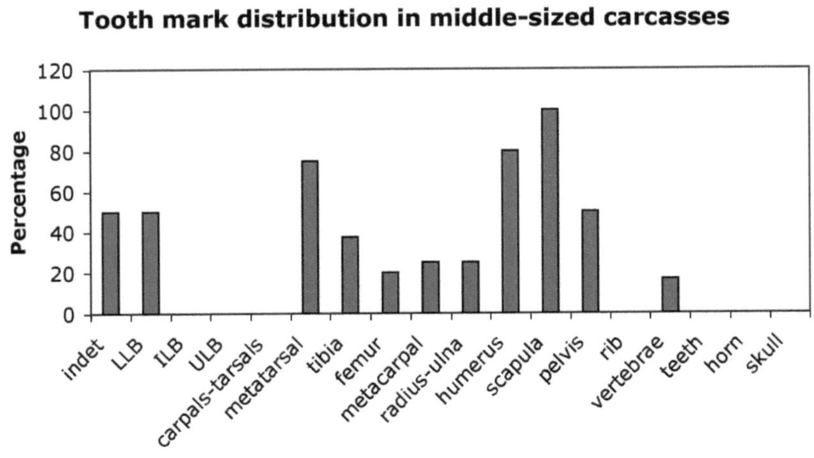

Figure 140. Tooth mark frequencies (as %NISP) in bones from middle-sized carcasses at FLK NN1. ULB = upper limb bone, ILB = intermediate limb bone, and LLB = lower limb bone.

the cancellous tissue are removed, is typical of felids. This can be observed on the only element from an Hippotragini: a complete tibia with furrowing in the proximal crest (Figure 141). Tooth marks are preferentially located on caudal condyles of humeri, on the proximal ends of the olecranea of ulnae, and on tibial crests. This pattern is typical of felid modification of bones (Domínguez-Rodrigo *et al.*, in press). The low proportion of tooth marks is distinct from the high frequency of tooth marks in hyaenid-modified assemblages (70%), and is much more similar to that reported by Selvaggio (1994) for felids (21%). The logical conclusion is that felids, not hyaenids, were the main agents in the consumption and accumulation of carcasses at the site.

Discussion and Conclusions

Leakey defined living floors as those sites where "the occupation debris is found on a paleosol or old land surface with a vertical distribution of only a few inches" (1971: 258). FLK NN1 and NN3 were included in this category, whereas NN2 was classified as a site with diffused material where "artifacts and faunal remains are found throughout a considerable thickness of clay or fine-grained tuff" (Leakey, 1971: 258). The information provided by Leakey (1971) about artifact densities suggests great variability among sites. The FC West site shows the highest density of lithic materials and FLK NN1 the lowest. However, in the FLK NN sequence, despite the discrete horizontal accumulations of archaeological remains in FLK NN1 and NN3 (10 cm) versus the more vertically dispersed FLK NN2 (27 cm), the density of bone remains in FLK NN1 is in fact very similar to that reported for FLK NN2 (Leakey, 1971: 260). The lack of flakes, the low density of remains, and the lack of any hominid-modified bone suggest that FLK NN1 can no longer be interpreted as a living floor.

The preservation of bones at FLK NN1 shows different weathering stages (following Behrensmeyer, 1978), clearly differentiated when focusing on long limb bones (Figure 142). Figure 143 shows the preservation of cortical surfaces. This representation of various weathering stages is suggestive of different depositional moments. Most bones from small carcasses show excellent cortical preservation with no weathering, suggesting rapid burial after deposition. By contrast, medium-sized carcasses show several weathering stages (0–4). Several elements show weathering stages 3–4. About 32% of elements from medium-sized carcasses show

LARGE CARCASSES

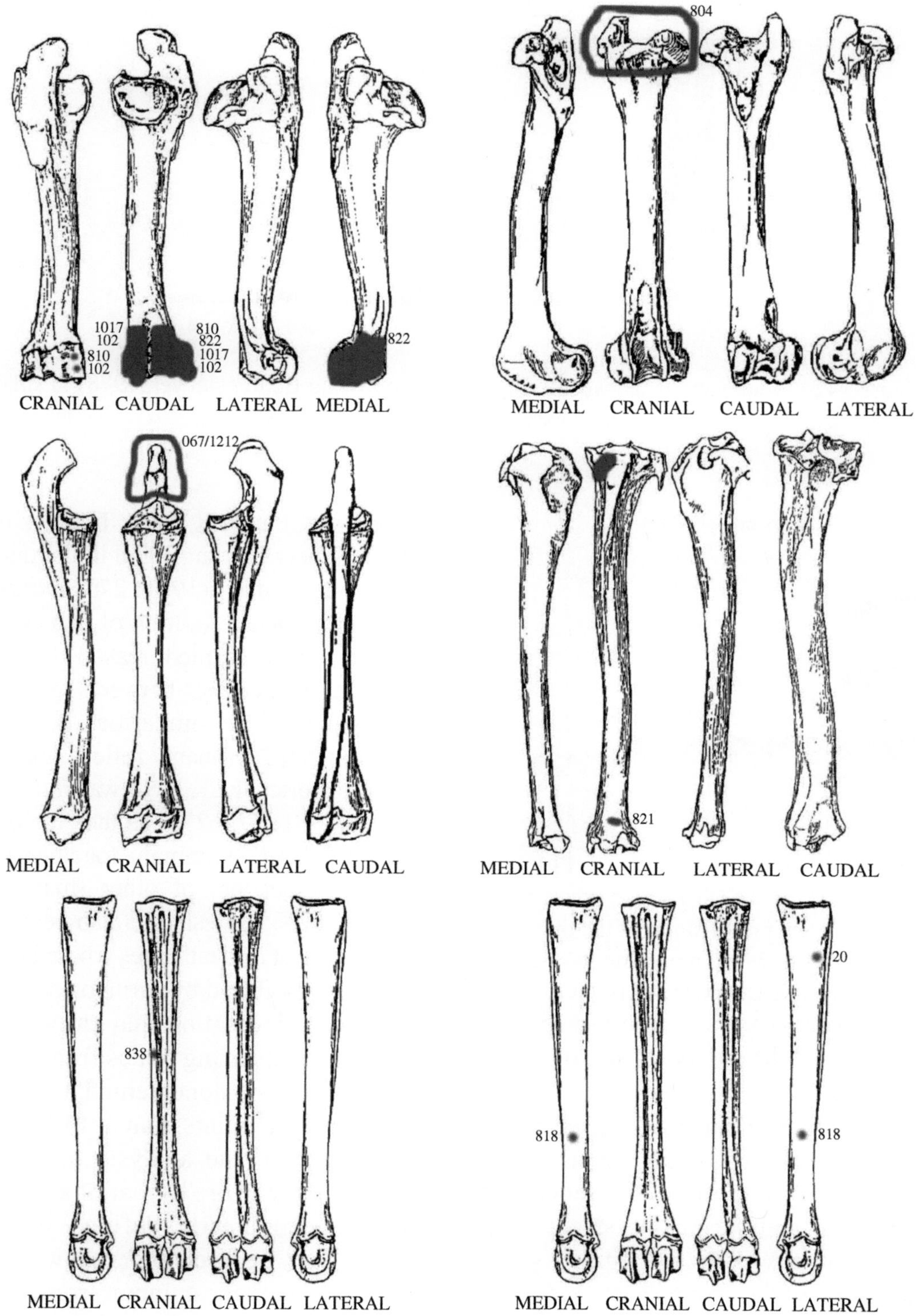

Figure 141. Anatomical distribution of tooth marks on long limb bones from large carcasses at FLK NN1. Letters and numbers are for the specimen number where these marks occur, where actual location of tooth mark could be ascertained. Shaded areas show intensive furrowing. Specimens bearing marks which could not be anatomically located were not incorporated into this figure. Bones are redrawn from Pales and Lambert (1971).

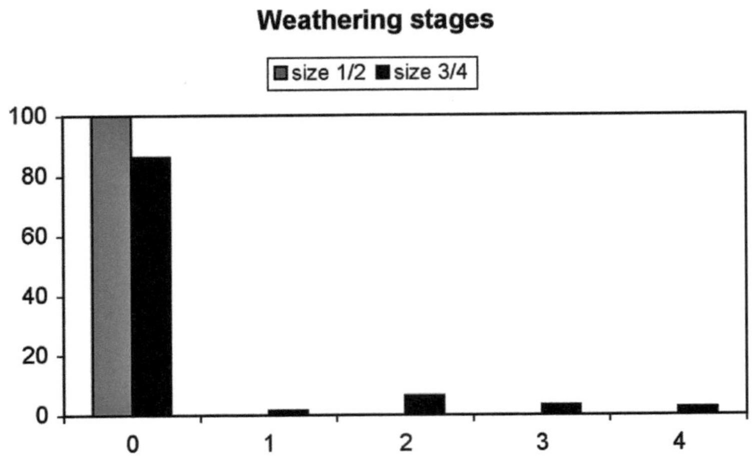

Figure 142. Frequency of the weathering stages (following Behrensmeyer, 1978) in small and large carcass elements at FLK NN1.

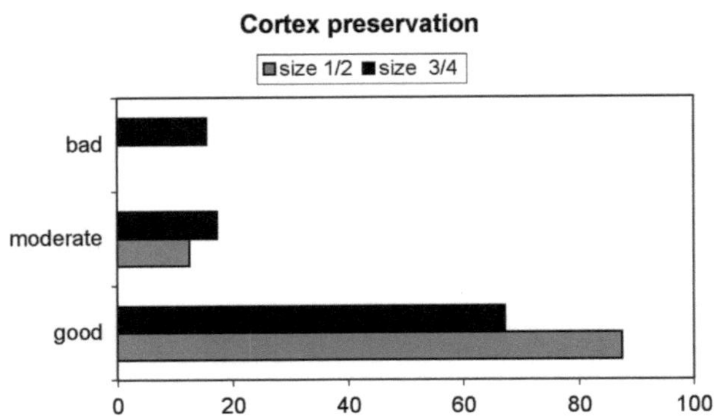

Figure 143. Cortical preservation in the FLK NN1 bone assemblage (as %NISP). Good preservation refers to complete preservation of the original cortical surface; moderate preservation refers to good cortex on at least two thirds of the surface and bad preservation refers to surfaces where less than two thirds of the surface is well-preserved.

poor cortical preservation, biochemical marks, and/or surfaces affected by corrosion. In sum, multiple depositional episodes are reflected in the preservation and weathering stages in the fauna, indicating a very prolonged time period in the formation of the assemblage.

The low frequency of tooth marks, the high number of complete elements, the high proportion of epiphyses, and the way bones are modified point to felids as the most likely agent in bone modification (and probably in bone accumulation) at FLK NN1. There is no functional link between fauna and lithic materials at the site. As Binford (1981: 273) indicated, the low integrity and resolution of this assemblage "provides no warranted reason for inferring behavioral connections between spatially associated things." The similarities between FLK NN 1 and NN 2 probably reflect similar depositional histories. FLK NN2 was interpreted by Binford (1981: 272–273) as the "the death site of several large mammals together with the background scatters of other (mainly bovid) remains." FLK NN1 should also be considered as a palimpsest of death sites where bones were moderately modified by carnivores.

A clearer indication that carnivores were actively accumulating bones from larger carcasses than those documented at FLK North and in larger amounts than at FLK NN 1 can be observed in the analysis of FLK NN2, where the "carnivore" agent of accumulation has not been clearly defined in previous studies. The next chapter shows that this level has the highest input from hyenas in bone modification among all the Bed I sites. The results of the present study also suggest that felids might have been able to accumulate multiple-sized animals in specific spots on the landscape.

13. Zooarchaeology and taphonomy of FLK North North 2

C.P. EGELAND

Introduction

Leakey (1971) originally reported no stone tools from FLK North North Level 2 (FLK NN2) and noted the presence of many complete bones and articulated skeletons. She classified FLK NN2 as a deposit with vertically dispersed material, unlike the classic Bed I "living floors" such as DK Level 3, FLK Level 22, and FLK North North Level 3 where artifacts and faunal material were concentrated in discrete horizons.

Both Bunn (1982) and Potts (1982, 1983) conducted early taphonomic analyses of the FLK NN2 assemblage, and the lack of associated lithic material and a dearth of hominid butchery damage, in addition to substantial evidence of carnivore damage suggested to both researchers that the site represented a carnivore accumulation. Citing distinctive patterns of bone breakage, both researchers surmised that hyenas were likely agents in bone accumulation. This renewed round of research supports and expands upon these conclusions. Specifically, the taphonomic data from FLK NN2 reflect a probable felid signal in addition to that of hyenas. Furthermore, ravaging and inferred levels of competition appear to be relatively low, in turn suggesting a closed microhabitat conducive to felid carcass transport.

The FLK North North 2 Faunal Assemblage

SITE DISTURBANCE AND INTEGRITY

As Figure 144 demonstrates, small fragments are underrepresented in the FLK NN2 assemblage. However, this is probably due to very low levels of fragmentation and not postdepositional disturbance. This is also evidenced by the high incidence of Bunn's (1982) circumference Type 3 (Figure 145).

TAXONOMIC AND SKELETAL REPRESENTATION

The FLK NN2 assemblage is dominated by bovids and suids (Table 66). The suid subassemblage consists mainly of juvenile remains. Bunn (1982) originally reported higher number of identified specimens (NISPs) and minimum number of individuals (MNIs) (Table 67) for the suids, which confirms that this study was not able to relocate some of the material in the National Museums of Kenya. Waterbuck (*Kobus sigmoidalis*) make up the majority of the bovid subassemblage, indicating the proximity of the site to a permanent water source.

M. Domínguez-Rodrigo, R. Barba, and C.P. Egeland, Deconstructing Olduvai: A Taphonomic Study of the Bed I Sites, 229–237.

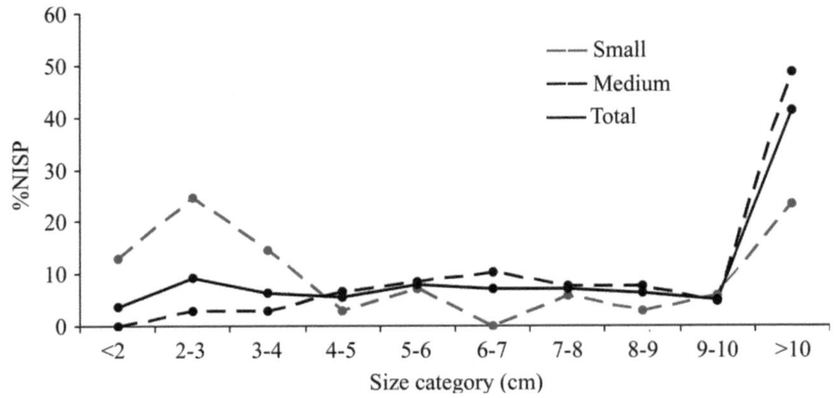

Figure 144. Percentage of specimens in each size range category at FLK NN2. Data are divided into specimens from small carcasses, medium carcasses, and combined.

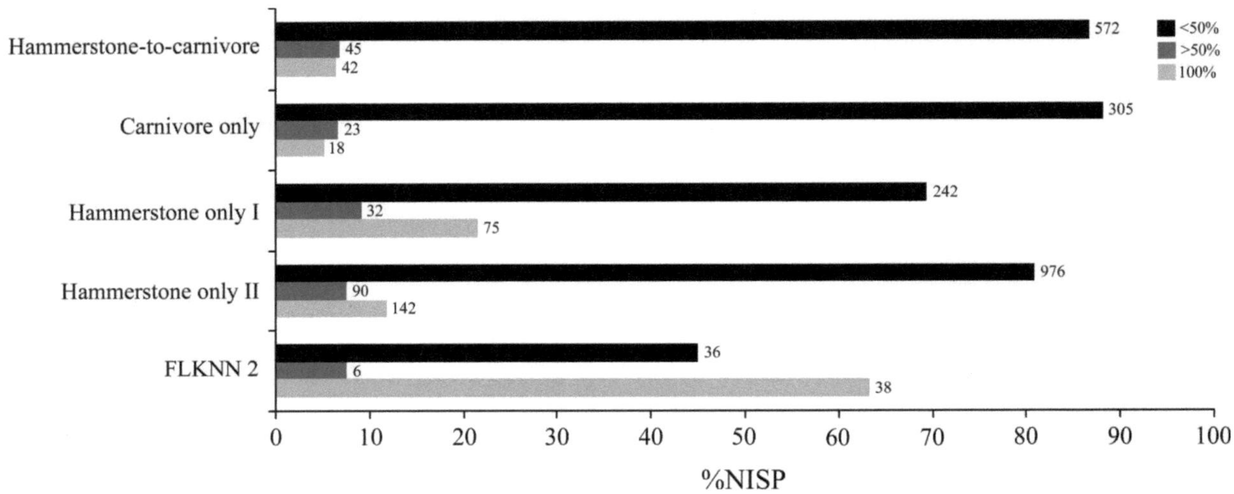

Figure 145. The percentage of Bunn's (1982) limb bone circumference types in the FLK NN2 assemblage compared to several experimental samples. Carnivore-only, hammerstone-to-carnivore, and hammerstone-only I samples from Marean *et al.* (2004). Hammerstone-only II sample from Pickering and Egeland (2006).

Table 66. Number of identified specimens (NISP) for major faunal groups at FLK NN2

Faunal group	NISP	%
Bovidae	175	51.8
Suidae	160	47.3
Equidae	1	0.3
Carnivora	2	0.6

Table 67. Minimum number of individuals (MNI) represented at FLK NN2

Taxon	MNI
K. sigmoidalis	8
Size 3a Alcelaphini	1
Size 3b Alcelaphini	1
Size 3 Tragelaphini	2
Size 3 Hippotragini	2
Kolpochoerus limnetes	3

It is likely that both bovids and suids were deposited as largely complete carcasses, as each group is represented by all parts of the skeleton (Table 68). Most of the bovid remains represent medium-sized (i.e., Size Class 3) animals; the few small carcasses are immature individuals.

Table 69 provides minimum number of element (MNE) estimates for the bovid subassemblage, and Table 70 summarizes MNE estimates for limb bones by portion (following Marean and Spencer, 1991). As in many of the Bed I assemblages, axial bones are poorly represented. However, given that axial representation is mediated by carnivore ravaging (Marean *et al.*, 1992; Capaldo, 1997, 1998b) and the fact that the limb bone MNE is almost exactly what would be predicted based on the number of crania, it is likely that most individuals were deposited as complete carcasses. Although ravaging is evident based on axial underrepresentation, midshaft portions of limb bones only significantly enhance MNE estimates for femora, which suggests a low overall level of ravaging (see later).

BONE SURFACE MODIFICATIONS

No hominid surface modifications were discovered in the FLK NN2 assemblage. (The cut-marked equid metacarpal identified by Bunn [1982:139] and Potts [1988:128] was not available for study.) However, almost 41% of specimens that were scored for surface preservation had poorly preserved cortices, which is among the highest of any Bed I assemblage. Table 71 summarizes tooth-mark incidence by skeletal part. The overall incidence of tooth-marking is 25%, and much of the damage clusters on the limb bones.

Figure 146 compares midshaft tooth-mark frequencies on medium-sized carcasses at FLK NN2 with experimental control samples. Midshaft tooth-mark frequencies fall within the range of variation for carnivore-only experiments, but are lower than Blumenschine's (1988, 1995) intensely ravaged samples, which are tooth-marked at an average rate of 86%. Nevertheless, at 50%, the rate of midshaft tooth-marking at FLK NN2 is among the highest of any Bed I

Table 68. Number of identified specimens (NISP) by major faunal group and skeletal element at FLK NN2

Element	Bovidae	Suidae
Cranium	12	2
Teeth	18	48
Mandible	12	7
Vertebrae	7	1
Ribs	3	4
Innominate	2	3
Scapula	10	2
Humerus	15	2
Radio-ulna	21	2
Carpals/Tarsals	4	5
Metacarpal	15	8
Femur	10	3
Tibia	16	3
Metatarsal	17	5
Patella	1	–
Phalanges	7	37
Sesamoids	3	13
Metapodial	2	2
Limb bone shaft	–	–

Table 69. Minimum number of element (MNE) estimates for small (Size Class 1 and 2) and medium (Size Class 3) carcasses at FLK NN2

Element	Small	Medium	Total
Cranium	2	6	8
Mandible	2	6	8
Vertebrae	1	10	11
Innominate	0	1	1
Ribs	2	13	15
Scapula	3	8	11
Humerus	2	13	15
Radius	1	12	13
Ulna	0	6	6
Carpals	1	2	3
Metacarpal	0	13	13
Femur	3	5	8
Patella	0	1	1
Tibia	3	9	12
Tarsals	0	1	1
Metatarsal	4	8	12
Phalanges	1	6	7
Sesamoids	0	3	3

Table 70. Minimum number of element (MNE) estimates for limb bones by bone portion for small (Size Class 1 and 2) and medium (Size Class 3) carcasses at FLK NN2

Element	Small	Medium	Total
Humerus			
PR	0	1	1
PRS	0	2	2
SH	1	13	14
DSS	2	13	15
DS	1	11	12
Radius			
PR	0	10	10
PRS	0	10	10
SH	1	12	13
DSS	1	9	10
DS	0	4	4
Metacarpal			
PR	0	11	11
PRS	0	11	11
SH	0	12	12
DSS	0	11	11
DS	0	10	10
Femur			
PR	0	0	0
PRS	2	3	5
SH	3	5	8
DSS	1	4	5
DS	0	2	2
Tibia			
PR	0	2	2
PRS	2	6	8
SH	3	9	12
DSS	3	7	10
DS	0	5	5
Metatarsal			
PR	0	8	8
PRS	4	8	12
SH	4	8	12
DSS	4	6	10
DS	0	1	1

Abbreviations: PR, proximal; PRS, proximal shaft; SH, midshaft; DSS, distal shaft; DS, distal

assemblage and therefore reflects a strong hyena signal relative to other assemblages.

Tooth-mark frequencies on epiphyseal and near-epiphyseal specimens at FLK NN2 also fall close to the 64–100% range produced in hyena-ravaged assemblages (Blumenschine, 1995; Capaldo, 1997, 1998b) (Table 72). In addition, tooth-mark frequencies are too high to be solely the result of felid processing (Figure 146). The sample of measurable tooth pits is too small to provide meaningful statistics; however, all measured pits are consistent with larger carnivores with more robust dentitions such as lions and hyenas.

BONE BREAKAGE

Diagenetic breakage has had little effect on the FLK NN2 assemblage (Figure 147). Given the high proportion of Bunn's (1982) Type 3 circumferences (Figure 145), the overall level of ravaging at FLK NN2 appears low. However, several factors indicate that small carcasses experienced higher levels of ravaging. For example, there are no complete limb bones among smaller carcasses (Table 73) while complete bones make up 28% of the medium carcass limb bone MNE, and, like FLK North 5 (see Chapter 10), bones appear broken in proportion to their nutritional content. Furthermore, the epiphysis-to-shaft fragment ratio for small carcasses is very low (0.08) compared to medium carcasses (1.03).

As both Bunn (1982) and Potts (1982) observed, limb bones are broken in a patterned way at FLK NN2 (e.g., Figure 148). Although the sample size of fracture planes for small carcasses is too small for comparison, fracture planes for medium carcasses fall within the carnivore range (Figure 149). The experimental sample of hammerstone-created notches (Capaldo and Blumenschine, 1994) is small and overlaps in range with carnivore notches; nonetheless, the dimensions of almost all notches in the FLK NN2 assemblage fall below even the range of carnivore notches (Figure 150). Unsurprisingly, micronotches, which are extremely common in carnivore-broken assemblages (Capaldo and Blumenschine, 1994; Egeland *et al.*, in preparation), are the dominant notch type (Figure 151).

Table 71. Tooth-mark frequencies for small (Size Class 1 and 2) and medium (Size Class 3) carcasses at FLK NN2

Element	Small	Medium	Total
Mandible	0/2 (0.0)	0/10 (0.0)	0/12 (0.0)
Vertebrae	0/4 (0.0)	0/14 (0.0)	0/18 (0.0)
Innominate	0/0 (0.0)	0/2 (0.0)	0/2 (0.0)
Ribs	0/3 (0.0)	1/39 (2.6)	1/42 (2.4)
Scapula	1/3 (33.3)	2/7 (28.6)	3/10 (30.0)
Humerus	1/2 (50.0)	6/14 (42.9)	7/16 (43.8)
Radius	1/1 (100.0)	7/12 (58.3)	8/13 (61.5)
Ulna	0/0 (0.0)	6/8 (75.0)	6/8 (75.0)
Carpals	0/1 (0.0)	0/2 (0.0)	0/3 (0.0)
Metacarpal	0/0 (0.0)	4/15 (26.7)	4/15 (26.7)
Femur	1/4 (25.0)	3/7 (42.9)	4/11 (36.4)
Patella	0/0 (0.0)	0/1 (0.0)	0/1 (0.0)
Tibia	1/3 (33.3)	8/18 (44.4)	9/21 (42.9)
Tarsals	0/0 (0.0)	0/1 (0.0)	0/1 (0.0)
Metatarsal	2/4 (50.0)	6/13 (46.2)	8/17 (47.1)
Phalanges	0/1 (0.0)	1/6 (16.7)	1/7 (14.3)
Sesamoids	0/0 (0.0)	0/3 (0.0)	0/3 (0.0)
Unidentified limb bone	1/4 (25.0)	1/6 (16.7)	2/10 (20.0)
Total	8/33 (24.2)	45/178 (25.3)	53/211 (25.1)

Numerator denotes number of marked specimens; denominator denotes total NISP for each skeletal element; percentage is in parentheses

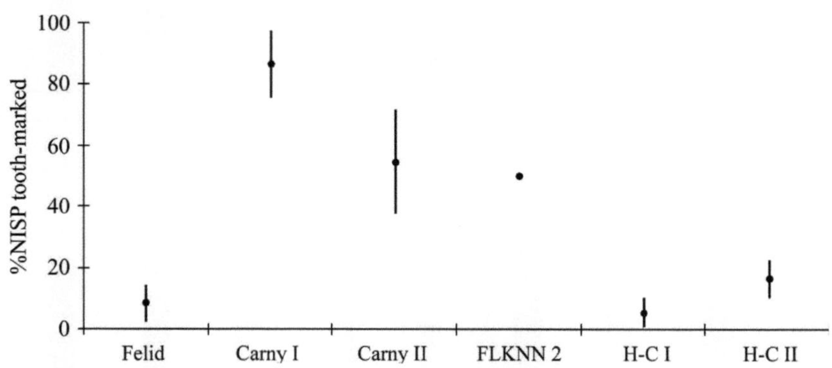

Figure 146. Incidence of tooth-marked midshaft fragments at FLK NN2 for medium-sized carcasses compared to the mean and 95% confidence intervals for actualistic control assemblages. Only fossil specimens with green breakage and good cortical surface preservation are considered. *Abbreviations*: Felid, felid-consumed carcasses; Carny, carnivore-only; H-C, hammerstone-to-carnivore. (Data sources: Blumenschine [1995]; Capaldo [1997, 1998a, 1998b]; Domínguez-Rodrigo *et al.* [in press].)

Discussion and Conclusions

This analysis agrees with previous studies (e.g., Bunn, 1982; Potts, 1982) that implicated hyenas as a significant agent in site formation at FLK NN2. In fact, unlike many other Bed I assemblages, midshaft tooth-mark frequencies at FLK NN2 match reasonably well with experimental controls that use hyenas as the carnivore bone modifier. Therefore, FLK NN2 reflects one of the strongest hyena signals within Bed I. However, the prevalence of complete limb bones (which were broken in direct proportion to their nutritional content),

Table 72. Percentage of epiphyseal, near-epiphyseal, and midshaft specimens bearing tooth marks at FLK NN2

Portion	EP	NEP	MSH	Total
Small carcasses				
NISP	0	3	0	3
NISP TM	0	3	0	3
%	0.0	100.0	0.0	100.0
Medium carcasses				
NISP	19	8	10	37
NISP TM	11	5	5	21
%	57.9	62.5	50.0	56.8

NISP counts include only those specimens with green breakage and good cortical surface preservation
Abbreviations: EP, epiphyseal; NEP, near-epiphyseal; MSH, midshaft

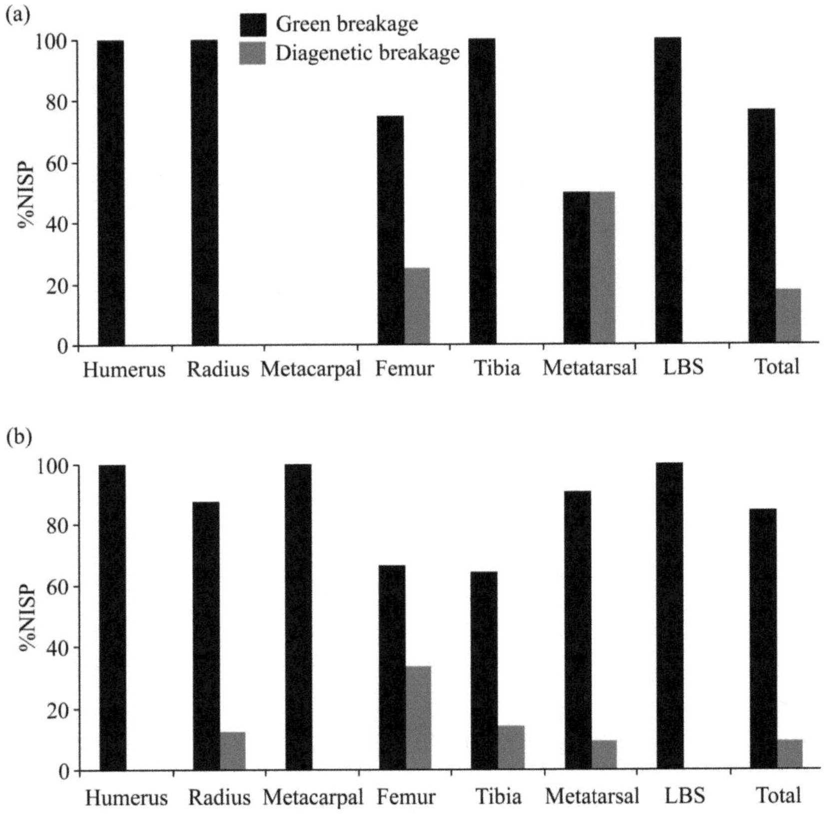

Figure 147. Incidence of green and diagenetic breakage on limb bones from FLK NN2. Percentages may not add to 100% as specimens with recent breakage are not considered. *Abbreviations*: LBS, unidentified limb bone shaft.

high epiphysis-to-shaft ratios (for medium carcasses), and a significant proportion of Type 3 circumferences all indicate that levels of hyena ravaging and competition were relatively low. The axial-to-limb bone and femur-to-tibia ratios place FLK NN2 within the lower range of Domínguez-Rodrigo and Organista's ravaging stage 2 (see Chapter 11). There appears to be

Table 73. Frequencies of complete limb bones for small (Size Class 1 and 2) and medium (Size Class 3) carcasses at FLK NN2

Element	Small	Medium	Total
Humerus	0/2 (0.0)	1/13 (7.7)	1/15 (6.7)
Radius	0/1 (0.0)	4/12 (33.3)	4/13 (30.8)
Metacarpal	0/0 (0.0)	8/13 (61.5)	8/13 (61.5)
Femur	0/3 (0.0)	1/5 (20.0)	1/8 (12.5)
Tibia	0/3 (0.0)	2/9 (22.2)	2/12 (16.7)
Metatarsal	0/4 (0.0)	2/8 (25.0)	2/12 (16.7)
Total	0/13 (0.0)	17/60 (28.3)	17/73 (23.3)

Numerator denotes number of complete bones; denominator denotes total MNE for each limb bone; percentage is in parentheses

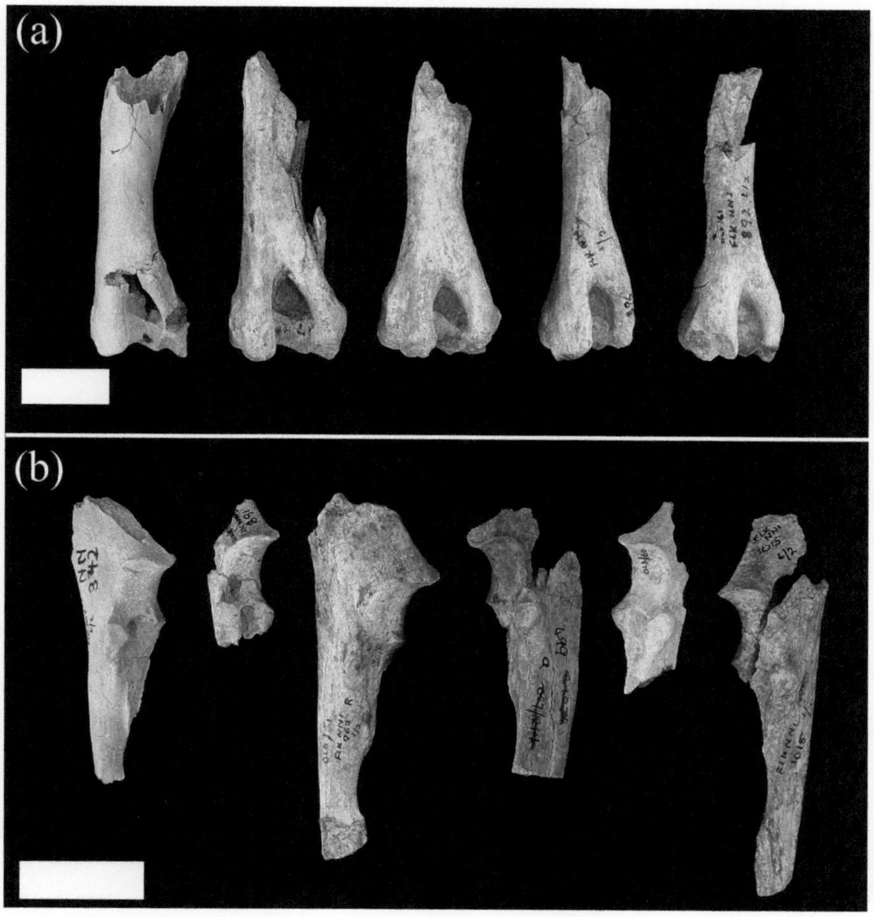

Figure 148. Examples of patterned bone breakage in the FLK NN2 assemblage. (A) Caudal views of right humeri showing patterned removal of proximal epiphysis and slight gnawing of distal epicondyles and (B) ulnae showing patterned gnawing of the olecranon process (scale bars = 5 cm).

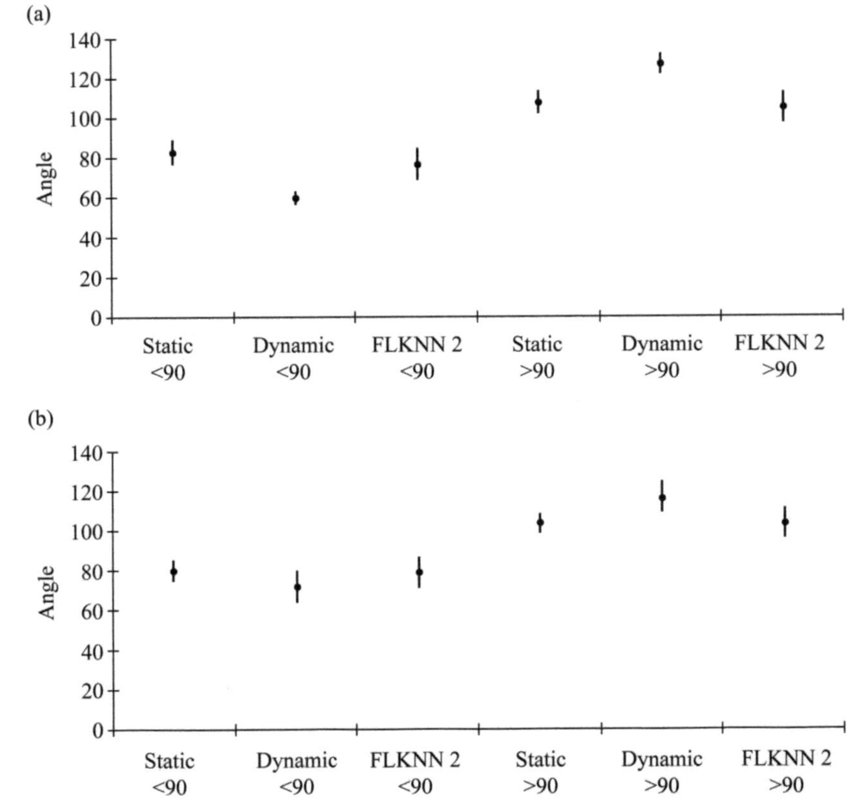

Figure 149. Mean and 95% confidence intervals of fracture angles from (A) oblique and (B) longitudinal planes on medium-sized carcasses at FLK NN2 compared to experimental samples.

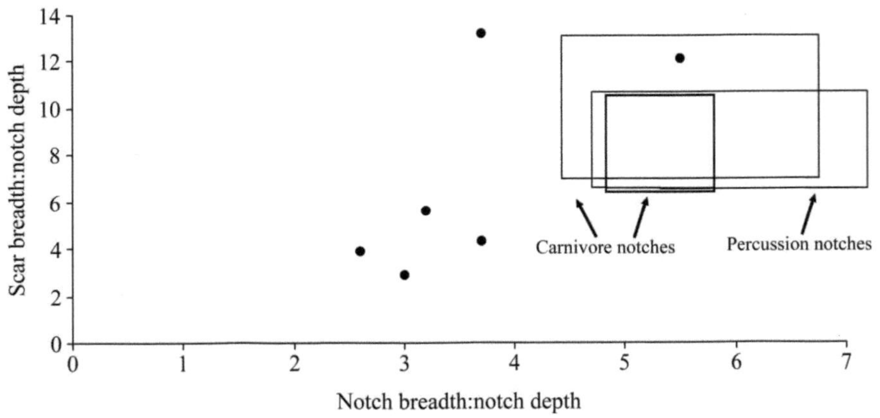

Figure 150. Notch dimensions on specimens from medium carcasses at FLK NN2. Boxes represent 95% confidence intervals of notch breadth:notch depth and scar breadth:notch depth ratios for experimental assemblages (Capaldo and Blumenchine, 1994) and a sample of notches from a hyena den in the Masaai Mara (Egeland *et al.*, in preparation). Solid circles represent individual notches from FLK NN2.

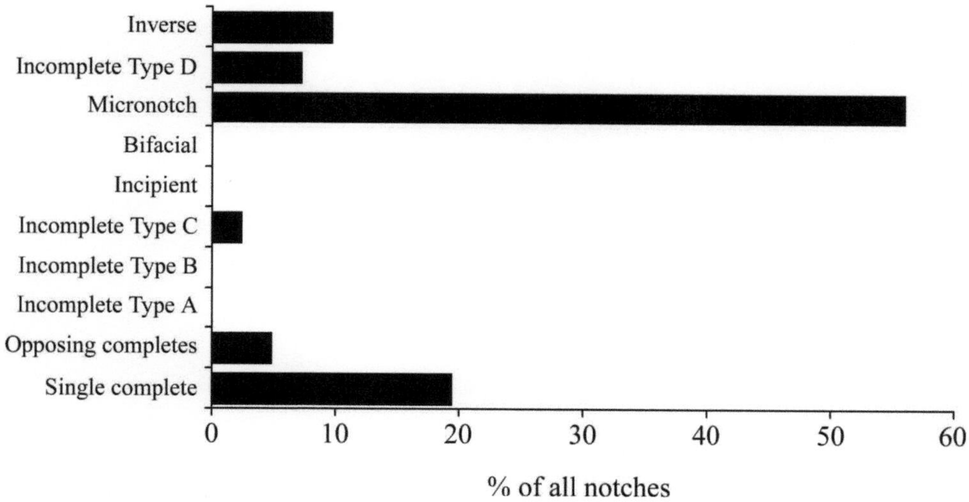

Figure 151. Incidence of notch types (modified from Capaldo and Blumenschine, 1994) from medium carcasses in the FLK NN2 assemblage.

only one case in which all the major elements were processed by carnivores. Interestingly, refitting and biometric data suggest that all these elements belong to a single individual. Therefore, among the 14 bovid carcasses that were deposited at the FLK NN2 locality, only one was completely processed.

Although midshaft tooth-mark frequencies at FLK NN2 match well with those recorded in lower competition hyena den assemblages (55%; Egeland *et al.*, in preparation), the lack of juvenile hyena remains, coprolites, or any stratigraphic evidence for den excavation argues against such an interpretation for FLK NN2. Therefore, the assemblage was probably deposited in close proximity to both permanent water and closed vegetation. As mentioned elsewhere, such environments are associated with lower degrees of competition (Blumenschine, 1986, 1987; Creel and Creel, 1996; Domínguez-Rodrigo, 2001). If felids can be implicated in the accumulation of carcasses at FLK NN2 as has been suggested for other assemblages in the FLK North North sequence (see Chapters 12 and 14), the size range of prey suggest a medium- to large-sized felid such as *Dinofelis*, *Megantereon*, *Homotherium*, or lion. Overall, it appears that FLK North North, like FLK North (see Chapters 8–10), served as a magnet for carcass transport and consumption by carnivores over a long period of time.

This interpretation is further supported by the results of the study of FLK North North 3 presented in the following chapter, where it is argued that felids played a more significant role than hyenas in carcass accumulation, further supporting the hypothesis that felids were also accumulating medium- and small-sized carcasses over long periods of time.

14. Reanalysis of FLK North North 3: yet another case of a palimpsest?

M. DOMÍNGUEZ-RODRIGO AND R. BARBA

Introduction

Level 3 of the FLK North North site (henceforth called FLK NN3) in Olduvai Bed I is situated on a land surface that rests on top of Tuff IB and it is therefore slightly older than the FLK *Zinj* level. It is the oldest site in Olduvai after the DK complex, discussed in Chapter 15 (Leakey, 1971). It is also one of the three archaeological levels in Bed I – the other two being DK-3 and FLK *Zinj* – that are limited to a thin horizon over a paleosol. FLK NN3 was located near the lake-margin deposits and the silty/clay sediments suggest that it was on the lacustrine plain, probably about 1 km from the lake margin (Hay, 1976). The site must have been situated in a moist environment, as suggested by the abundant fish and turtle remains. The fauna found at the site suggested a closed environment, between closed woodland and subtropical forest, with no modern counterpart (Kappelman, 1984; Potts, 1988; Plummer and Bishop, 1994; Fernández-Jalvo *et al.* 1998; Andrews and Humphrey, 1999).

Binford (1981) interpreted the site as the result of hominids having transported carcass parts from animal kills that had already been utilized (i.e., defleshed) by carnivores. Potts (1988:140) noted that such an interpretation was at odds with the fact that a skeleton of a *Kobus sigmoidalis* was seemingly transported complete to the site; some of the bones of this skeleton (vertebrae and tarsal elements) appeared articulated. Potts' (1988) study of FLK NN3 remains the only complete hands-on analysis of the site to date. He concluded that hominids played an important role in the accumulation of bones at this level. The number of manuports at the site and their spatial relationship to the fauna suggested to him that the site was a "stone cache" where carcasses were processed by hominids. Although Potts used some of the analytical variables which we include in the physical attribute approach (Chapter 3), most of his supporting arguments stemmed from the application of the "eco-taphonomic" variant of the paleontological approach, discussed in Chapters 3 and 7.[1] The density of bones (both in number of specimens and number of individuals represented), together with taxonomic range and skeletal element representation, contrasted with those found in natural savanna environments. In this chapter, we present the results of a taphonomic study of the site from a strictly physical approach, results which contradict previous assertions about the formation of the site.

Sample and Method

Potts (1988) showed that the site had a diverse taxonomic representation consisting of primates (NISP = 39), carnivores (NISP = 54), equid, suid, and bovid (NISP = 252) remains. In the

[1] Most of the analytical tools described in the "Physical attribute" approach were not available when Potts (1982) collected his data.

M. Domínguez-Rodrigo, R. Barba, and C.P. Egeland, Deconstructing Olduvai: A Taphonomic Study of the Bed I Sites, 239–252.

assemblage, the following taxa are represented: *Equus oldowayensis, Kolpochoerus limnetes, Parmularius altidens, Kobus sigmoidalis*, indeterminate antilopini and indeterminate tragelaphini. Potts (1988) also identified indeterminate Cercopithecinae and the following carnivore taxa: *Hyaena hyaena, Mungus minutes, Galerella primitivus,* and *Protocyon recki*. In our study, we did not study carnivores and primates, whose presence at the site did not seem to be related to hominid behavior (Potts, 1988). We also did not include any equids, since they were not available in the National Museums of Kenya during our study. Given the small equid sample size (Potts identified 10 specimens), their absence in our analysis does not make any significant difference.

Table 74 shows the distribution of specimens according to faunal group and sample size. A total of 373 specimens were analyzed. Most of them (NISP = 302) belong to bovids. Potts (1988) identified 34 individuals, which, if excluding carnivores and primates make up a total of 24, of which 18 are bovids (in our study, we only identified 15 bovid individuals). Suids are mostly represented by skulls, dentition, and foot elements. Elements from the entire skeleton are only found in bovids. Estimates of MNE significantly differ depending on whether epiphyses or shafts are used for element identification (Table 75). In large carcasses, MNE is almost double when using shafts as indicators. In small carcasses (including an abundant number of elements from size 3 juvenile carcasses), the difference is almost seven times higher. This is an indicator of intervening taphonomic processes biasing the original number of elements, probably related to carnivore ravaging.

In the present study, we will analyze the distribution of specimen size to detect any preservation bias introduced by physical agents. An evaluation of the cortical surfaces will be made, followed by an analysis of bone surface modifications: cut marks, tooth marks, percussion marks and natural marks (i.e., biochemical and abrasion marks). Marks were identified by using hand lenses under strong direct light following the methodological and mark diagnostic criteria specified in Blumenschine (1988, 1995) and Blumenschine and Selvaggio (1988) for tooth and percussion

Table 74. Number of identified specimens (NISP) at FLK North North 3

	Bovid	Suid
Horn	0	0
Skull	5	2
Teeth	1	28
Mandible	6	5
Vertebra	17	1
Ribs	50	0
Pelvis	9	0
Scapula	9	1
Humerus	27	4
radius-ulna	33	1
Carpals	5	15
Metacarpal	19	0
Femur	24	0
Tibia-fibula	32	0
Tarsals	5	8
Metatarsal	5	0
Patella	4	0
Phalanges	18	6
Other	10	0
Indeterminate	23	0
Total	302	71

Table 75. Minimum number of elements (MNE) represented at FLK North North 3, estimated using epiphyses and shafts. Small carcasses are Bunn's (1983) sizes 1 and 2 (including size 3 infantiles and early juveniles). Large carcasses are size 3 and higher

	Small carcasses		Large carcasses	
	Epiphysis	Shaft	Epiphysis	Shaft
Humerus	1	5	4	12
Radius	0	3	5	13
Metacarpal	1	1	10	10
Femur	0	5	0	5
Tibia	1	4	2	4
Metatarsal	0	2	2	2
Total	3	20	23	46

marks, and Bunn (1981) and Domínguez-Rodrigo (1997a, 2002) for cut marks (see Chapter 3 for details on methods).

Breakage patterns were analyzed from three different and complementary perspectives, as discussed in Chapter 3. A study of a randomly selected number of oblique breakage planes larger than 4 cm was carried out, measuring the angle that they form with respect to the cortical surface. The physical principle is the same that applies for notches. Dynamic loading (i.e., hammerstone percussion) creates more acute or obtuse angles than static loading (i.e., carnivore gnawing), which creates more right angles (Alcantara *et al.*, 2006; Pickering *et al.*, 2005a).

Given that carnivores are known to delete certain bones and bone portions from any faunal assemblage, a study of skeletal part distribution including appendicular bone portions is included. Brain (1967, 1981) reported how dogs ravaged Hottentot bone assemblages and documented that proximal epiphyses of humeri and tibiae, together with both ends of femora, were underrepresented. Experimental studies with hyenas have shown a similar pattern (Marean *et al.*, 1992; Capaldo, 1995, 1997; Pickering *et al.*, 2003). Therefore, these bone portions are sensitive to carnivore intervention in any assemblage.

Finally, a study of the lithic assemblage is included. The goal is to test the possibility of a functional association between lithics and fauna. If the fauna were exploited by hominids, a functional link would be expected in the form of cutting tools (i.e., flakes) proportionate to carcass remains.

Results

All the specimen size ranges are represented at the site (Figure 152). The small specimens are as highly represented as the larger ones, suggesting that there was no significant preservation bias according to specimen size. Specimens

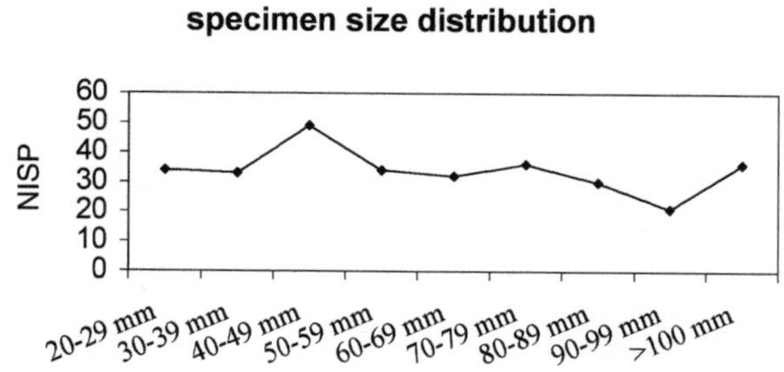

Figure 152. Specimen size frequencies at FLK NN 3.

larger than 4 cm are represented as expected. Those smaller than 4 cm are represented in a slightly smaller frequency than expected according to experimental assemblages (Blumenschine, 1988, 1995). This might be suggestive of minor postdepositional disturbance, very likely by hydraulic processes. No bones showing the typical abrasion and polishing caused by water transport were documented. Therefore, the bone accumulation at the site seems to be autochthonous and minimally disturbed by physical agents.

Skeletal part profiles (Table 74) show that axial specimens are abundant. According to Potts, FLK NN3 has the second-highest number of axial bones (46%) of the Olduvai Bed I sites. However, a look at the number of elements that those specimens represent and the number of individuals represented shows that the actual number of complete axial bones represented is far smaller than that (Table 76 and Figure 153). When plotted against the expected number of elements for to the number of individuals represented, it can be observed that all elements are represented in small numbers. Most of them are represented by less than one-tenth of the expected number. Axial bones are scarcely represented and suggest intensive deletion, probably by carnivores.

This is further supported by the differential representation of appendicular bone epiphyseal portions. Proximal ends of humeri and tibiae are far less represented than distal ends. Proximal ends of radii are more represented than distal ends, as was the case in the ravaged

Table 76. Minimum number of bovid skeletal elements documented at FLK North North 3

	Bovid MNE
Skull*	3
Vertebrae	14
Ribs	14
Pelvis	4
Scapula	9
Humerus prox.	3
Humerus dist.	7
Radius prox.	4
Radius dist.	3
Carpals	5
Metacarpal prox.	4
Metacarpal dist.	5
Femur prox.	2
Femur dist	1
Tibia prox.	1
Tibia dist	3
Tarsals	5
Metatarsal prox.	2
Metatarsal dist.	1
Phalanges	17

*including mandible

assemblages documented by Brain (1967, 1981). Femora are the most underrepresented limb elements. Scapulae are represented by the glenoid fossa and acromium. This indicates that the less dense elements and bone epiphyseal portions are less represented at the site than the denser ones. Additionally, compact and small limb bones such as carpals, tarsals, and phalanges are very marginally represented, as would be expected if hyenas ravaged the assemblage (Capaldo, 1995). The anatomical representation indicates, therefore, significant biases in the preservation of elements and portions introduced by carnivore postdepositional bone ravaging.

Further support for this claim lies in the analysis of bone breakage. The ratio of shafts (86) to epiphyses (36, excluding complete bones) is 2.3. The representation of shaft circumference types (following Bunn 1982) has no clear equivalent (especially given the lack of percussion marks) in experimental

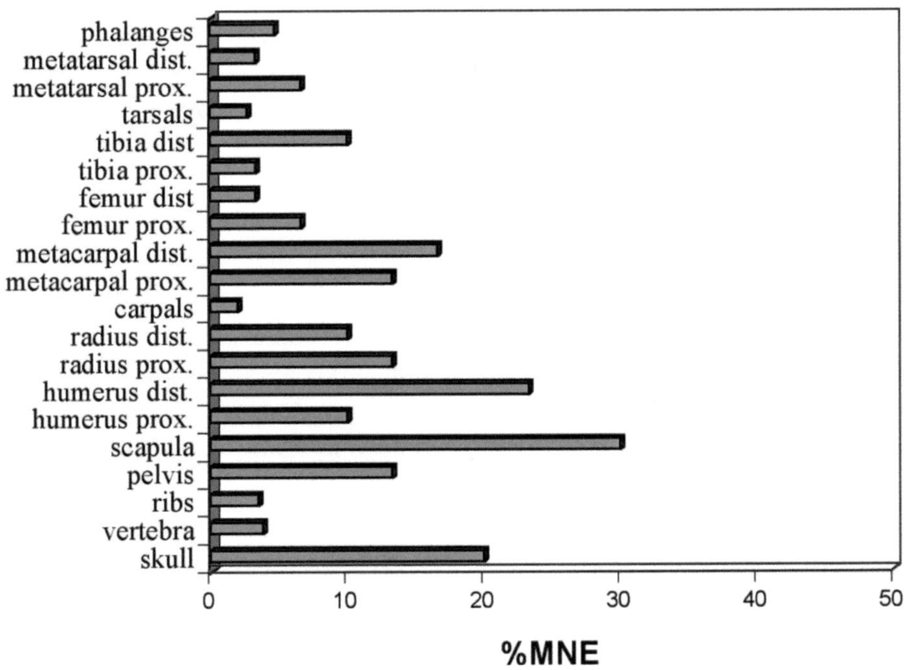

Figure 153. Frequencies of skeletal elements at FLK NN 3 (as %MNE), taking into account the minimum number of individuals represented.

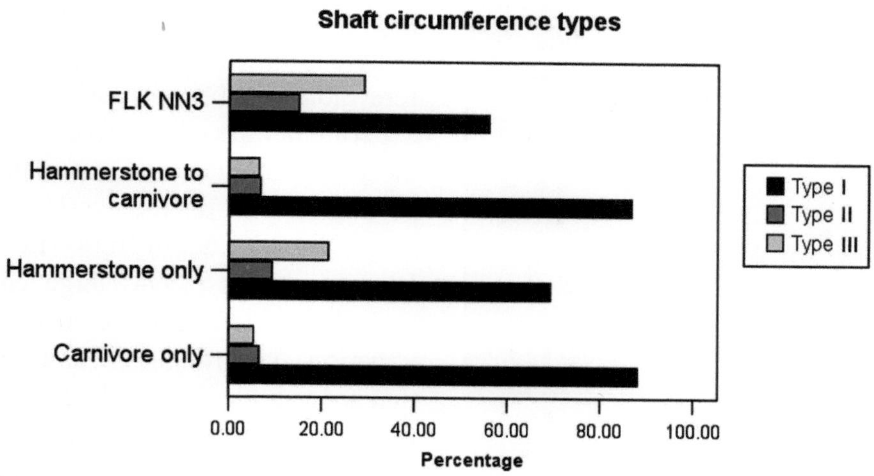

Figure 154. Distribution of the long bone shaft circumference types (following Bunn, 1982) in experimental assemblages and at FLK NN 3. (Data for experimental assemblages are from Marean and Spencer [1991] and Marean *et al.* [2004]).

assemblages modeling either carnivore first or human-carnivore interactions (Figure 154). Types 2 and 3 are almost as abundant as Type 1, signifying limited bone breakage (Bunn, 1982, 1983a; Marean *et al.*, 2004; see also Chapter 12). This suggests that if hyenas modified the assemblage, they did so in a very marginal way and not as intensively as has been documented in experiments.

This limited bone breaking is also not documented in human-first assemblages, where marrow extraction would have produced a higher number of Type 1 shafts. If a taphonomic bias is suggested as an explanation for this distribution of shaft types, the high presence of Types 2 and 3 argues against marrow exploitation by hominids followed by hyenas. This is supported by the fact that the FLK NN3 has one of the highest number of complete shafts in Olduvai Bed I. Twenty-one complete shafts have been documented, in several states of preservation. Eleven complete bones have been found, that is, over 40% of the total number of long limb bones at the site. Complete bones are not restricted to lower limb bones, as they are in other Olduvai sites. A total of five humeri (with furrowing on the epiphyses), one radius, one femur, three metacarpals and one metatarsal have been found complete. The high number of complete shafts and the missing epiphyses are undoubtedly the result of hyena ravaging. This hypothesis is further strengthened by the presence of some digested bone at the site (Figure 155).

Figure 155. Example of digested bone at FLK NN 3 (scale = 1 cm).

Capaldo and Blumenschine (1994) suggested that carnivore (static loading) and percussion (dynamic loading) notches created during bone breakage could be differentiated by using a combination of five variables (see Chapter 3). We observed some notches that Capaldo and Blumenschine called incipient

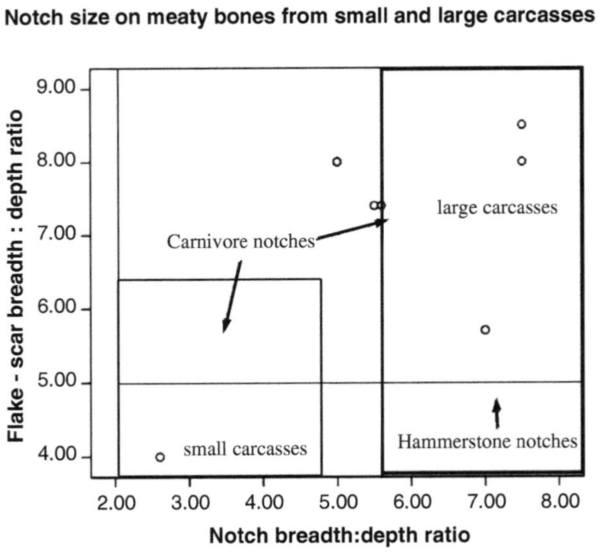

Figure 156. Notch measurements for the FLK NN 3 specimens, showing notch length:breadth ratio from the cortical surface and flake scar length:breadth ratio from the medullary surface. Boxes represent the ranges of variation (95% confidence intervals) for these ratios as documented in experimental assemblages produced by static (carnivore) and dynamic (hammerstone percussion) loading (Capaldo and Blumenschine, 1994). The box with the thinnest outline at the top represents the range of variation for carnivore notches on large carcasses. The box with a moderately thick outline, on the left, represents the range of variation for carnivore notches on small carcasses. The box with the thickest outline, on the right, represents the range of variation for percussion notches on large carcasses. The range of variation for percussion notches on small carcasses falls to the extreme right, outside the graph.

notches and incomplete notches Types A–C. However, since our intention was to test the heuristics of the combination of the five variables, we only measured complete notches in which all five measurements could be taken (Figure 156). Very few complete notches were observed ($n = 8$), given the partial completeness of the limb bone sample. The notch breadth: depth ratio and the flake-scar breadth:depth ratio show that most of the measurements cluster around the ranges of variation reported for carnivore notches in Capaldo and Blumenschine (1994). It is important to stress that for small carcasses, their experimental sample with carnivores is far larger, and therefore more reliable. Sample sizes for carnivore experiments are about four times larger than those for hammerstone experiments. This may account in part for the overlap and the narrow notch breadth:depth ratio in Capaldo and Blumenschine's (1994) large carcass sample, since it may not have included all the possible variation. Nevertheless, it is important to note that not a single notch at FLK NN3 falls outside the range reported for carnivores (Figure 156). This is also the case for the measurement of the angle of the medullary flake scar relative to the cortical surface (Figure 157).

The degree of diagenetic breakage is minimal. Only 16 specimens (4%) in the whole assemblage showed traces of breakage during fossil diagenesis. The freshly broken bones were also measured. All oblique breakage planes larger than 3 cm (n = 34) were measured. The distribution of their angles also differs from ranges of variation in experimental models (Figure 158). This may be due to the fact that experiments have not reproduced all the possible ranges of variations. A large number of bones in Capaldo and Blumenschine's (1994) experimental sample, as well as in the sample in Alcántara *et al.* (2006), were broken by applying artificial pressure. This models most carnivore bone-breaking behavior, but not all of it. It is valid for carnivores like canids but not necessarily for all the behavioral variation shown by hyenas. As discussed in Chapter 4, hyenas have sometimes been observed to break open bones by intentionally opening their jaws widely and letting both mandible and maxilla hit the bone before breaking it, creating notches and breakage planes that are often similar to those generated by hammerstone percussion. Some of the aforementioned experimental studies have not modeled this behavior, and this may account for their narrower variation ranges.

If the planes falling outside the carnivore range and within the hammerstone range were broken by hominids, this would be reflected in the presence of percussion marks, especially

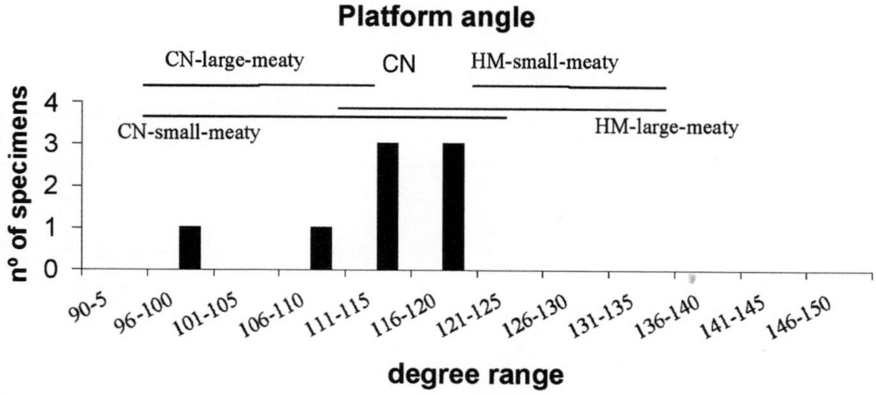

Figure 157. Distribution of notches from FLK NN 3 according to platform angle. Ranges on top of bars are for carnivore-made notches (CN) and hammerstone-made notches (HM) from Capaldo and Blumenschine (1994).

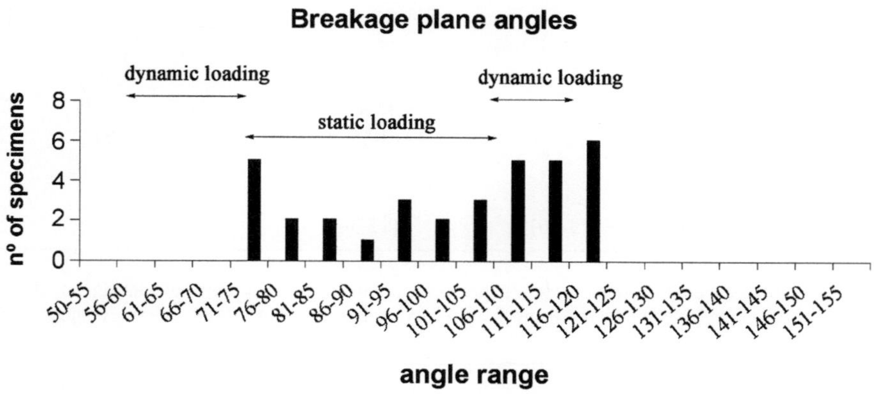

Figure 158. Mean values and 1 SD values for experimental assemblages (Alcántara *et al.*, 2006) on longitudinal and oblique planes modeling dynamic loading (percussion) and static loading (carnivore) for small and large carcasses, and for the FLK NN 3 breakage planes.

given the pristine preservation of the cortex on bones from FLK NN3 (only 17 specimens (4.5%) had bad cortex). Not a single specimen has a clear percussion mark. This supports the hypothesis that hominids did not break open any bones at the site. Additionally, we did not find a single convincing cut mark. A small percentage of bones (n = 15; 4%) showed traces of abrasion, probably due to trampling. None of these marks could be mistaken for cut marks.

Only 62 specimens (16.3%) are tooth-marked. Tooth marks appear on ribs, pelves, phalanges, and scapulae, but they are most abundant on limb bones (Table 77). Distribution of tooth marks according to bone portion on the appendicular skeleton is shown following Blumenschine's (1988, 1995) method of dividing long limb specimens into epiphyseal, near-epiphyseal, and midshaft portions (Figure 159). Given the high quality of the cortical surfaces, we are certain that we did not miss any tooth mark; only eight specimens (2%) showed biochemical marks that could mimic inconspicuous tooth marks (see Domínguez-Rodrigo and Barba, 2006). Tooth

Table 77. Mean percentages of tooth-marked specimens in relation to the total number of specimens in experimental assemblages and at FLK North North 3, according to bone portion: epiphysis (EPI), near-epiphysis (NEP) and midshaft (MSH)

Small-sized carcasses		EPIPH	NEP	MSH	ALL
Carnivore only*					
	Mean %	100	75	69.1	70.8
	95% C. I.	(–)	(0–100)	(47.5–90.7)	(33.1–100)
Hammerstone-to-carnivore*					
	Mean %	50	60.7	15.9	21.9
	95% C. I.	(0–100)	(29.3–92.1)	(2.6–29.2)	(5.1–38.7)
Hammerstone-to-carnivore**					
	Mean %	–	31.4	14.9	19.4
	95% C. I.	–	(19.9–42.9)	(11.2–18.6)	(15.5–23.3)
Large-sized carcasses		**EPIPH**	**NEP**	**MSH**	**ALL**
Carnivore only*					
	Mean %	100	90.5	86.5	87.7
	95% C. I.	(–)	(76.5–100)	(75.7–97.3)	(78.5–96.9)
Hammerstone-to-carnivore*					
	Mean %	85.7	39.5	5.1	16.4
	95% C. I.	(35.6–100)	(8.4–70.6)	(0.2–10)	(10.5–22.3)
Hammerstone-to-carnivore**					
	Mean %	–	48.4	16.2	27
	95% C. I.	–	(40.5–56.3)	(10.4–22.8)	(22.5–31.5)
FLK North North 3					
	Percent	25	40	43.2	37.5
	N° marked/total	(9/36)	(12/30)	(29/67)	(50/133)

Analysis of 95% CI (confidence intervals) for sets of experiments are included. They were calculated using the *t* distribution, where *t*0 • 025 is the critical value of *t* with *n*-1 degrees of freedom. * Blumenschine (1988, 1995),** (Capaldo, 1995, 1998b)

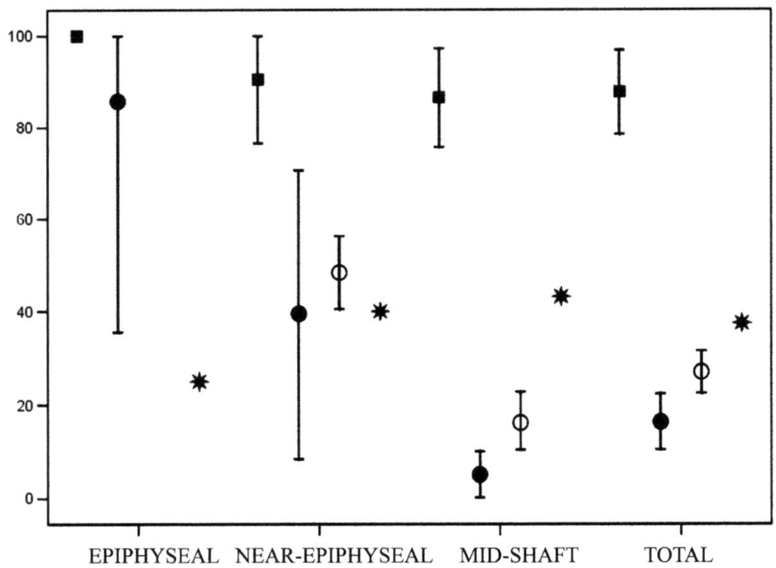

Figure 159. Distribution of the 95% CI (confidence intervals) for the frequency of tooth-marked specimens for each bone portion from large carcasses in experimental assemblages and at FLK NN 3. ■, Carnivore only; ●, Human (hammerstone)-to-carnivore (Blumenschine, 1995); o, Human (hammerstone)-to-carnivore (Capaldo, 1997); ✱, FLK NN 3.

mark frequency and distribution are unlike those reported for carnivore-first and human (hammerstone)-to-carnivore experiments. The surviving epiphyses are very rarely tooth-marked, even below the mean expected for human-to-carnivore scenarios. Near-epiphyseal fragments are ambiguous, since they fall within the range of variation reported for carnivore-only and human-to-carnivore models for small-sized carcasses, although they are unlike the carnivore-only models for large carcasses.

Blumenschine emphasized that midshaft portions were most clearly diagnostic as to the number and order of agents in carcass modification. Tooth mark frequencies reported for midshafts at FLK NN3 are in between carnivore-only and human-to-carnivore experimental scenarios and fall outside the ranges of variation for both scenarios for small- and large-sized carcasses. Large carcasses are more highly tooth-marked than smaller ones. Tooth mark distribution is most abundant on upper limb bones (i.e., femora and humeri), which are also the most affected by ravaging and whose epiphyses are most destroyed. Tooth-marking on the shafts of intermediate bones is more reduced (Figure 160). The intermediate frequency of tooth marks and their distribution according to section and element, including abundance of complete shafts and complete bones, suggest that hyenas ravaged the assemblage in a much milder way than they have done in experimental assemblages, and that they were not the primary collectors of the carcasses accumulated at the site (Figure 161). Otherwise a higher degree of bone tooth-marking and fragmentation would be expected (Brain, 1981; Blumenschine, 1988).

Analysis of the Lithic Assemblage

The purported stone tool assemblage at FLK NN3, according to Leakey (1971), is composed of 24 manuports (14 on angular blocks and 10 on cobbles), five utilized materials (blocks), one complete core, eight core fragments, two choppers, 19 broken flakes, and seven flakes. Potts' (1988) revision of the assemblage was very similar. The most abundant category is manuports. Figure 162 shows that most of them are similar in size. A careful analysis of their surfaces show that none of them was used in any activity (e.g., battering) that might have left inconspicuous traces. An analysis of the quality of raw material indicates that most of them are not apt for flaking purposes. All the angular manuports are lava pieces except one quartz piece. All the cobble manuports are also lava. This contrasts with most of the debitage and flakes, which are made of quartz. Most of these manuports have abundant porosity which renders them inappropriate for knapping (Figure 163). This, and the fact that none of them show traces of having been used for any activity, could support a nonanthropogenic origin of these pieces.

Recently, de la Torre and Mora (2005) argued that most Olduvai manuports could be ecofacts; that is, natural clasts and cobbles deposited on the ground by natural processes rather than by hominids. They documented the existence of different-sized clasts in low-energy contexts in modern lacustrine floodplains, which are similar in size and quality to most manuports unearthed at Olduvai. In this chapter, we suggest that the FLK NN3 manuports are also ecofacts. This would explain their similarity in size (Figure 162), and their presence at the site could be the result of physical forces of size selection, such as the hydraulic jumbles or streams so common in the transition between alluvial plains and floodplains, which nowadays show selection of the clasts that they transport (personal observation).

Nonetheless, hominid presence is indicated by the presence of the other stone tools. There is one complete basalt core and eight more fragments of quartz and lava cores. The lava fragments are extremely small and show the

LARGE AND SMALL CARCASSES

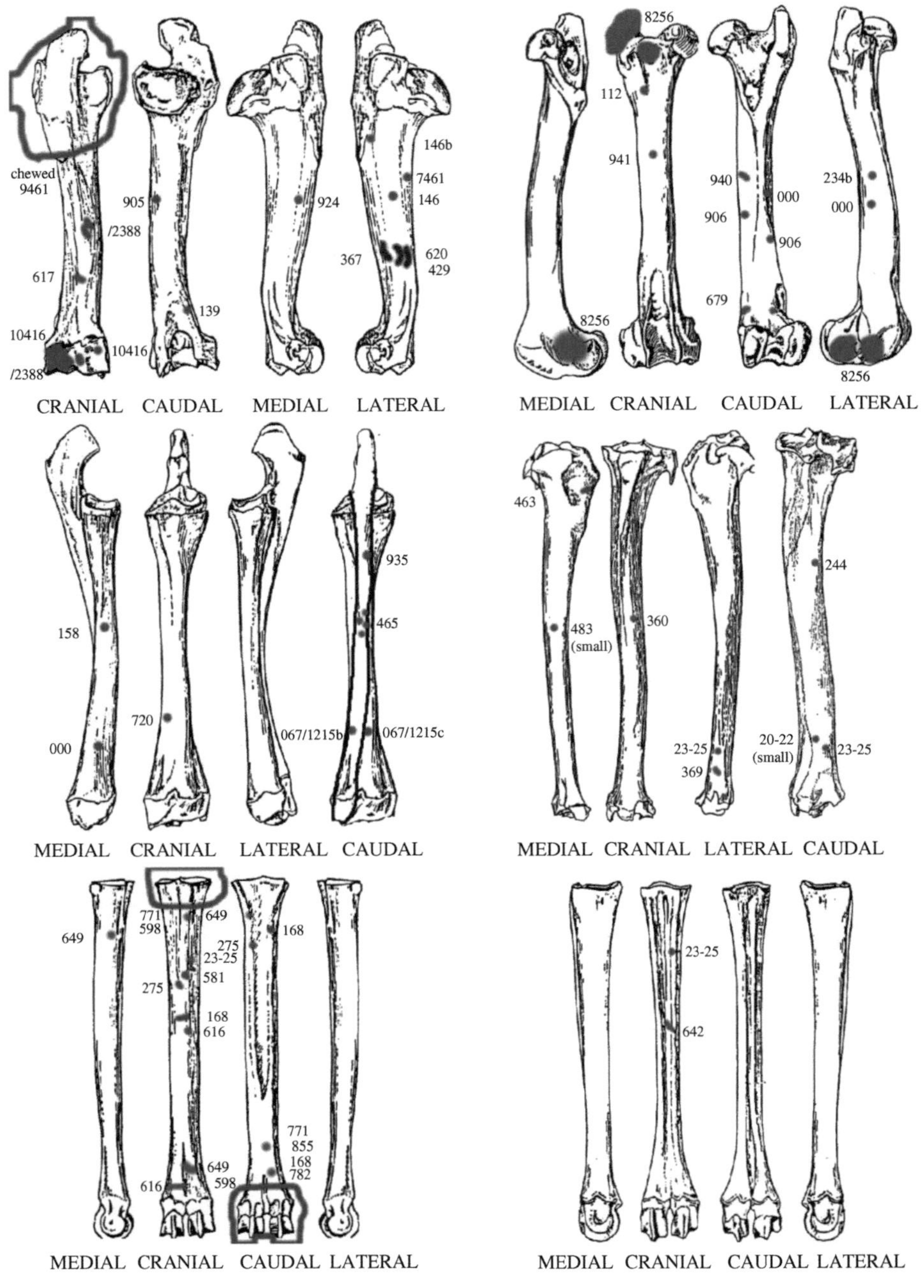

Figure 160. Anatomical distribution of tooth marks on humerus, femur, radio-ulna, and tibia from small and large carcasses. The letters and numbers indicate specimen numbers from the FLK NN 3 assemblage. Specimens bearing marks which could not be anatomically located were not incorporated into this figure. Bones are redrawn from Pales and Lambert (1971).

last stages of core exploitation, which is probably why they were discarded. This would explain why most flake and flake fragments are so small (Figure 164). A total of 36 cm of edge length has been measured combining all flakes together. This limited amount of edge, together with the small number of flakes present at the site, suggests that the stone tools at FLK NN3 are insufficient to account for the exploitation of the faunal resources associated

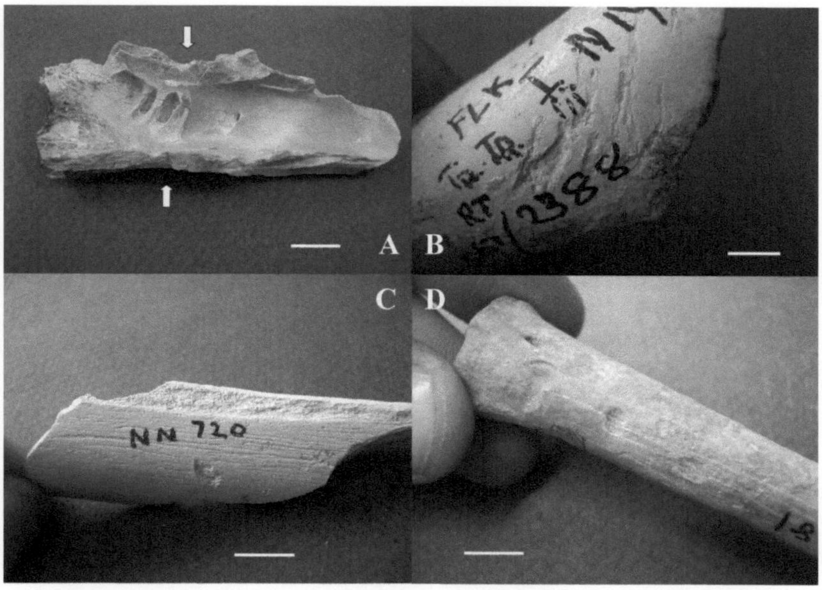

Figure 161. Some examples of tooth-marked specimens. (A) double-opposing and overlapping notch in a humerus caused by carnivore bone breakage. (B) shaft fragments bearing intensive tooth-marking. (C) example of tooth pit on a shaft. (D) Several conspicuous pits on the shaft of a metapodial (scale = 1 cm).

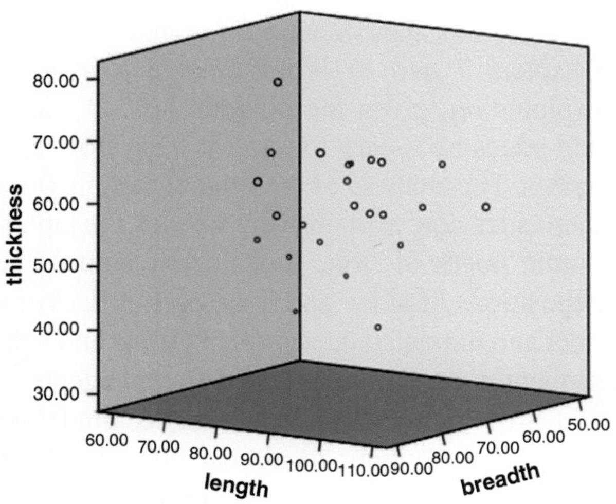

Figure 162. Length (A), breadth (B), and thickness (C) of "manuports" at FLK NN 3.

with them. As mentioned in previous chapters, an average of 15–20 flakes is necessary to fully deflesh one carcass (work in progress). This means that the number of flakes at the FLK NN3 and their small size cannot even account for the exploitation of a single carcass.

Together these data suggest that the deposition of these few lithics might have occurred independently of the deposition of the faunal remains. Hominids may have visited the site very briefly, according to the number of lithic artifacts discarded, and faunal remains may have been accumulated by natural processes (see below) either before or after stone tool discard took place. This would not be an extraordinary event. Other contemporaneous Bed I sites such as FLK North 6 seem to have been formed the same way (see Chapter 7).

Discussion

The lack of taphonomic indicators associating faunal remains with hominid activities at sites such as FLK NN 2, together with the lack of artifacts, led researchers to conclude that these were natural accumulations created by carnivores (Bunn, 1982; Potts, 1988). The same lack of taphonomic indicators of carcass exploitation by hominids has been documented at other

Figure 163. Some examples of "manuports," showing typical porous structure.

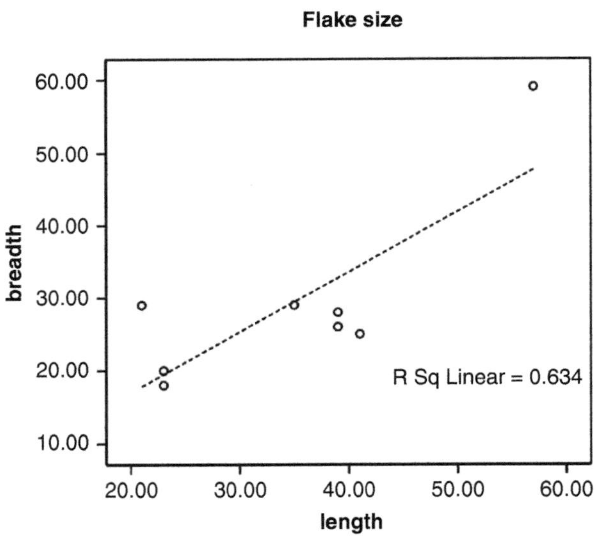

Figure 164. Length and breadth of the few flakes (n = 8) documented at FLK NN 3.

Bed I sites (see previous chapters) and here at FLK NN 3. The only criterion linking the faunal accumulations at these sites to hominid activities was the presence of stone tools. However, hominid contribution to these sites can now be better understood, whether by showing that most of "stone artifacts" are ecofacts (as in the present chapter) or by discovering that their function (i.e., battering; see Chapters 7 and 8) is not related to carcass exploitation, given the obvious lack of flakes and edges necessary to remove flesh from carcasses. The analysis of the lithic materials from these sites now explains why we lack the taphonomic traces of stone tool use on fauna. The depositions of stone and bone each reflect distinct and unrelated depositional events and subsequently, are attributable to different agents.

FLK NN3 shows an important accumulation of bones belonging to almost 20 individuals (16 of which are bovids), which has no counterpart in accumulations created by natural processes in the absence of bone transport (Behrensmeyer and Boaz, 1980; Behrensmeyer, 1983; Hill and Behrensmeyer, 1984). Contrary to sites where individuals are mostly represented by teeth (Chapter 7), individuals at FLK NN3 are represented by several postcranial bones. This indicates that such a large number of animals must have been accumulated at the site as the result of transport. Given the lack of taphonomic evidence for hominid involvement with carcasses, only carnivores can be claimed responsible for the assemblage. There is no doubt that hyenas visited the assemblage, as suggested by the paucity of axial elements and some epiphyses, and by the presence of digested bone. However, the presence of several non-metapodial complete bones and the low percentage of tooth marks on bones argues against hyenas being the main carcass-accumulation agents.

Domínguez-Rodrigo *et al.* (in press) showed that felid modification of carcasses resulted in very low frequencies of tooth-marked bone. Complete bones are very frequent in felid-consumed carcasses. The analysis of tooth marks at the FLK NN3 shows similarities with slight modification of small carcasses by felids. The intermediate values of tooth marks obtained on midshaft specimens at the site could be the result of both felid modification of carcasses and of hyena post-ravaging, which, according to the timing of access to bone remains accumulated by felids, could have resulted in extensive or moderate to low ravaging, depending on whether bones were recently discarded or not. The latter scenario, with bones long-since discarded by the time hyenas arrived, would have produced very few tooth marks. If the site was formed by multiple hyena visits to the same place or by just one visit during which carcasses showed different degrees of bone freshness, ravaging would very likely have produced the frequencies of tooth-marking reported. If this hypothesis is valid, the felids responsible for the accumulation are more likely to have been middle-sized for the following reasons:

1. Large felids (*Homotherium, Panthera leo*) would have been in the upper part of the trophic pyramid, with probably not enough competitors to motivate carcass transport. This assertion can be contradicted if inter-specific competition between lions and *Homotherium* is accepted, which would result in the accumulation of larger carcasses. However, no remains belonging to *Homotherium* have been found in Bed I, whereas lion remains have been found.
2. The predatory range represented is more likely to have been that of a middle-sized felid (e.g., *Dinofelis, Panthera pardus*), since nearly all the carcasses from the site were under 150 kg. *Parmularius altidens* is the most abundant species and it is a size 3A (Bunn, 1982). In addition, almost half of the carcasses were juveniles (which are even smaller, and therefore easily transported in a similar way to that observed in leopards nowadays (Brain, 1981; Cavallo, 1998; de Ruiter and Berger, 2000). FLK NN3 shows the highest frequency of immature animals from all Olduvai Bed I sites (see Potts, 1988).

Felids have been reported to use certain spots, usually where trees are present, to transport carcasses that were obtained nearby, thus creating an accumulation of several individuals (Tappen, 1992; Cavallo, 1998). The accumulations made by felids are smaller than the one reported at FLK NN3, but this could be due to several factors, limited time of felid observations being one of them. The site could also have acted as a preferential depositional environment as is common in modern alluvial environments (Chapter 7), thus creating a bone accumulation to which some carnivores (i.e., felids) might have contributed.

Conclusions

Whether or not felids were the primary bone-accumulating agents at FLK NN3, there is no doubt from a taphonomic point of view that carnivores played the most relevant role in carcass modification and very likely, carcass accumulation, at the site. This chapter has shown that only a taphonomic approach can pinpoint responsibility for carcass accumulation and the presence and possible interaction of agents at any given site. Site integrity and resolution can therefore only be tackled through taphonomic techniques. Faunal analyses must also be reconciled with lithic analyses to understand site formation processes and the extent of taphonomic biases and hominid contribution.

A study of FLK NN1 showed no evidence of hominid modification (Chapter 12). Given that FLK NN2 has been interpreted as a carnivore accumulation, it should be no surprise that FLK NN3 is also a natural accumulation, probably formed by carnivores. This location must have been a magnet for carnivores for prolonged time periods, resulting in successive levels with evidence of carnivore activity.

Site formation processes at most Plio-Pleistocene sites are still poorly understood. Each site needs to be addressed independently. Interpretations of sites that have not undergone rigorous taphonomic study are usually based on the assumption that the spatial association of stone tools and bone remains is the result of hominid behavior. Following the logic of an "eco-taphonomic" approach, discussed in Chapter 3 and 7, the density of faunal materials at Olduvai "Type C" sites cannot be observed in natural scatters in modern savannas, based on the number of remains, MNE, MNI, skeletal part profiles, and the ecological niches of the taxa represented. These sites were therefore thought to be the result of hominid transport, given their association with stone tools (Potts, 1988). The present work has shown this is not the case. After all this time, when archaeologists were confident that the site formation histories of Olduvai Bed I were well known, we realize that the formation of several of those sites, and the contribution of hominids to them, is far from understood.

An example of the complexity of site formation processes at Bed I sites is shown in the following chapter with the taphonomic study of the DK site. Hominid behavior at this site is less well-understood than carnivore behavior. However, hominid–carnivore interactions are better understood at this site: each agent used the site in alternation, with minimal overlap or interdependence in bone accumulation and modification. This site supports claims made in previous chapters that there was overlap in the use of the same ecological lacustrine habitats by hominids, felids, and hyenids. Chapter 15 will present DK as a natural death arena, similar to the depositional scenario inferred for FLK North 6, but probably with a higher degree of carnivore input in the accumulation of carcasses.

15. Zooarchaeology and taphonomy of the DK Site

C.P. EGELAND

Introduction

The DK site complex preserves one of the larger faunal assemblages in Bed I and is particularly well known for the stone circle encountered at the base of Level 3, which like FLK NN3 from the previous chapter, was found on top of a weathered paleosol. Leakey (1971) noted the resemblance of the stone circle to the supports of modern living structures among the Okombambi people of southwest Africa. Coupled with the co-occurrence of faunal and lithic material within a relatively discrete horizon, DK Level 3 was interpreted by Leakey (1971) as the oldest of the living floors within the Gorge.

Potts' (1982, 1983, 1984, 1986, 1988) seminal analyses of the DK assemblages, which acknowledged and drew extensively upon the available taphonomic data, interpreted the site as a primarily hominid accumulation representing repeated episodes of carcass part transport, perhaps over several years. However, these interpretations are now questioned because of their heavy reliance on skeletal part frequencies, which have been shown to be ambiguous indicators of hominid transport behavior (Marean *et al.*, 1992; Capaldo, 1997, 1998b). Although Potts (1988: 23) did examine all of the larger mammalian fauna, his analyses of skeletal part frequencies and bone surface modifications are based almost exclusively on taxonomically identifiable pieces. Such a focus eliminates a large amount of potentially informative material, including the majority of the limb bone shaft fragments. The inclusion of these fragments at all levels of analysis has implications not only for estimates of skeletal part frequencies but also for previous analyses of the cut marks and tooth marks (Potts and Shipman, 1981; Shipman, 1983, 1986).

This renewed investigation calls into question the inference of a high level of hominid involvement in the accumulation of the DK fauna. The site is interpreted here as a complex palimpsest due largely to the activities of large carnivores, with only minor and sporadic input from hominids. Although DK consists of three discrete levels, the faunal assemblage from Level 1 is very small and therefore only the samples from Levels 2 and 3 are considered in this report.

The DK Faunal Assemblage

SITE DISTURBANCE AND INTEGRITY

Fragments from all size ranges are represented from both Levels 2 and 3 (Figure 165). However, fragments <4 cm are present in lower frequencies than expected when compared to experimental assemblages (Blumenschine, 1995; Pickering and Egeland, 2006). Fragments

M. Domínguez-Rodrigo, R. Barba, and C.P. Egeland, Deconstructing Olduvai: A Taphonomic Study of the Bed I Sites, 253–268.

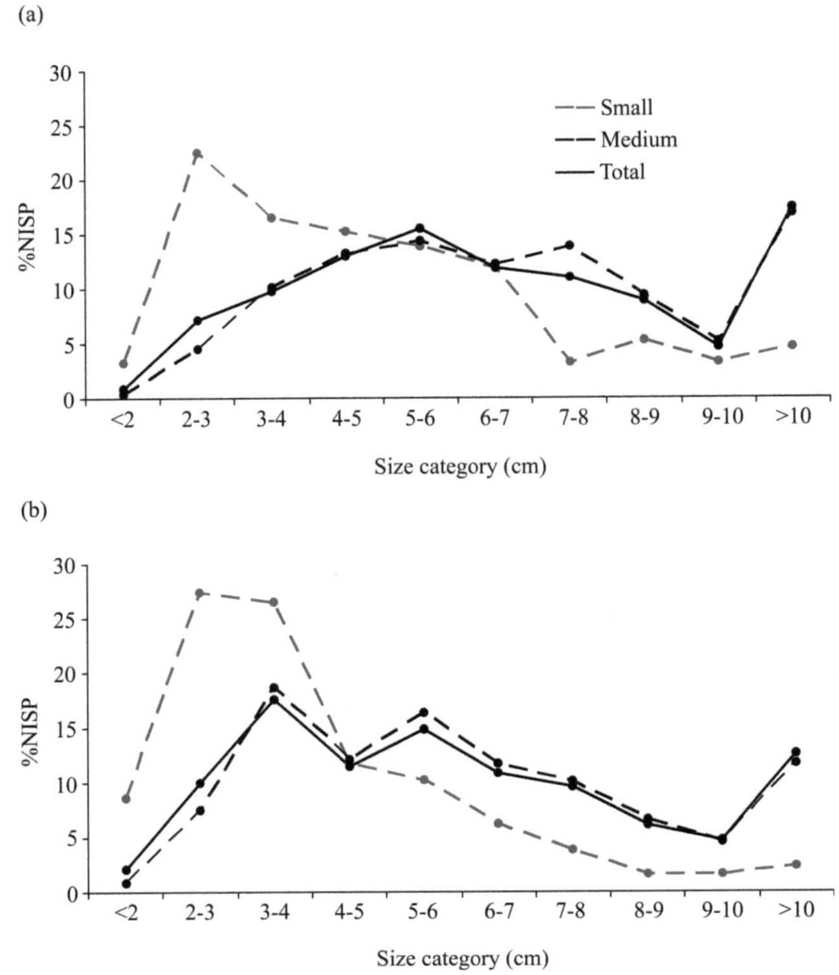

Figure 165. Percentage of specimens in each size range category for (A) Level 2 and (B) Level 3 at DK. Data are divided into specimens from small carcasses, medium carcasses, and combined.

<4 cm are relatively more abundant in Level 3. Only ten specimens (0.6% of total number of identified specimens [NISP]) from Level 2 and eight specimens (0.6% of total NISP) from Level 3 display the polishing characteristic of long-distance water transport. Overall, this suggests minor levels of postdepositional disturbance by physical agents, although Level 2 appears to have been more severely affected by water activity.

Bunn's (1982) circumference types show a predominance of Type 1 (<50% of original diaphyseal circumference) (Figure 166), suggesting both complete retention of material and minimal site disturbance (Marean *et al.*, 2004).

TAXONOMIC AND SKELETAL REPRESENTATION

Table 78 summarizes the taxonomic representation of larger mammals for Levels 2 and 3. As at many classic "Type C" sites throughout Bed I, a diverse fauna is represented at DK. Bovids dominate the assemblages both in terms of minimum number of individuals (MNI) (Table 78) and NISP (Table 79). Relative to the other faunal groups, which are represented largely by teeth, bovids are represented by fragments from all parts of the skeleton (Table 80), suggesting that they experienced a different taphonomic history from the remainder of the assemblage.

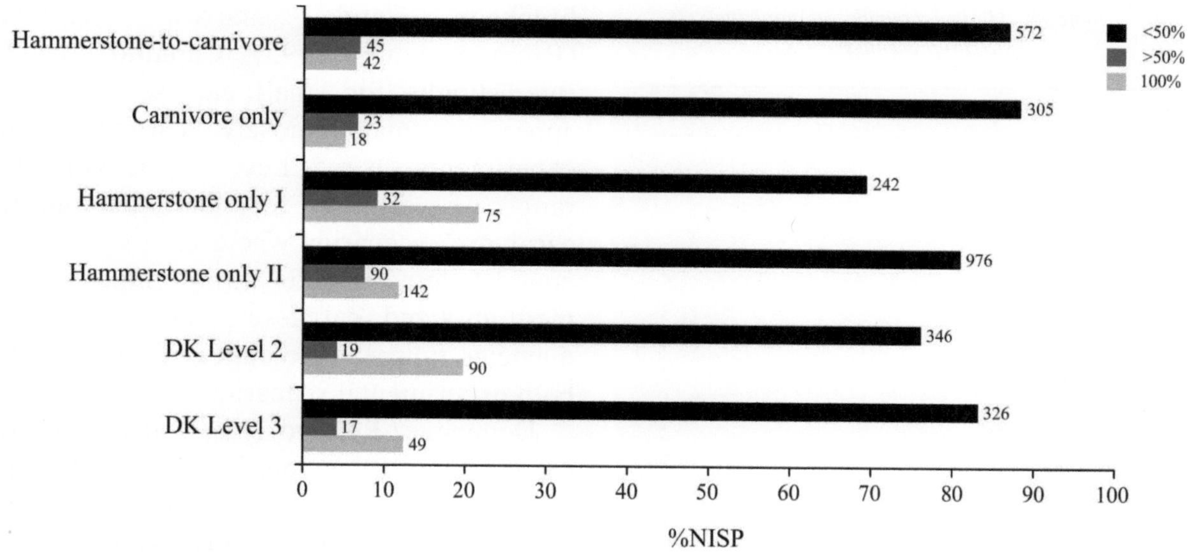

Figure 166. The percentage of Bunn's (1982) limb bone circumference types in the DK assemblages compared to several experimental samples. Carnivore-only, hammerstone-to-carnivore, and hammerstone-only I samples from Marean *et al.* (2004). Hammerstone-only II sample from Pickering and Egeland (2006).

Table 81 presents minimum number of element (MNE) estimates for three body size aggregations: small (Size Class 1 and 2), medium (Size Class 3a and 3b), and large (Size Class 4 and larger). Following Marean and Spencer (1991), MNEs for each of the six major limb bones were calculated for five separate bone portions (proximal epiphysis, proximal shaft, midshaft, distal shaft, distal epiphysis) in order to examine the influence of shaft inclusion on MNE estimates and to gauge the intensity of epiphyseal destruction. In Level 2, midshaft sections provide the highest MNE estimate for all limb bones in small and medium carcasses. In Level 3, 83.3% of small and medium carcass limb bones are best represented by midshafts (Table 82).

BONE SURFACE MODIFICATIONS

Tables 83 and 84 summarize raw surface-mark frequencies for Level 2 and 3, respectively. Cortical surface preservation at DK is moderate to good. In both Level 2 and 3, about 30% of specimens that were scored for surface preservation were observed to have poorly preserved cortices. Therefore, over two thirds of each assemblage is amenable to confident surface-mark identifications. Although tooth marks, cut marks, and percussion marks appear in relatively low frequencies, carnivore damage is far more abundant than hominid damage, both of which cluster on the appendicular skeleton. One Size Class 2 metatarsal from Level 2 preserves tooth marks overlying cut marks, indicating that carnivore bone breakage followed hominid butchery (likely skinning) in at least one instance.

Figure 167 provides tooth-mark frequencies on midshaft fragments (following Blumenschine's [1988, 1995] bone portion definitions) from DK and a number of actualistic samples (see also Table 85). Experimental work demonstrates that tooth-mark frequencies on epiphyses and near-epiphyses are ambiguous indicators of carnivore access to carcasses, as these portions contain nutritious tissues (i.e., grease) before and after hominid hammerstone

Table 78. Minimum number of individuals (MNI) represented at DK

Taxon	MNI
Level 2	
Parmularius altidens	2
Size 3a Alcelaphini	4
Size 3b Alcelaphini	4
Size 1 Antilopini	5
Size 3b Tragelaphini	5
Size 3 Reduncini	4
Size 3 Hippotragini	2
Size 4 Bovini	1
Proboscidean indet.	2
Metridiochoerus modestus	1
Kolpochoerus limnetes	2
Hippopotamus gorgops	1
Libytherium sp.	1
Giraffa sp.	1
Pseudocivetta ingens	1
Panthera pardus	1
Level 3	
P. altidens	1
Size 3a Alcelaphini	11
Size 3b Alcelaphini	3
Size 1 Antilopini	4
Size 3 Tragelaphini	4
Size 3 Reduncini	3
Size 3 Hippotragini	1
Size 4 Bovini	2
M. modestus	1
K. limnetes	2
Deinotherium sp.	1
H. gorgops	2
Giraffa sp.	1
Crocuta crocuta	1

Table 79. Number of identified specimens (NISP) for major faunal groups at DK

	Level 2		Level 3	
Faunal group	NISP	%	NISP	%
Bovidae	824	77.6	516	78.1
Suidae	116	10.9	53	8.0
Proboscidean	20	1.9	20	3.0
Hippopotamidae	13	1.2	21	3.2
Carnivora	9	0.8	9	1.4
Giraffidae	8	0.8	4	0.6
Rhinocerotidae	4	0.4	10	1.5
Equidae	18	1.7	23	3.5
Primates	50	4.7	5	0.8

Equid numbers include 17 specimens from Level 2 and 22 from Level 3 listed in Potts (1988: Table A.1)
Primate numbers from Potts (1988: Table A.1)

breakage and marrow removal (Blumenschine, 1995; Capaldo, 1997). Interestingly, for small carcasses Level 2 appears to match closely "carnivore-first" experiments whereas Level 3 falls within the range of variation for "hammerstone-to-carnivore" scenarios where carnivores scavenged demarrowed limb bones. For medium-sized carcasses, tooth-mark frequencies fall near the edges of variation for both experimental scenarios.

Domínguez-Rodrigo *et al.* (in press) have recently argued that equifinality in tooth-mark frequencies can arise if felids rather than hyenas (as in previous actualistic studies) are used as the primary agent of bone modification. Specifically, they show that felid consumption followed by bone breakage can mimic the tooth-mark frequencies produced in "hammerstone-to-carnivore" experiments (see Figure 167). Therefore, it is not possible to distinguish either carnivore or hominid access to carcasses with absolute confidence based simply on midshaft tooth-mark frequencies. In addition, one must assume when applying these experimental tooth-mark frequencies that all carcasses were available to and processed by carnivores. The fact that the tooth-mark data do not match actualistic samples in every case suggests that DK, as many of the Bed I sites at Olduvai, is a more complex palimpsest that has not yet been accurately modeled experimentally.

Figure 168 summarizes tooth pit dimensions on limb bone diaphyses at DK. Only the sample of pits on medium-sized carcasses from Level 2 is large enough to provide reasonable ranges of variation; therefore, only these data appear in Figure 168. Pit length and breadth are similar to those created by carnivores such as lions and hyenas with larger and more robust dentitions. Although not pictured and despite the small sample size, the remaining data are also consistent with this interpretation.

Further examination of tooth-mark frequencies may help distinguish between hyenas and

Table 80. Number of identified specimens (NISP) by major faunal group and skeletal element at DK

	Bovidae	Suidae	Hippopotamidae	Giraffidae	Rhinocerotidae	Proboscidean
Level 2						
Cranium	26	5	–	–	–	–
Teeth	120	19	9	1	2	4
Mandible	36	6	–	1	–	–
Vertebrae	26	2	–	–	–	–
Ribs	13	3	–	–	–	6
Innominate	15	–	–	1	2	–
Scapula	15	–	–	–	–	–
Humerus	31	1	–	–	–	–
Radio–ulna	66	4	1	–	–	–
Carpals/tarsals	26	3	2	–	–	1
Metacarpal	50	–	–	–	–	–
Femur	40	1	–	–	–	–
Tibia	69	–	1	3	–	–
Metatarsal	50	1	–	–	–	–
Patella	4	–	–	–	–	–
Phalanges	65	4	–	2	–	–
Sesamoids	28	–	–	–	–	1
Metapodial	20	–	–	–	–	–
Limb bone shaft	1	–	–	–	–	–
Level 3						
Cranium	13	4	–	–	–	–
Teeth	69	7	15	2	7	9
Mandible	22	5	–	–	1	–
Vertebrae	12	4	3	–	–	–
Ribs	17	1	–	–	–	7
Innominate	8	–	–	–	–	–
Scapula	5	–	–	–	–	–
Humerus	27	2	–	–	–	–
Radio–ulna	35	1	–	–	1	–
Carpals/tarsals	47	1	–	–	–	2
Metacarpal	27	–	–	2	–	–
Femur	30	–	–	–	–	–
Tibia	54	2	–	–	–	–
Metatarsal	46	1	–	–	1	–
Patella	–	–	–	–	–	–
Phalanges	26	2	1	–	–	–
Sesamoids	12	–	1	–	–	1
Metapodial	23	–	1	–	–	–
Limb bone shaft	1	–	–	–	–	1

large felids. In "carnivore-only" experiments where hyenas are the primary carcass consumers, between 64 and 100% of epiphyseal and near-epiphyseal fragments are tooth-marked (Blumenschine, 1995; Capaldo, 1997, 1998b). However, at DK epiphyseal and near-epiphyseal tooth-mark frequencies are relatively low, especially in Level 3 (Table 85). Lion consumption results in epiphyseal and near-epiphyseal tooth-mark frequencies of 33.3% and 23.1%, respectively (Domínguez-Rodrigo *et al.*, in press). Leopards and cheetahs tooth mark these bone portions at rates of about 15.9% and 16.8% (Domínguez-Rodrigo *et al.*, in press). These data are more consistent with a felid pattern of tooth-marking along with minor to intermediate levels of postdepositional hyena ravaging.

Table 81. Minimum number of element (MNE) estimates for small (Size Class 1 and 2), medium (Size Class 3) and large (Size Class 4–6) carcasses at DK

Element	Small	Medium	Large	Total
Level 2				
Cranium	7	15	3	25
Mandible	9	24	3	36
Vertebrae	12	38	6	56
Innominate	6	15	2	23
Ribs	7	7	11	25
Scapula	3	13	2	18
Humerus	4	25	0	29
Radius	8	21	2	31
Ulna	3	10	2	15
Carpals	9	17	1	26
Metacarpal	11	20	0	31
Femur	8	23	1	32
Patella	4	1	0	5
Tibia	10	30	3	43
Tarsals	16	23	1	40
Metatarsal	6	23	0	29
Phalanges	28	31	2	61
Sesamoids	9	20	4	33
Level 3				
Cranium	3	19	3	25
Mandible	5	13	3	21
Vertebrae	6	17	5	28
Innominate	2	7	2	11
Ribs	3	8	9	20
Scapula	1	6	2	9
Humerus	5	18	1	24
Radius	2	13	1	16
Ulna	1	5	2	8
Carpals	2	11	1	14
Metacarpal	6	9	1	16
Femur	6	17	2	25
Patella	0	0	0	0
Tibia	6	22	5	33
Tarsals	12	21	2	35
Metatarsal	7	16	0	23
Phalanges	17	7	2	26
Sesamoids	4	9	2	15

BONE BREAKAGE

Green fractures predominate for both small and medium carcasses (Figure 169). About 8.5% of limb bones are complete in both levels, although complete limb bones from small-sized carcasses are more abundant in Level 3 (Table 86). The distribution of circumference types matches more closely those experiments where ravaging hyenas did not play a role in carcass consumption (see Figure 166).

Fracture plane analysis indicates that most bone breakage can be attributed to the static loading characteristic of carnivore consumption. The sample of fracture planes from small carcasses is too small for meaningful comparisons, but the data from medium-sized carcasses is more informative (Figure 170). For both levels acute angles from both oblique and longitudinal planes reflect a mixture of hominid and carnivore breakage, but obtuse angles are consistent solely with static loading. The few measurable notches on small carcasses fall within the range of variation for carnivore notches (Figure 171).

Most of the notches on medium-sized carcasses are probably the result of carnivore breakage, although a few may have been created through hammerstone percussion (see, e.g., Figure 172). The frequencies of Capaldo and Blumenschine's (1994) notch types are also suggestive of carnivore breakage (Figure 173). Carnivore assemblages tend to show relatively high frequencies of overlapping (Incomplete Type C; e.g., Figure 174) and opposing complete notches (Figure 175) in addition to abundant micronotches (Capaldo and Blumenschine, 1994; Egeland *et al.*, unpublished data).

Discussion and Conclusions

There is an increase in hominid involvement with carcasses at DK relative to other stone tool-bearing assemblages such as FLK North. This makes some sense in light of the DK lithic assemblage, which contains many more supposed butchering tools such as flakes than those from the FLK North sequence (de la Torre, 2006). In the near absence of percussion marks, the fracture angle data provide the best evidence for hominid bone-breaking, although the preponderance of the data demonstrate that hominids played a very marginal role in the

Table 82. Minimum number of element (MNE) estimates for limb bones by bone portion for small (Size Class 1 and 2), medium (Size Class 3) and large (Size Class 4–6) carcasses at DK

Level 2					Level 3				
Element	Small	Medium	Large	Total	Element	Small	Medium	Large	Total
Humerus					*Humerus*				
PR	1	1	0	2	PR	0	2	0	2
PRS	1	10	0	11	PRS	1	7	0	8
SH	3	24	0	27	SH	0	16	1	17
DSS	2	22	0	24	DSS	2	15	0	17
DS	2	12	0	14	DS	3	10	0	13
Radius					*Radius*				
PR	1	13	2	16	PR	1	8	1	10
PRS	3	11	2	16	PRS	1	6	1	8
SH	8	21	0	29	SH	2	11	0	13
DSS	4	11	0	15	DSS	0	3	0	3
DS	1	10	0	11	DS	2	1	0	3
Metacarpal					*Metacarpal*				
PR	4	14	0	18	PR	2	6	0	8
PRS	4	14	0	18	PRS	2	5	1	8
SH	8	19	0	27	SH	4	9	1	14
DSS	2	13	0	15	DSS	1	7	1	9
DS	1	8	0	9	DS	1	5	1	7
Femur					*Femur*				
PR	4	3	0	7	PR	2	2	0	4
PRS	1	8	0	9	PRS	4	6	0	10
SH	6	23	1	30	SH	6	16	1	23
DSS	3	16	0	19	DSS	2	6	0	8
DS	1	5	0	6	DS	3	5	2	10
Tibia					*Tibia*				
PR	1	3	0	4	PR	2	3	0	5
PRS	2	12	2	16	PRS	1	10	1	12
SH	7	29	3	39	SH	5	22	5	32
DSS	7	19	1	27	DSS	5	12	3	20
DS	6	16	0	22	DS	3	5	0	8
Metatarsal					*Metatarsal*				
PR	1	15	0	16	PR	4	9	0	13
PRS	1	19	0	20	PRS	4	10	0	14
SH	6	22	0	28	SH	6	16	0	22
DSS	2	9	0	11	DSS	2	5	0	7
DS	1	5	0	6	DS	4	3	0	7

Abbreviations: PR, proximal; PRS, proximal shaft; SH, midshaft; DSS, distal shaft; DS, distal

modification of the assemblage, which can thus be attributed largely to carnivores. It is possible that hominids transported some carcass parts from the site for further processing elsewhere. Unlike some of the assemblages from the FLK North sequence, which, for example, likely represent accumulations of a specialized felid predator, the diversity of bovid taxa and body sizes at DK indicates a slightly more complicated taphonomic history.

The low representation of axial elements and the deletion of limb bone epiphyses are consistent with carnivore ravaging. Several lines of data can be used to reconstruct levels

Table 83. Surface–mark frequencies for small (Size Class 1 and 2), medium (Size Class 3), and large (Size Class 4–6) carcasses for DK Level 2

	Small			Medium			Large		
Element	TM	CM	PM	TM	CM	PM	TM	CM	PM
Mandible	2/11 (18.2)	0/11 (0.0)	0/11 (0.0)	6/37 (16.2)	0/37 (0.0)	0/37 (0.0)	0/4 (0.0)	0/4 (0.0)	0/4 (0.0)
Vertebrae	1/15 (6.7)	0/15 (0.0)	0/15 (0.0)	6/47 (12.8)	0/47 (0.0)	0/47 (0.0)	0/7 (0.0)	0/7 (0.0)	0/7 (0.0)
Innominate	0/6 (0.0)	0/6 (0.0)	0/6 (0.0)	6/25 (24.0)	0/25 (0.0)	0/25 (0.0)	0/4 (0.0)	0/4 (0.0)	0/4 (0.0)
Ribs	1/19 (5.3)	0/19 (0.0)	0/19 (0.0)	4/32 (12.5)	0/32 (0.0)	0/32 (0.0)	3/67 (4.5)	0/67 (0.0)	0/67 (0.0)
Scapula	0/6 (0.0)	0/6 (0.0)	0/6 (0.0)	3/21 (14.3)	0/21 (0.0)	0/21 (0.0)	0/6 (0.0)	0/6 (0.0)	0/6 (0.0)
Humerus	2/6 (33.3)	1/6 (16.7)	0/6 (0.0)	13/44 (29.5)	1/44 (2.3)	0/44 (0.0)	0/0 (0.0)	0/0 (0.0)	0/0 (0.0)
Radius	5/10 (50.0)	0/10 (0.0)	0/10 (0.0)	8/43 (18.6)	0/43 (0.0)	0/43 (0.0)	0/2 (0.0)	0/2 (0.0)	0/2 (0.0)
Ulna	1/3 (33.3)	1/3 (33.3)	0/3 (0.0)	3/16 (18.8)	0/16 (0.0)	0/16 (0.0)	0/5 (0.0)	0/5 (0.0)	0/5 (0.0)
Carpals	0/9 (0.0)	0/9 (0.0)	0/9 (0.0)	2/17 (11.8)	0/17 (0.0)	0/17 (0.0)	0/1 (0.0)	0/1 (0.0)	0/1 (0.0)
Metacarpal	2/14 (14.3)	0/14 (0.0)	1/14 (7.1)	11/37 (29.7)	1/37 (2.7)	0/37 (0.0)	0/0 (0.0)	0/0 (0.0)	0/0 (0.0)
Femur	1/14 (7.1)	0/14 (0.0)	0/14 (0.0)	16/44 (36.4)	0/44 (0.0)	0/44 (0.0)	0/1 (0.0)	0/1 (0.0)	0/1 (0.0)
Patella	0/4 (0.0)	0/4 (0.0)	0/4 (0.0)	0/1 (0.0)	0/1 (0.0)	0/1 (0.0)	0/1 (0.0)	0/1 (0.0)	0/1 (0.0)
Tibia	5/17 (29.4)	0/17 (0.0)	0/17 (0.0)	14/73 (19.2)	0/73 (0.0)	1/73 (1.4)	0/5 (0.0)	1/5 (20.0)	0/5 (0.0)
Tarsals	0/16 (0.0)	0/16 (0.0)	0/16 (0.0)	1/12 (8.3)	0/12 (0.0)	0/12 (0.0)	0/2 (0.0)	0/2 (0.0)	0/2 (0.0)
Metatarsal	3/10 (30.0)	1/10 (10.0)	0/10 (0.0)	13/41 (31.7)	0/41 (0.0)	1/41 (2.4)	0/0 (0.0)	0/0 (0.0)	0/0 (0.0)
Phalanges	2/30 (6.7)	0/30 (0.0)	0/30 (0.0)	1/35 (2.9)	0/35 (0.0)	0/35 (0.0)	0/2 (0.0)	0/2 (0.0)	0/2 (0.0)
Sesamoids	0/9 (0.0)	0/9 (0.0)	0/9 (0.0)	1/20 (5.0)	0/20 (0.0)	0/20 (0.0)	0/4 (0.0)	0/4 (0.0)	0/4 (0.0)
Unidentified limb bone	1/23 (4.3)	0/23 (0.0)	0/23 (0.0)	18/109 (16.5)	1/109 (0.9)	2/109	1/24 (4.2)	0/24 (0.0)	0/24 (0.0)
Total	25/216 (11.6)	3/216 (1.4)	1/216 (0.5)	126/654 (19.3)	3/654 (0.5)	4/654 (0.6)	4/120 (3.3)	1/120 (0.8)	0/120 (0.0)

Numerator denotes number of marked specimens; denominator denotes total NISP for each skeletal element; percentage is in parentheses
Abbreviations: TM, tooth mark; CM, cut mark; PM, percussion mark

Table 84. Surface–mark frequencies for small (Size Class 1 and 2), medium (Size Class 3), and large (Size Class 4–6) carcasses for DK Level 3

	Small			Medium			Large		
Element	TM	CM	PM	TM	CM	PM	TM	CM	PM
Mandible	2/6 (33.3)	0/6 (0.0)	0/6 (0.0)	2/27 (7.4)	0/27 (0.0)	0/27 (0.0)	0/6 (0.0)	0/6 (0.0)	0/6 (0.0)
Vertebrae	0/8 (0.0)	0/8 (0.0)	0/8 (0.0)	2/25 (8.0)	0/25 (0.0)	0/25 (0.0)	0/6 (0.0)	0/6 (0.0)	0/6 (0.0)
Innominate	1/2 (50.0)	0/2 (0.0)	0/2 (0.0)	5/12 (41.7)	0/12 (0.0)	0/12 (0.0)	1/2 (50.0)	0/2 (0.0)	0/2 (0.0)
Ribs	0/16 (0.0)	0/16 (0.0)	0/16 (0.0)	4/26 (15.4)	0/26 (0.0)	0/26 (0.0)	0/26 (0.0)	0/26 (0.0)	0/26 (0.0)
Scapula	0/2 (0.0)	0/2 (0.0)	0/2 (0.0)	2/15 (13.3)	0/15 (0.0)	0/15 (0.0)	0/4 (0.0)	0/4 (0.0)	0/4 (0.0)
Humerus	1/5 (20.0)	0/5 (0.0)	0/5 (0.0)	7/33 (21.2)	0/33 (0.0)	0/33 (0.0)	0/1 (0.0)	0/1 (0.0)	0/1 (0.0)
Radius	1/4 (25.0)	0/4 (0.0)	0/4 (0.0)	6/26 (23.1)	0/26 (0.0)	0/26 (0.0)	0/1 (0.0)	0/1 (0.0)	0/1 (0.0)
Ulna	0/1 (0.0)	0/1 (0.0)	0/1 (0.0)	2/9 (22.2)	0/9 (0.0)	0/9 (0.0)	1/2 (50.0)	0/2 (0.0)	0/2 (0.0)
Carpals	0/2 (0.0)	0/2 (0.0)	0/2 (0.0)	0/11 (0.0)	0/11 (0.0)	0/11 (0.0)	0/1 (0.0)	0/1 (0.0)	0/1 (0.0)
Metacarpal	2/8 (25.0)	1/8 (12.5)	0/8 (0.0)	1/20 (5.0)	0/20 (0.0)	0/20 (0.0)	1/2 (50.0)	0/2 (0.0)	0/2 (0.0)
Femur	0/8 (0.0)	0/8 (0.0)	0/8 (0.0)	6/35 (17.1)	0/35 (0.0)	1/35 (2.9)	0/4 (0.0)	0/4 (0.0)	0/4 (0.0)
Patella	0/0 (0.0)	0/0 (0.0)	0/0 (0.0)	0/0 (0.0)	0/0 (0.0)	0/0 (0.0)	0/0 (0.0)	0/0 (0.0)	0/0 (0.0)
Tibia	3/11 (27.3)	0/11 (0.0)	0/11 (0.0)	5/48 (10.4)	0/48 (0.0)	0/48 (0.0)	0/8 (0.0)	0/8 (0.0)	0/8 (0.0)
Tarsals	0/12 (0.0)	0/12 (0.0)	0/12 (0.0)	4/21 (19.0)	0/21 (0.0)	0/21 (0.0)	0/2 (0.0)	0/2 (0.0)	0/2 (0.0)
Metatarsal	1/11 (9.1)	0/11 (0.0)	0/11 (0.0)	12/35 (34.3)	0/35 (0.0)	0/35 (0.0)	0/0 (0.0)	0/0 (0.0)	0/0 (0.0)
Phalanges	0/17 (0.0)	0/17 (0.0)	0/17 (0.0)	1/8 (12.5)	0/8 (0.0)	0/8 (0.0)	0/2 (0.0)	0/2 (0.0)	0/2 (0.0)
Sesamoids	0/4 (0.0)	0/4 (0.0)	0/4 (0.0)	0/9 (0.0)	0/9 (0.0)	0/9 (0.0)	0/2 (0.0)	0/2 (0.0)	0/2 (0.0)
Unidentified limb bone	2/32 (6.3)	0/32 (0.0)	0/32 (0.0)	3/126 (2.4)	0/126 (0.0)	0/126 (0.0)	1/7 (14.3)	0/7 (0.0)	0/7 (0.0)
Total	13/149 (8.7)	1/149 (0.7)	0/149 (0.0)	62/486 (12.8)	0/486 (0.0)	1/486 (0.2)	4/76 (5.3)	0/76 (0.0)	0/76 (0.0)

Numerator denotes number of marked specimens; denominator denotes total NISP for each skeletal element; percentage is in parentheses
Abbreviations: TM, tooth mark; CM, cut mark; PM, percussion mark

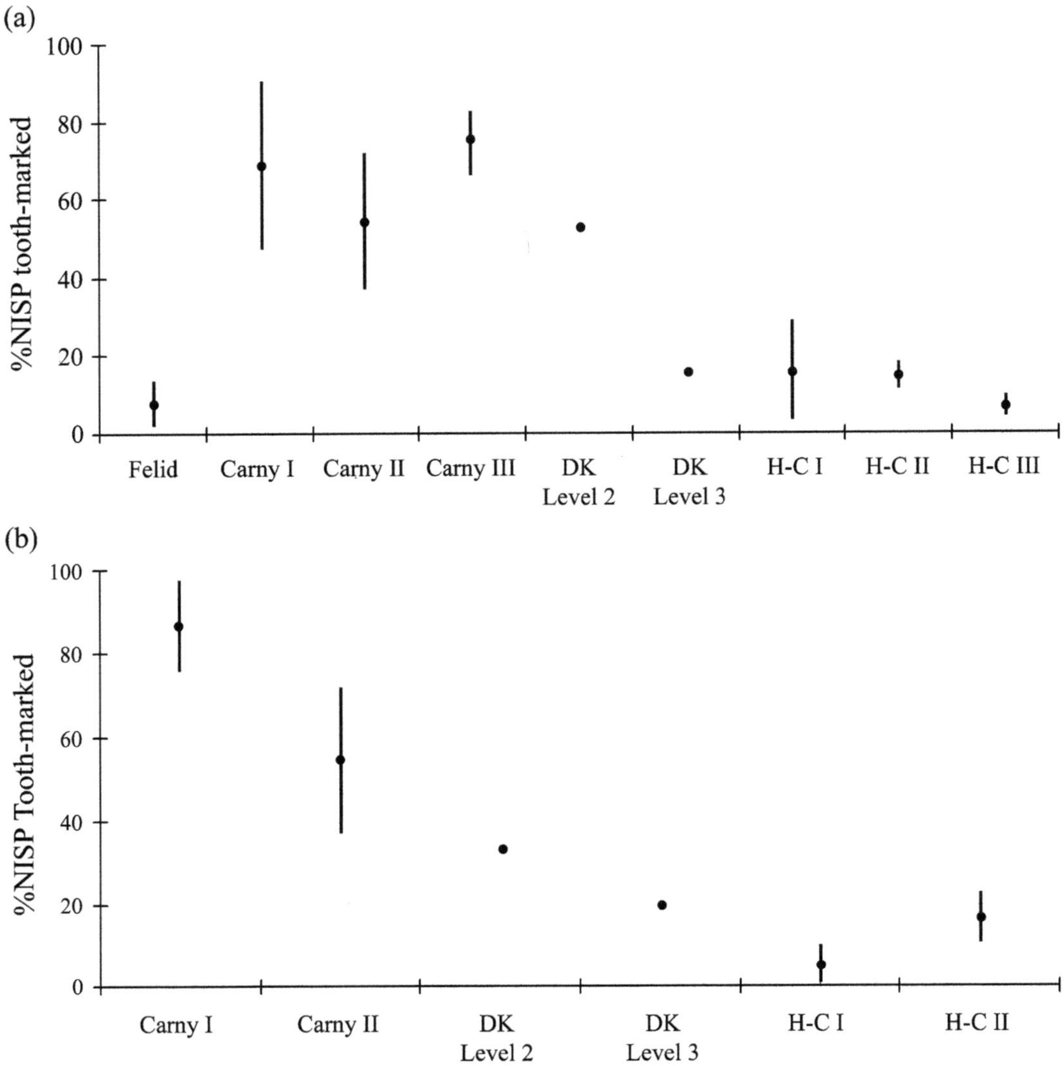

Figure 167. Incidence of tooth-marked midshaft fragments at DK for (A) small-sized carcasses and (B) medium-sized carcasses compared to the mean and 95% confidence intervals for actualistic control assemblages. Only fossil specimens with green breakage and good cortical surface preservation are considered. *Abbreviations*: Felid, felid-consumed carcasses; Carny, carnivore-only; H-C, hammerstone-to-carnivore. (Data sources: Blumenschine [1995]; Capaldo [1997, 1998a, 1998b]; Marean *et al.* [2000]; Domínguez-Rodrigo *et al.* [in press].) *Note*: The range of variation from Marean's experiments ("Carny III" and "H-C III") are somewhat smaller because confidence intervals were calculated by bootstrapping a single sample (Marean *et al.*, 2000: Table 3).

of on-site competition for carcasses at DK. The presence of complete limb bones (about 8.5% of the total limb bone MNE) indicates that some within-bone nutrients were not processed. Epiphysis-to-shaft fragment ratios for Levels 2 and 3 (0.24 and 0.21, respectively) are somewhat higher than carnivore-only assemblages reported on by Blumenschine and Marean (1993: Figure 16.5) and far higher than Capaldo's (1998b) carnivore-only experiments. The axial-to-limb bone and femur-to-tibia ratios place both Levels 2 and 3

Table 85. Percentage of epiphyseal, near–epiphyseal, and midshaft specimens bearing tooth marks at DK

	EP	NEP	MSH	Total
Level 2				
Small carcasses				
NISP	6	6	17	29
NISP TM	2	4	9	15
%	33.3	66.7	52.9	51.7
Medium carcasses				
NISP	24	31	84	139
NISP TM	12	17	28	57
%	50.0	54.8	33.3	41.0
Level 3				
Small carcasses				
NISP	8	4	19	31
NISP TM	3	0	3	6
%	37.5	0.0	15.8	19.4
Medium carcasses				
NISP	11	19	87	117
NISP TM	1	5	17	23
%	9.1	26.3	19.5	19.7

NISP counts include only those specimens with green breakage and good cortical surface preservation
Abbreviations: EP, epiphyseal; NEP, near–epiphyseal; MSH, midshaft.

within Domínguez-Rodrigo and Organista's ravaging stage 1 (see Chapter 11). The DK site itself was probably located within a few hundred yards of the lake shore and very near to closed vegetation (Figure 176), a suggestion supported by both geological and paleoecological reconstructions. In modern savanna mosaic environments, such habitats tend towards low degrees of competition given the low visibility and lack of visual cues to carcass location (e.g., vultures) (Blumenschine, 1986, 1987; Creel and Creel, 1996; Domínguez-Rodrigo, 2001).

The almost complete lack of cut marks and percussion marks and other evidence for hominid bone-breaking mean that the low overall incidence of tooth marks in both levels cannot be explained solely by carnivore ravaging of hominid food refuse. It also seems that some carcasses were initially modified by felids (which damage bones at much lower frequencies than hyenas) and subsequently ravaged by hyenas and/or many carcasses simply were not discovered to be consumed. In addition, it is certainly possible that the behaviorally derived fauna from both levels is mixed with background scatters of bones that were deposited independently. All of these processes would depress surface-mark frequencies below those documented in actualistic samples. The location of DK near a semipermanent water source and intermittent streams would have made an ideal arena for natural deaths as well as serial predation by felids. In addition, low visibility probably prevented many carcasses from being discovered by scavengers. Although about 50% of the bovid individuals represented in both levels fall within the size range of leopard prey (Size Classes 1, 2, and 3a), tooth pit dimensions suggest that large felids and hyenas were responsible for modifying most of the carcasses regardless of body size.

The few hominid surface modifications that are preserved provide an interesting picture of hominid–carnivore interactions at DK. The frequency of specimens showing both hominid and carnivore damage has been argued to reflect the level of hominid–carnivore interdependence in assemblage formation (Marean and Kim, 1998; Marean *et al.*, 2000; Egeland *et al.*, 2004). Although the overall incidence of these specimens in Level 2 is low (0.7% of total small + medium carcass NISP), it is interesting that 63.6% of hominid-modified specimens also preserve carnivore tooth marks whereas only 7.3% of carnivore-modified specimens also preserve hominid damage. This demonstrates that when hominids were utilizing the site, carnivores were almost always in the vicinity at the same time or within days of hominid abandonment. This also speaks to the only intermittent use by hominids of the site for carcass-processing activities.

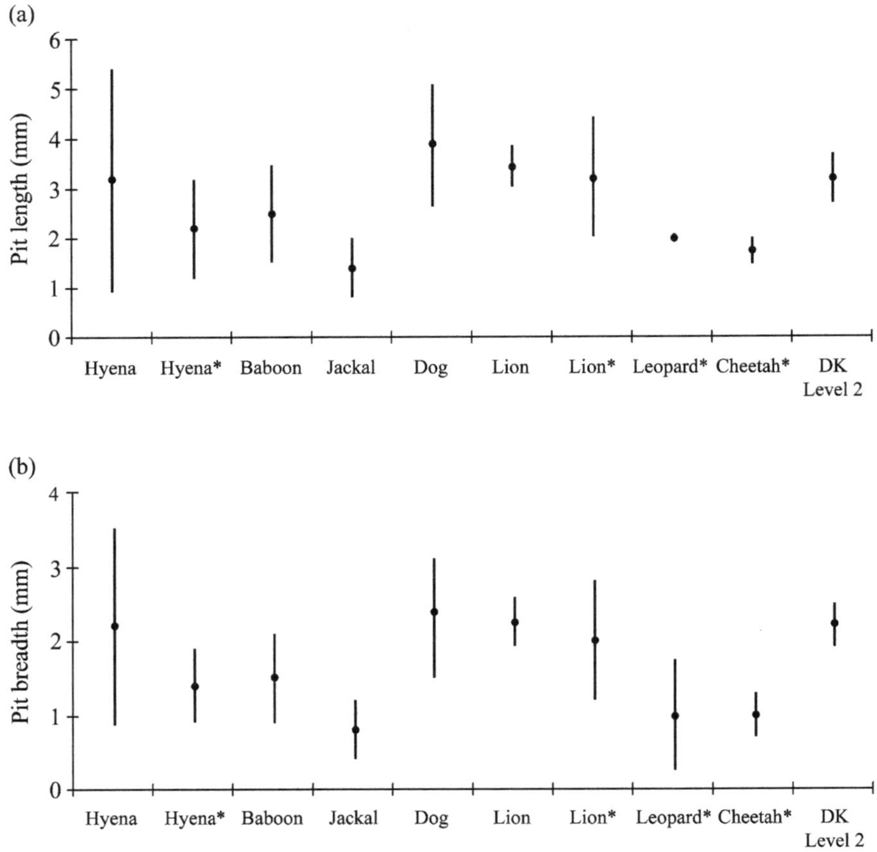

Figure 168. Mean and 95% confidence intervals for tooth pit (A) lengths and (B) breadths on the limb bone diaphyses of medium-sized carcasses at DK Level 2 compared to actualistic samples. (Data indicated with an asterisk (*) are from Selvaggio [1994]; all other modern carnivore data are from Domínguez-Rodrigo and Piqueras [2003].)

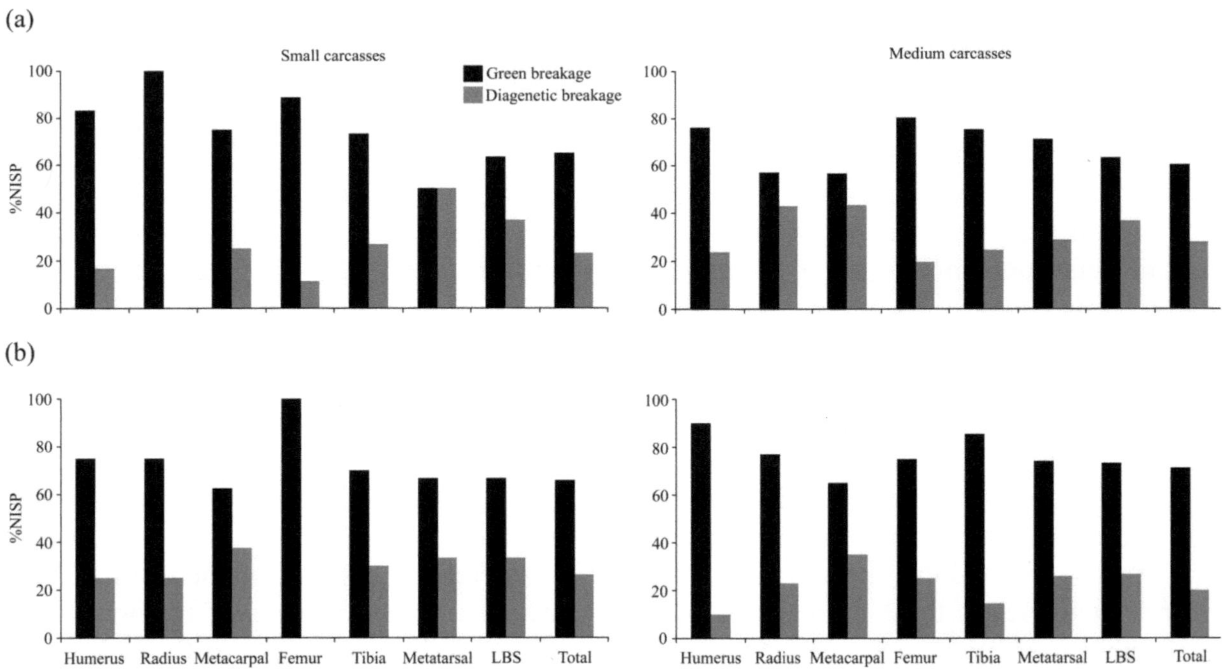

Figure 169. Incidence of green and diagenetic breakage on limb bones in (A) Level 2 and (B) Level 3 at DK. Percentages may not add to 100% as specimens with recent breakage are not considered. *Abbreviations*: LBS, unidentified limb bone shaft.

Table 86. Frequency of complete limb bones for small (Size Class 1 and 2) and medium (Size Class 3) carcasses at DK

Element	Small	Medium	Total
Level 2			
Humerus	0/4 (0.0)	1/25 (4.0)	1/29 (3.4)
Radius	1/8 (12.5)	2/21 (19.0)	3/29 (10.3)
Metacarpal	0/11 (0.0)	5/20 (25.0)	5/31 (16.1)
Femur	0/8 (0.0)	1/23 (4.3)	1/31 (3.2)
Tibia	0/10 (0.0)	1/30 (3.3)	1/40 (2.5)
Metatarsal	0/6 (0.0)	3/23 (13.0)	3/29 (10.3)
Total	1/47 (2.1)	15/142 (10.6)	16/189 (8.5)
Level 3			
Humerus	0/5 (0.0)	1/18 (5.6)	1/23 (4.3)
Radius	0/2 (0.0)	0/13 (0.0)	0/15 (0.0)
Metacarpal	0/6 (0.0)	0/9 (0.0)	0/15 (0.0)
Femur	2/6 (33.3)	1/17 (5.9)	3/23 (13.0)
Tibia	1/6 (16.7)	3/22 (13.6)	4/28 (14.3)
Metatarsal	1/7 (14.3)	2/16 (12.5)	3/23 (13.0)
Total	4/32 (12.5)	7/95 (7.4)	11/127 (8.7)

Numerator denotes number of complete bones; denominator denotes total MNE for each limb bone; percentage is in parentheses

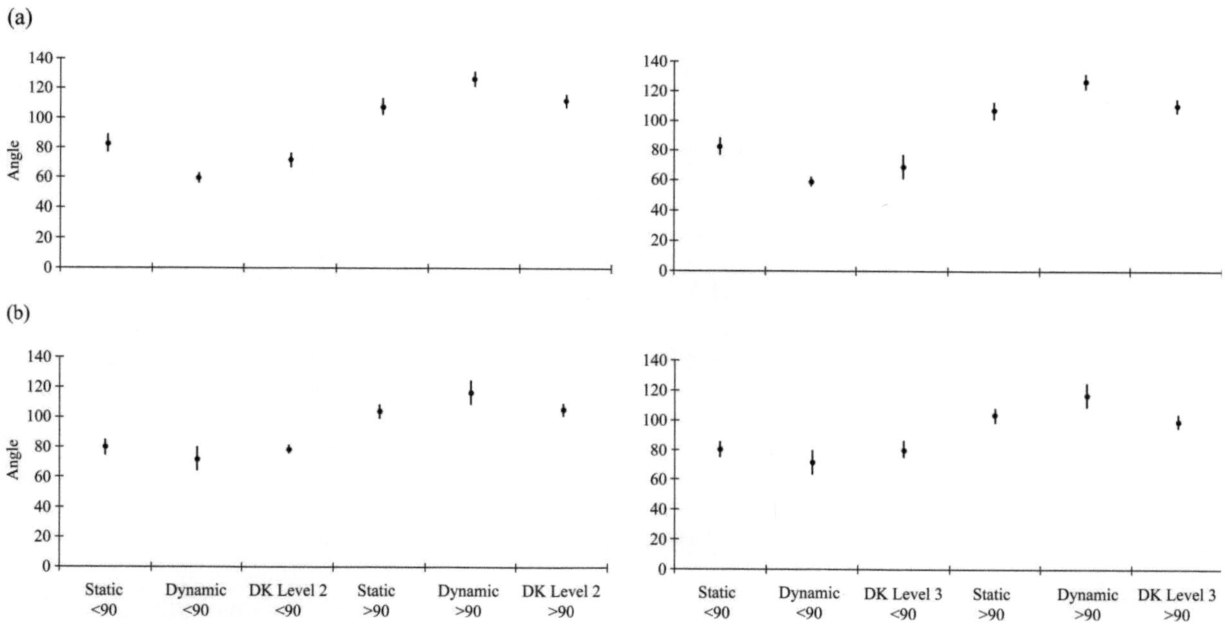

Figure 170. Mean and 95% confidence intervals of fracture angles from (A) oblique and (B) longitudinal planes on medium-sized carcasses at DK compared to experimental samples.

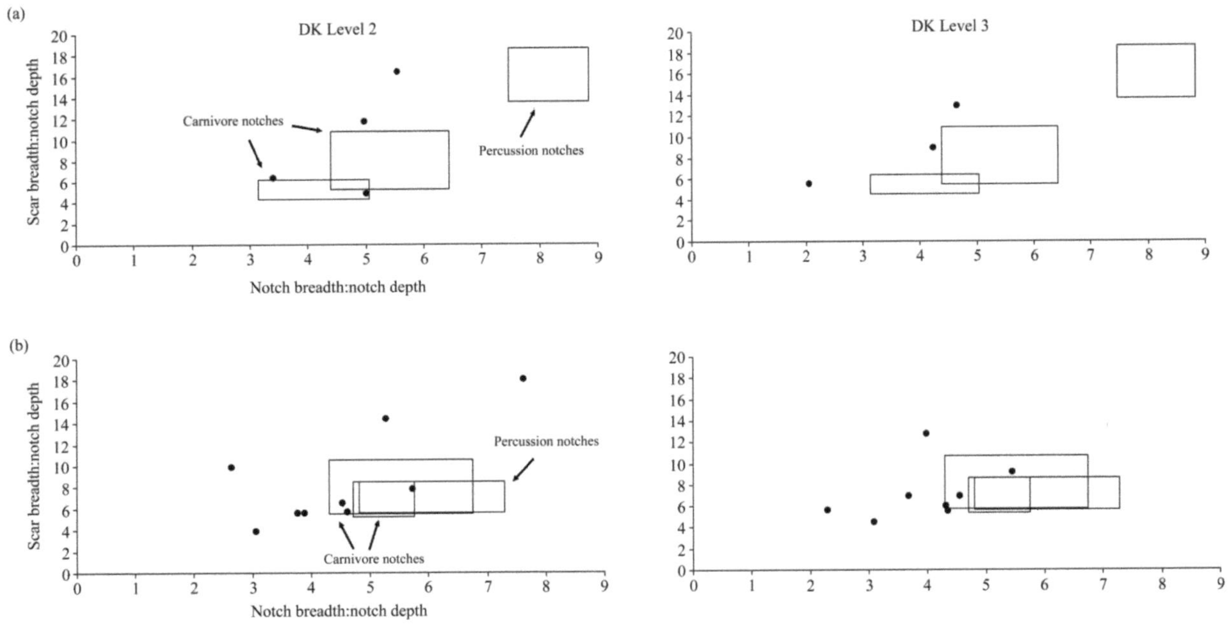

Figure 171. Notch dimensions on specimens from (A) small and (B) medium carcasses at DK. Boxes represent 95% confidence intervals of notch breadth:notch depth and scar breadth:notch depth ratios for experimental assemblages (Capaldo and Blumenchine, 1994) and a sample of notches from a hyena den in the Masaai Mara (Egeland *et al.*, unpublished data). Solid circles represent individual notches from DK.

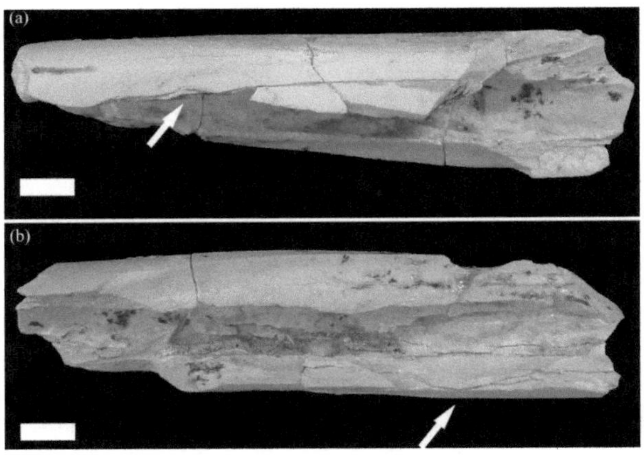

Figure 172. (A) Cortical and (B) medullary views of an incipient notch with a partially detached flake (arrows) on a Size Class 3a tibia from DK Level 2. This notch type is commonly associated with breakage by hammerstone percussion. Scale bar = 1 cm.

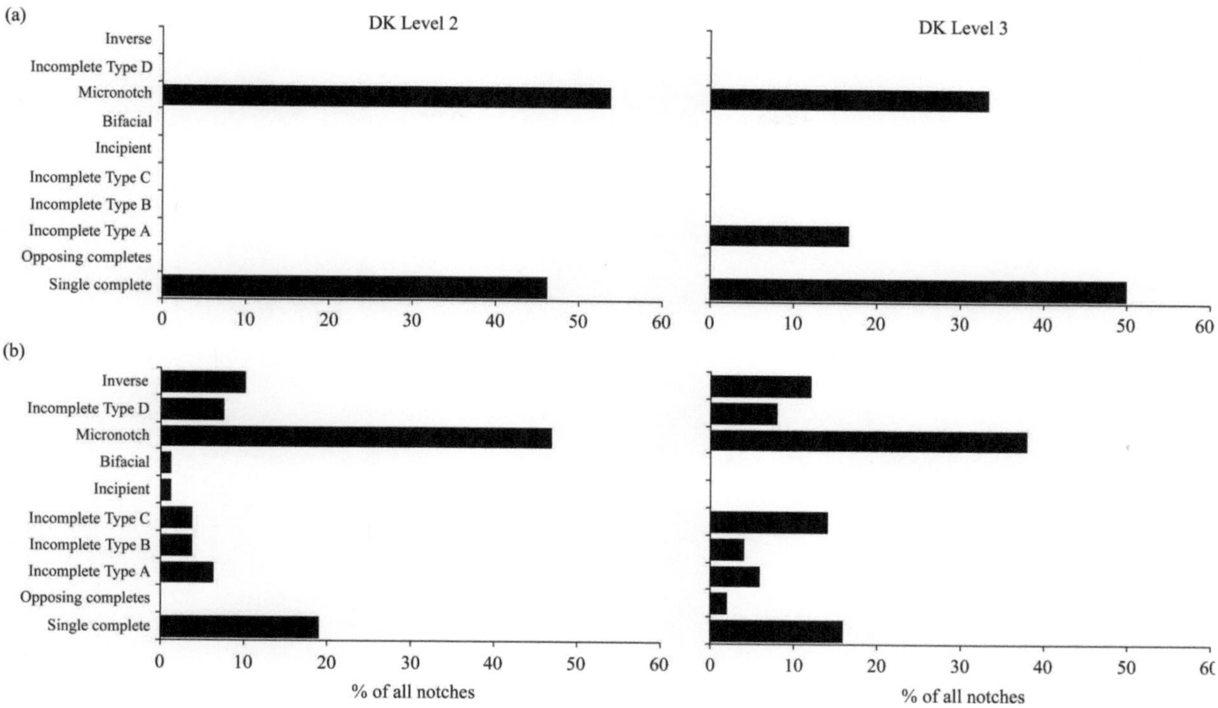

Figure 173. Incidence of notch types (modified from Capaldo and Blumenschine, 1994) from (A) small and (B) medium carcasses in the DK assemblage.

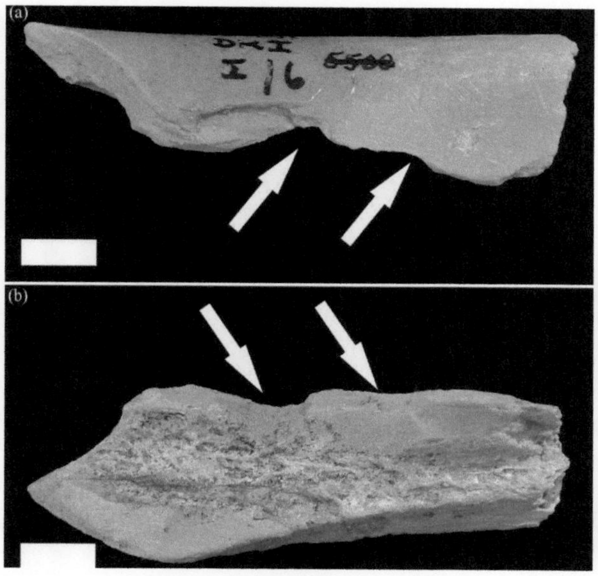

Figure 174. (A) Cortical and (B) medullary views of an Incomplete Type C notch (arrows) on a Size Class 3 intermediate (i.e., radius or tibia) limb bone from DK Level 2. These notches are common in carnivore-broken assemblages and result from adjacent tooth cusps creating multiple, closely spaced impact points. Scale bar = 1 cm.

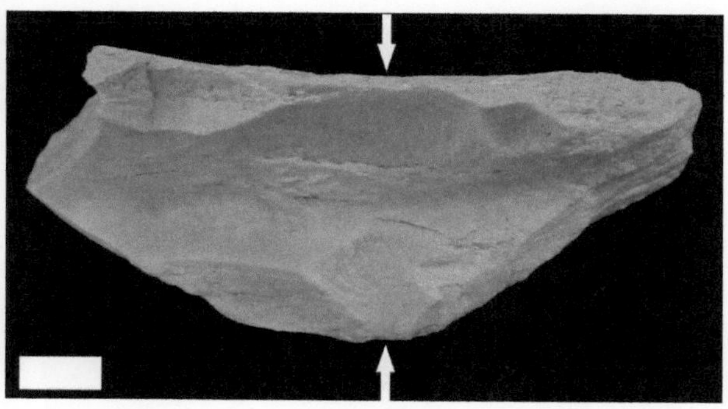

Figure 175. Cortical view of opposing notches (arrows) on a Size Class 3 femur from DK Level 2. These notches are the result of opposite loading points created largely, though not exclusively (e.g., hammerstone and anvil), by the opposing forces of the upper and lower dentition of carnivores. Scale bar = 1 cm.

Figure 176. View of stream running through broken woodlands and eventually into a lake at Nakuru National Park, Kenya. Lake Nakuru is approximately 200 yards behind the photographer. The DK site probably looked much like this ca. 1.8 million years ago (photo by C.P. Egeland).

Conclusions

M. DOMÍNGUEZ-RODRIGO, C.P. EGELAND, AND R. BARBA

The present study provides new evidence that FLK *Zinj* was created as a result of hominids selecting a specific spot (i.e., a central place) on the landscape, presumably with the presence of trees, to which they transported carcasses, and raw materials that were knapped, used, and discarded after butchery. The new results provide compelling evidence of primary access to carcasses by hominids, and therefore of meat-eating as the main goal of carcass acquisition. However, this interpretation does not apply to any other Bed I site. This is not due to the type of ecological settings where the Olduvai sites are located.

The southwestern lacustrine plain of the former Olduvai lake, where all the Bed I archaeological sites occur, seems to have been an important area of attraction for animals, including hominids. The presence of carnivores such as lions, leopards, jackals, hyenas, and saber-toothed felids (*Megantereon*, *Dinofelis*), whose remains were found in the same area, suggests that part of the herbivore biomass must have used the floodplain edaphic grasslands at least on a seasonal basis, and that carnivores must have been fairly active in the area. This situation might also have attracted hominids to the same place, probably at the same time. Competition among carnivores could have prompted carcass transportation by solitary hunters (i.e., medium-sized felids) and in some cases by larger felids to safe spots where bone remains were repeatedly accumulated. Lions, leopards, and *Dinofelis* are likely candidates for the carnivore-generated accumulations discovered in sites that were previously believed to be made by hominids. FLK North North 1 (Chapter 12) and FLK North North 3 (Chapter 14) are clearly carnivore-made accumulations. Tooth mark frequencies, a very low degree of bone fragmentation and an abundance of complete elements seem to support a non-bone cruncher as the main agent responsible for the accumulation.

FLK North North 2 (Chapter 13) has previously been reported as a bone accumulation of carnivore origin (Bunn, 1986; Potts, 1988). However, the type of carnivore agent was never specified. Bunn (1986) identified the carnivore as large and probably a scavenger (hyena), given the large size (3B) of most of the carcasses accumulated. However, the high frequency of fairly complete long limb bones, together with a frequency of tooth-marking much lower than is reported for hyena dens, suggests that hyenas were not a major taphonomic agent. Our data indicate that a medium to large-sized felid could have been responsible for the accumulation, with hyenas acting as intermittent scavengers. Given that the under- and overlying levels provide sustained support for this hypothesis, the FLK North North site could in fact have resulted from repeated use of the same space by a similar kind of carnivore. This site suggests a surrounding environment of very closed vegetation. This offers support for the possibility of felids storing carcasses in trees, which could account

M. Domínguez-Rodrigo, R. Barba, and C.P. Egeland, Deconstructing Olduvai: A Taphonomic Study of the Bed I Sites, 269–278.

for bone accumulation on the ground and for moderate modification of discarded bones by hyenas. Almost no hominid signature has been detected on the bone accumulations at FLK North North, except for two cut-marked bones. The few flakes found at FLK North North 1 and FLK North North 3 could be accounted for by brief visits by hominids, during which no activity related to carcass-processing took place. Most of the purported artifacts in both levels (Leakey, 1971; Potts, 1988) are most likely natural ecofacts rather than "manuports," as was previously argued by de la Torre and Mora (2005).

In contrast to the relatively straightforward, single-agent formational scenarios of the FLK North North faunas, many of the other Bed I sites are more complicated palimpsests, with a predominance of carnivore participation over that of hominids. Hominids seem to have been more actively involved in the formation of the various levels at FLK North than at FLK North North. At the former, they performed a slightly higher number of activities that required stone knapping and especially battering (de la Torre, 2006). None of these activities were linked to carcass-processing, let alone to the bone accumulation at FLK North 6 (Chapter 7). The same can be said of FLK North 1–2, FLK North 3, FLK North 4, and FLK North 5 (Chapters 8–10), although in these levels, some marginal contribution by hominids to the carnivore-made bone accumulation has been observed. Most bones from these sites seem to be background scatters or accumulations resulting from predation over time or from carcass-storing or -transport by felids. The co-occurrence of bones and stone tools in these levels seems to represent several independent, unrelated depositional events. The predominant activity inferred from the study of stone tools (battering) is not related to bone demarrowing, since neither percussion marks nor green fractures caused by dynamic loading have been documented. Most bones appear fairly complete, proving that carnivores also did not demarrow most of them. In summary, the superposition of different archaeological levels, in which faunal remains seem to be natural background scatters and accumulations, suggests that sites were situated in an ecologically favorable place to which herbivores, carnivores, and hominids were attracted, creating independent depositional events.

The contribution of hominids seems to be marginally more significant in the FLK North 1–2 level, though the overall taphonomic signature suggests that carnivores were responsible for most of the accumulations. This is a perfect example of a palimpsest. Given the carcass sizes that dominate the bone assemblages at FLK North 1–2, 3, and 4 (*Antidorcas* and *Parmularius*), leopards or *Dinofelis* could account for most of the assemblage. A specialized hunter of size 1–3A carcasses (following Bunn, 1982) fits well with the type of animals upon which modern leopards prey. Alternatively, leopards are not known to store size 3A carcasses in trees, and a somewhat larger felid could be responsible for the *Parmularius* portion of the assemblage on the ground. In this case *Dinofelis* may be a more likely agent, and remains of this felid have been found at the site. The same can probably be said of the DK locality, which preserves a moderate hominid signal that is (very distantly) third to FLK *Zinj* among the Bed I assemblages (FLK North 1–2 being second). The presence of many flakes, cut marks, percussion marks, and some evidence for dynamically induced bone fractures show that hominids performed carcass manipulation at DK. In addition, the co-occurrence of hominid and carnivore damage on the same specimens indicates a level of formational interdependence not seen at many Bed I sites. Nevertheless, the hominid contribution to the bone assemblage that resulted largely from natural deaths and carnivores is negligible overall.

In summary, there are four major issues to be considered:

1. Felids were accumulating carcasses in the Olduvai lacustrine environment at rates not observed today. Given the predominance of small carcasses in the FLK North levels, one could argue for leopard and *Dinofelis* as the main accumulators at FLK North 1–4 and FLK North North 3. The larger carcass size represented at FLK North North 1, FLK North North 2, and DK3 (Chapter 15) suggests that a larger felid, probably a lion, was accumulating these carcasses. This scenario suggests much higher intraspecific competition among felids during the formation of Bed I, compared to modern savannas in which half of those species are now extinct. A larger and more diverse predatory guild may have prompted several felids to transport carcasses to low-competition areas. Therefore, modern savannas could be proxies of limited value to account for trophic dynamics and the resulting bone accumulation and bone scattering in Plio-Pleistocene savannas.
2. Accordingly, the lacustrine environments at Olduvai must have been ecologically different from modern lacustrine habitats, and competition at that time must have been higher. This would have prevented closed-vegetation habitats from being extremely low-competition settings, as can be observed in modern savannas, and therefore from being a place where carcasses would have been available for scavenging by hominids. The presence of so many felids, and (in most instances) of assemblages modified by hyenas, indicates that both felids and hyenas were very active in the lacustrine environment and that chances for carcasses to remain undisturbed were very low. Therefore while passive scavenging of felid-abandoned remains could be feasible (to a limited extent) in modern savannas, it seems highly unlikely for the Olduvai Bed I ecosystem.
3. Sites formed as palimpsests through the independent deposition of bones and stone tools. Hominids visited the same places as did carnivores, but they carried out activities that were mostly unrelated to carcass-processing. The study of Olduvai Bed I sites proves that the spatial association of stone tools and bones should never been taken as direct evidence of functional relatedness. The lack of flakes (or the very small number of them) underscores the lack of a functional relationship between lithics and bones at these sites.
4. It has been traditionally argued that Olduvai Bed I sites were situated in a lacustrine floodplain not far from the lake shoreline (Leakey, 1971; Hay, 1976; Blumenschine and Masao, 1991). However, most of the space in a floodplain is open and is subjected to the same degree of intense competition on a seasonal basis. If the Olduvai sites are the result of carnivores and hominids repeatedly transporting carcasses to the same spots, those spots could not have been in open spaces. A new look at the Olduvai "manuports" yields further information in this regard. If the abundance of "manuports" is due to natural processes which created small patches of ecofacts in the landscape, as suggested by de la Torre and Mora (2005), these processes can only be observed in specific spots in modern savannas (Figure 177). In modern savannas, no stream energy can bring cobbles into the proximal lacustrine floodplain in the vicinity of the lake shoreline. In these completely flat plains, the rivers have insufficient energy to do so. We

Figure 177. Lake Manyara (Tanzania) floodplain and lacustrine forest. Arrows show connection between floodplain and alluvial plain (forested ecotone) in the upper photographs. Here large numbers of ecofacts were found resting on a clay surface, as a result of mudflow washed into the floodplain during the rainy season (lower photographs).

conducted a survey of natural processes bringing cobbles into low-energy depositional environments in the lacustrine floodplains of Natron, Ndutu, and Manyara in Tanzania. Our results showed that the only place where cobbles could be observed overlying the clayish floodplain sediments was in the ecotone-transition between the alluvial plain and the lacustrine floodplain. There, stream energy is still high enough to drag cobbles through the alluvial plain and discard them in the form of fans over the connecting lake floodplain (Figure 1).

The ecological landmark of this transitional zone is that it sustains a closed-vegetation habitat: the lacustrine forest. Today this is a very low-competition habitat (Blumenschine, 1986; Domínguez-Rodrigo, 2001). It probably was also a relatively low-competition habitat during the formation of Bed I, although with higher trophic dynamics and competition than similar modern settings because of a more diverse predatory guild. This type of vegetation cover might have prompted carnivores and hominids to obtain their prey nearby and transport it to safer consumption places located away from the lake shoreline and either inside the lacustrine forest or in the wooded transitional area between the forest and the open floodplain containing a biomass-rich grassland (Figures 178 and 179). Domínguez-Rodrigo (1994b) divided carnivore carcass transport strategies into three types:

1. Lack of transport in favor of in situ consumption, which is carried out by carnivores in a high position in the trophic order or by any carnivore in an open habitat without access to nearby shelter.
2. Peripheral transport of prey, which only occurs when the nearby shelter is less

Figure 178. Ecotone between the dense alluvial forest and the lake floodplain, next to Lake Nakuru (Kenya).

Figure 179. Ecotone between the dense alluvial forest and the lake floodplain next to Lake Nakuru (Kenya) associated with a small stream channel, usually documented in several Olduvai Bed I sites.

than 300 m away, and is a common strategy of solitary carnivores like cheetah or leopard.

3. Systematic transport, which could be carried out over short or longer distances (as is done by hyenas), resulting

in accumulations of bones from multiple carcasses.

If Bed I sites are the results of these strategies, especially of peripheral or systematic transport, open landscapes must have been very close to the place where sites were formed, suggesting that the semiclosed ecotone marked by the transition of the lacustrine forest into the open grassland might have been the chosen place for felids to accumulate carcasses.

Further proof that sites were never created in the open floodplain comes from the presence, in most of the assemblages, of moderate to high frequencies of biochemical marks caused by fungi and bacteria (Chapter 5). Fungi need certain humidity and shade conditions to grow during the prolonged time needed to develop macroscopic marks on bone surfaces (Marchiafava *et al.*, 1974; Hackett, 1981; Piepenbrink, 1984; Child, 1995; Greenlee, 1996; Sharmin *et al.*, 2003). These conditions are almost never met in open floodplain environments and can only be explained if it is accepted that sites were created in wooded environments, which would have provided the necessary shade and would have been close to water sources.

Recent results of the study of soil phytoliths and macroplant remains from Olduvai Bed II also show that wooded and bushy vegetation was important in the *loci* where sites occur (Albert *et al.*, 2006; Bamford *et al.*, 2006). The FLK site shows evidence of a landscape similar to the modern Manyara wooded savanna. The HWK site showed a more densely covered habitat than modern analogs, with an overwhelming dominance of palm trees (Figure 180).

To what extent are these paleoecological interpretations supported by other available data? Paleoenvironmental reconstructions based on macro and microfaunal remains

Figure 180. Palm tree alluvial forest in the north of Lake Eyasi, showing the ecotones marked by the presence of palm trees in the proximal section of the floodplain. This landscape could be used for a proxy for some of the areas in Bed II where phytoliths have been recovered.

(Kappelman, 1984; Plummer and Bishop, 1994; Fernández-Jalvo *et al.*, 1998; Andrews and Humphrey, 1999) suggest that lower Bed I, where the DK and FLK North North sites occur, was very moist, wooded (probably including tropical forest), and embodied a larger diversity of bovids (including a diversity of tribes) than at the top of Bed I (Kappelman, 1984; Plumer and Bishop, 1994; Fernández-Jalvo *et al.*, 1998, Andrews and Humphrey, 1999). During middle Bed I, represented by FLK and by lower levels of FLK North, there is evidence for increasing aridity towards the upper part of the sequence. FLK *Zinj*, at the bottom of middle Bed I, showed wetter conditions than FLK North, which is stratigraphically situated above it. Bovids at FLK *Zinj* included alcelaphini (such as *Parmularius*), and antilopini (such as *Antidorcas*), representing a more open environment, in smaller proportions to other bovids such as reduncini (*Kobus*) and tragelaphini, suggesting a mixed habitat, in which we can infer more woodland and perhaps more edaphic grassland than was present during the formation of upper FLK North (Plumer and Bishop, 1994; Fernández-Jalvo *et al.*, 1998). FLK North 1–4, in upper Bed I, represent more arid conditions and more open-vegetation environments. The overwhelming predominance of *Antidorcas* and *Parmularius* supports this interpretation. The populations of tragelaphines and reduncines must have been much smaller during the formation of FLK North than during that of FLK *Zinj*.

If we link the paleoecological information derived from the analysis of micro- and macromammals to the present taphonomic study, some agreement can be found. In Chapter 11, it was shown that the upper FLK levels (above FLK *Zinj*) and the FLK North levels showed more intensive stages of carnivore ravaging (DK can also be included in this group), as would be expected with an increasingly open landscape, using modern savannas as a referential framework. The least-ravaged assemblages are from FLK *Zinj* and FLK North North, at the lowest part of the stratigraphic sequence, deposited in the more closed environments that have been inferred for Bed I (see above).

This concurrence between data sets does not seem to be accidental, and its implications are far-reaching. These data support the assertion that, despite greater diversity of the carnivore and herbivore paleoguilds during Bed I times, trophic dynamics seem to have been similar overall to modern savannas; that is, open habitats underwent greater carnivore competition than closed habitats. However, some significant differences with modern savannas should be stressed. The higher diversity of bovids (including large-bodied, closed-habitat herbivores) explains the higher diversity of the carnivore paleoguild. The existence of saber-toothed felids and extinct hyaenids (*Chasmaporthetes*) suggests that predatory opportunities might have been greater than in modern savannas. It has been argued that *Chasmaporthetes* was more hunter than scavenger, although it has not been documented in Olduvai during Bed I times (Lewis, 1997). The presence of saber-toothed felids also supports the hypothesis that hunting in closed-vegetation riparian environments was more intense during Bed I times than in modern savannas. However, if inferences drawn from the taphonomic study of the Bed I sites are valid, which imply the participation of saber-toothed felids in several of the bone accumulations from this time period, it could be argued that the bone modifications observed are exactly the same as those reported for modern felids (from cheetahs to lions) (Domínguez-Rodrigo *et al.*, in press). This would suggest that carcass defleshing by these felids would also have been similar to that observed in modern felids – supporting Marean and Ehrhardt's (1995) arguments for thorough carcass defleshing by *Homotherium*. This would also support the hypothesis that saber-toothed felids did not leave more flesh on bones than modern felids do, and that experimental frameworks modeling hominid primary or secondary access to carcasses, based on observations of carcass consumption by felids, are warranted.

As further support, it is interesting to note that the bone-modifying behavior by felids repeatedly documented in the Bed I sites is not observed during Bed II times; rather, most Bed II sites seem to have been modified almost exclusively by hyenas, coinciding with the increasing disappearance of saber-toothed felids from Olduvai, and also from the East African Plio-Pleistocene record (Domínguez-Rodrigo and Egeland, in preparation). If modern felids were exclusively responsible for those taphonomic modifications attributed to felids in the Bed I sites, why would the same felid accumulations not be documented during Bed II times, when all modern felids were well-represented? A similar prey size to that of modern felids has been inferred for *Dinofelis* and *Megantereon*, the only saber-toothed felids securely identified at Olduvai during Bed I times (Lewis, 1997).

Therefore, competition in Bed I times must have been stronger and distinct from competition in modern savannas, making resources for scavenging in riparian closed-habitats scarcer than can be observed today in this type of environment. However, flesh and other nutrient availability after felid prey consumption was probably the same as is observed in modern savannas. The higher degree of ravaging of bone assemblages by hyenas documented in Bed I, and especially during Bed II times, supports the interpretation that competition in riparian environments (at least in Olduvai) was stronger than is documented in modern savannas. This supports Monahan's (1996) arguments for limited availability of scavengeable resources during the formation of Bed II. The environment in lower Bed I has no modern counterpart, which renders our inferences of carnivore competition less solid. However, middle and upper Bed I find good parallels in modern savanna environments, both in terms of taxonomic representation of animals (Plummer and Bishop, 1994), taphonomic processes affecting them (Fernández-Jalvo *et al.*, 1998; the present study), and paleobotanical evidence (Bonnefille, 1984).

Despite the evidence which points to most Olduvai Bed I bone assemblages being accumulated by carnivores, there is one site where the bone accumulation could be of almost exclusive anthropogenic origin: FLK *Zinj* (level 22). Cut mark, percussion mark, and tooth mark frequencies and distribution, together with the analysis of breakage planes and notches, show that hominids were the main accumulators of bones at the site, and that bones came from carcasses that were butchered and demarrowed by hominids. These carcasses were acquired by primary access. This strongly supports primary access to carcasses and revives the hunting hypothesis.

This book has shown that most of what we believed we knew of early human behavior was based on a poor understanding of site formation. Presently, we only have one site (FLK *Zinj*; see Chapters 5 and 6) with which we can understand human behavior in the first million years of the earliest archaeological records associated with human evolution, that is, from 2.5 Ma to 1.5 Ma. Most other sites outside Olduvai belonging to this time period either lack fauna associated with stone tools, or the preservation of faunal remains is so poor that this link cannot be confidently established.

The abundance of Pliocene sites with pristinely preserved stone tools but no faunal remains, such as at Gona (Ethiopia), combined with meager evidence of butchering traces at Gona and Bouri (Ethiopia), and archaeological records in derived positions at Omo (Ethiopia) and Senga (D.R. Congo), do not warrant claims of functional links between lithics and entire bone assemblages, regardless of their spatial association, at any Pliocene site during the first 600 Ka of the archaeological record (Harris *et al.*, 1990; Semaw *et al.*, 1997; de Heinzelin *et al.*, 1999; de la Torre, 2004; Domínguez-Rodrigo *et al.*, 2005). The present taphonomic revision of the Olduvai sites not only demonstrates that incidental associations of stone tools and bones were possible during the Plio-Pleistocene, but that they were fairly common. The paleosurfaces where stone tools were discarded at these sites were exposed for prolonged time periods, during which successive unrelated depositional events by non-hominid biotic agents might have generated the spatial association that archaeologists unearthed. A similar palimpsest effect was previously stressed by Binford (1981).

Time-averaging combined with poor preservation also prevents us from determining a functional link between stones and bones in most sites from Koobi Fora. However, at some sites such as FxJj 82, taphonomic analyses suggest a natural depositional history of bones unrelated to the deposition of stone tools (unpublished study by Domínguez-Rodrigo and Egeland). There are also cut-marked bones found not associated with stone tools (Bouri and Koobi Fora) that suggest that butchery was performed at various *loci* (Bunn, 1994; de Heinzelin *et al.*, 1999).

All this evidence is suggestive of meat and marrow consumption by hominids. However, this marginal sample, including the cut-marked bone from Gona and Bouri and the few cut-marked elements from the Olduvai sites other than FLK *Zinj,* is all we have over one million years of archaeological record. At no site older than 1.5 Ma, other than at FLK *Zinj*, have taphonomic analyses been able to detect a clear functional association of stone tools and bones that could support the interpretation that the faunal assemblage was accumulated by hominids as a result of their carcass exploitation activities. A complete taphonomic review of most Plio-Pleistocene sites, including a list of arguments supporting these claims appears in detail in Domínguez-Rodrigo (in press).

Relying only on FLK *Zinj* to understand site functionality and hominid carcass acquirement strategies is equivalent to looking at a 360° landscape through the hole of a static straw. Through that small peephole, assuming that FLK *Zinj* is representative of the main behavior exhibited by hominids at that time, we can see that passive scavenging strategies were not used by hominids, who seemed to have regular primary access to the resources they accumulated at the site. Furthermore, it also seems that if the spot was selected in a low-competition environment, the repeated transport of carcasses suggests that the site was not a near-kill location. The fauna accumulated by hominids at FLK *Zinj* included local taxa like waterbucks (*Kobus*) and open-vegetation species brought into the site from further away. The near-kill location model, discussed in Chapter 1, was built upon observations of the Hadza using a riverine environment that drew both humans and herbivores together, whereas the location of FLK *Zinj* suggests otherwise: part of the fauna accumulated did not need to visit the site or nearby area, since water would be accessible in the floodplain and in major streams pouring into it. Those animals were intentionally and repeatedly brought into the site by hominids. That suggests that the site may have had a sociosubsistential function beyond that observed in Hadza near-kill locations.

We have also seen that sites could not have been stone caches or mere refuges (also discussed in Chapter 1). FLK *Zinj* was probably selected because it afforded shelter and low competition for hominids. However, had hominids behaved like extant apes or competing carnivores, the repeated carcass transport and accumulation to the same spot would be hard to explain. The available evidence suggests that FLK *Zinj* acted as a "central place" to which hominids repeatedly brought food surpluses. This behavior was probably triggered by intentional food-sharing.

The study presented in this book took seven months of intensive work. Here we summarize the results obtained:

1. Looking at dense "site" faunal assemblages, it could be argued that the degree of competition in modern savannas (Domínguez-Rodrigo, 2001) is distinct from that in Olduvai Bed I times. A higher degree of competition forced several types of felids to accumulate carcasses in the landscape at frequencies unobservable today. This has enabled the reconstruction of competition in a way not observed in modern savannas.
2. Faunal assemblages and stone tool assemblages were deposited in the same spaces with a virtual lack of functional relationship. Independent depositional histories show that there was none of the hominid–carnivore interaction exhaustively modeled and suggested by some researchers (e.g., Selvaggio, 1994; Blumenschine, 1995; Capaldo, 1995, 1997, 1998a, 1998b).
3. Ravaging intensity has been modeled based on a set of variables, all of them experimentally shown to be sensitive to postdepositional ravaging by carnivores. This approach has enabled the differentiation of various stages of ravaging in the assemblages at each of the studied sites. This in turn has informed us about the *loci* in which these sites might have formed.
4. We have not observed a single bone specimen with marks similar to those documented by Njau and Blumenschine (2006) among modern crocodiles in our analysis of the FLK *Zinj*, nor in any of the other sites analyzed. Crocodiles may have been a important agent in bone accumulation or modification in Lowermost Bed II (and this remains to be well demonstrated, given the reported evidence of only two bones), but they certainly did not play any significant role in the formation of Bed I sites.
5. The only nearly completely anthropogenic site in Bed I (FLK *Zinj*) was formed by the repeated transport of carcasses, obtained through primary access, by hominids to a selected spot. No passive scavenging was documented in any of the Bed I sites.

In sum, the "site" approach, in which we focus on the specific taphonomic histories of individual sites, has yielded an abundant amount of information about trophic dynamics and competition from a landscape perspective, and offers clues to the correct location of sites, often have often been erroneously portrayed as formed in the middle of a barren lake floodplain. Why would carnivores and hominids accumulate carcasses in such a location? We have seen that the alluvial plain/floodplain ecotone is a much better candidate and explains perfectly the carnivore and hominid behaviors unraveled through our taphonomic analyses at these sites.

We finish as we began: by emphasizing that nothing can be understood of early sites without a complete taphonomic analysis. This deconstruction of Olduvai is intended to be positive rather than negative: it places in front of archaeologists the challenge of reconstructing it, and this is essential to know how many of our interpretive tales can be scientifically supported, rather than being just-so stories. Archaeologists have been less rigorous in the study of other African sites, believing that Olduvai had provided the clues to reconstruct much of the picture. That is simply not true: our interpretations now stand naked. We know less of the first million years of human behavior with an archaeological record (i.e., 2.5–1.5 Ma) than was previously thought. Every single site from this time period with good bone preservation should therefore be considered unique and treated with utmost care. This is one of the main challenges for archaeologists in the twenty-first century.

References

Albert, R. M., Bamford, M. K., Cabanes, D. 2006. Taphonomy of phytoliths and macroplants in different soils from Olduvai Gorge (Tanzania) and the application to Plio-Pleistocene palaeoanthropological samples. Quaternary International 148, 78–94.

Alcalá, L. 1994. Macromamíferos neógenos de la fosa de Alfambra-Teruel. Monografías del Instituto de Estudios Tulolenses, Teruel.

Alcántara, V., Barba, R., Barral, J., Crespo, A., Eiriz, A., Falquina, A., Herrero, S., Ibarra, A., Megías, M., Pérez, M., Pérez, V., Rolland, J., Yravedra, J., Vidal, A., Domínguez-Rodrigo, M. 2006. Determinación de procesos de fractura sobre huesos frescos: un sistema de análisis de los ángulos de los planos de fracturación como discriminador de agentes bióticos. Trabajos de Prehistoria 63(1), 25–38.

Andrews, P., Cook, J. 1985. Natural modifications to bones in a temperate setting. Man 20, 675–691.

Andrews, P., Fernández-Jalvo, Y. 1997. Surface modifications of the Sima de los Huesos fossil humans. Journal of Human Evolution 33, 191–217.

Andrews, P., Groves, C. P., Horne, J. F. M. 1975. The ecology of the lower Tana River flood plain. Journal of East African Natural History Society 151, 1–31.

Andrews, P., Humphrey, L. 1999. African Miocene environments and the transtition to early hominines. In Bromage, T., Schrenk, F. (eds.) African Biogeography, Climate Change, and Early Hominid Evolution. Oxford University Press, Oxford, pp. 282–300.

Auffenberg, W. 1981. The fossil turtles of Olduvai Gorge, Tanzania, Africa. Copeia 3, 509–522.

Auguste, P. 1991. Chasse ou charognage au Paléolithique moyen: l'apport du gisement de Biache-Saint-Vaast (Pas de Calais). Bulletin de la Societé Préhistorique Française, 88, 1–167.

Auguste, P. 1995. Chasse ou charognage au Paléolithique moyen: l'apport du gisement de Biache-Saint-Vaast (Pas de Calais). Bulletin de la Societé Préhistorique Française 92, 155–167.

Bamford, M. K., Albert, R. M., Cabanes, D. 2006. Plio-Pleistocene macroplant fossil remains and phytoliths from lowermost Bed II in the eastern pleolake margin of Olduvai Gorge. Quaternary International 148, 95–112.

Bar-Yosef, O. 1989. Upper Pleistocene cultural stratigraphy in southwest Asia. In Trinkaus, E. (ed.) The Emergence of Modern Humans. Biocultural Adaptations in the Later Pleistocene. Cambridge University Press, Cambridge, pp. 159–180.

Bar-Yosef, O. 1994. The Lower Paleolithic of the Middle East. Journal of World Prehistory 8, 211–265.

Barba, R., Domínguez-Rodrigo, M. 2005. The taphonomic relevance of the analysis of long limb bone shaft features and their application to implement element identification: study of bone thickness and the morphology of the medullary cavity. Journal of Taphonomy 3, 111–124.

Barba, R., Domínguez-Rodrigo, M. (in press). Nueva aproximación al estudio de las marcas de corte: definición de "zonas calientes" como indicador de un consumo inicial de las carcasas por parte de los homínidos. Aplicación al conjunto de FLK Zinj (Olduvai). Trabajos de Prehistoria.

Bartram, L. E. 1993. An ethnoarchaeological analysis of Kua San (Bostwana) bone food refuse, Ph.D. Thesis. University of Wisconsin, Madison.

Bartram, L. E., Kroll, E. M., Bunn, H. T. 1991. Variability in camp structure and bone refuse patterning at Kua San hunter-gatherer camps. In Kroll, E., Price, D. (eds.) The Interpretation of Archaeological Spatial Patterning. Plenum Press, New York, pp. 77–148.

Bartram, L. E., Marean, C. W. 1999. Explaining the "Klasies Pattern": Kua ethnoarchaeology, the

Die Kelders Middle Stone Age archaeofauna, long bone fragmentation and carnivore ravaging. Journal of Archaeological Science 26, 9–29.

Behrensmeyer, A. K. 1978. Taphonomic and ecologic information from bone weathering. Paleobiology 4, 150–162.

Behrensmeyer, A. K. 1983. Patterns of natural bone distribution on recent land surfaces: implications for archaeological site formation. In Clutton-Brock, J., Grigson, C. (eds.) Animals and Archaeology: Hunters and their Prey, B.A.R. International Series, 163, pp. 93–106.

Behrensmeyer, A. K. 2005. Changes over three decades in skeletal part survival and bone modification in the Amboseli ecosystem, southern Kenya. In Pickering, T. R., Toth, N., and Shick, K. (eds.) African Taphonomy: A Tribute to the Career of C.K. Brain. CRAFT Press, Bloomington, Indiana.

Behrensmeyer, A. K., Boaz, D. 1980. The recent bones of Amboseli Park, Kenya, in relation to East African paleoecology. In Behrensmeyer, K. and Hill, A. (eds.) Fossils in the Making. University of Chicago Press, Chicago, pp. 72–92.

Behrensmeyer, A. K., Todd, N. E., Potts, R., McBrinn, G. E. 1997. Late Pliocene faunal turnover in the Turkana Basin, Kenya and Ethiopia. Science 278, 1589–1594.

Bender, M. M. 1971. Variations in the $^{13}C/^{12}C$ ratios of plants in relation to the pathway of photosynthetic carbon dioxide fixation. Phytochemistry 10, 1239–1245.

Binford, L. R. 1978. Nunamiut Ethnoarchaeology. Academic Press, New York.

Binford, L. R. 1981. Bones: Ancient Men, Modern Myths. Academic Press, New York.

Binford, L. R. 1984. Faunal Remains from Klasies River Mouth. Academic Press, New York.

Binford, L. R. 1985. Human ancestors: changing views of their behavior. Journal of Anthropological Archaeology 4, 292–327.

Binford, L. R. 1986. Reply to Bunn and Kroll. Current Anthropology 27, 444– 446.

Binford, L. R. 1988a. Fact and fiction about the Zinjanthropus floor: data, arguments and interpretations. Current Anthropology 29, 123–135.

Binford, L. R. 1988b. The hunting hypothesis, archaeological methods and the past. Yearbook of Physical Anthropology 30, 1–9.

Binford, L. R., Ho, C. K. 1985. Taphonomy at a distance: Zhoukoudian, "the cave-home of Beijing Man"? Current Anthropology 26, 413–439.

Binford, L. R., Mills, L. G. L., Stone, N. M. 1988. Hyena scavenging behavior and its implications for the interpretation of faunal assemblages from FLK 22 (the Zinj floor) at Olduvai Gorge. Journal of Anthropological Archaeology 7, 99–135.

Blasco Sancho, F. 1995. Hombres y Fieras. Estudio Zooarqueológico y Tafonómico del Yacimiento del Paleolítico Medio de la Cueva de Gabasa 1 (Huesca). Monografías de la Universidad de Zaragoza, Zaragoza.

Blumenschine, R. J. 1986. Early Hominid Scavenging Opportunities. Implications of Carcass Availability in the Serengeti and Ngorongoro Ecosystems. B.A.R. International Series, 283, Oxford.

Blumenschine, R. J. 1987. Characteristics of an early hominid scavenging niche. Current Anthropology 28, 383–407.

Blumenschine, R. J. 1988. An experimental model of the timing of hominid and carnivore influence on archaeological bone assemblages. Journal of Archaeological Science 15, 483–502.

Blumenschine, R. J. 1989. A landscape taphonomic model of the scale of prehistoric scavenging opportunities. Journal of Human Evolution 18, 345–371.

Blumenschine, R. J. 1991. Hominid carnivory and foraging strategies, and the socio-economic function of early archaeological sites. Philosophical Transactions of the Royal Society (London) 334, 211–221.

Blumenschine, R. J. 1995. Percussion marks, tooth marks and the experimental determinations of the timing of hominid and carnivore access to long bones at FLK Zinjanthropus, Olduvai Gorge, Tanzania. Journal of Human Evolution 29, 21–51.

Blumenschine, R. J., Bunn, H. T. 1987. On theoretical framework and tests of early hominid meat and marrow adquisition. A reply to Shipman. American Anthropologist 89, 444 – 448.

Blumenschine, R. J., Cavallo, J. A., Capaldo, S. D. 1994. Competition for carcasses and early hominid behavioral ecology: a case study and a conceptual framework. Journal of Human Evolution 27, 197–213.

Blumenschine, R. J., Madrigal, T. C. 1993. Long bone marrow yields of some African ungulates. Journal of Archaeological Science 20, 555–587.

Blumenschine, R. J., Marean, C. W. 1993. A carnivore's view of archaeological bone assemblages. In Hudson, J. (ed.) From Bones to Behavior: Ethnoarchaeological and Experimental Contributions to the Interpretations of Faunal Remains. Southern Illinois University, Illinois, pp. 271–300.

Blumenschine, R. J., Marean, C. W., Capaldo, S. D. 1996. Blind tests of inter-analyst correspondence and accuracy in the identification of cut marks, percussion marks, and carnivore tooth marks on bone surfaces. Journal of Archaeological Science 23, 493–507.

Blumenschine, R. J., Masao, F. T. 1991. Living sites at Olduvai Gorge, Tanzania? Preliminare lanscape archaeology results in the basal Bed II lake margin zone. Journal of Human Evolution 21, 451–462.

Blumenschine, R. J., Masao, F. 1995. Landscape ecology and hominid land use in the lowermost Bed II Olduvai basin. Proposal submitted to the National Science Foundation.

Blumenschine, R. J., Peters, C. R. 1998. Archaeological predictions for hominid land use in the paleo-Olduvai Basin, Tanzania, during lowermost Bed II times. Journal of Human Evolution 34, 565–607.

Blumenschine, R. J., Peters, C. R., Capaldo, S. D., Andrews, P., Njau, J. K., Pobiner, B. L. in press. Vertebrate taphonomic perspectives on Oldowan hominin land use in the Plio-Pleistocene Olduvai basin, Tanzania. In Pickering, T., Schick, K., and Toth, N. (eds.) African Taphonomy: A Tribute to the Career of C. K. "Bob" Brain. CRAFT Press (Indiana University), Bloomington, Indiana.

Blumenschine, R.J., Peters, C.R., Masao, F.T., Clarke, R.J., Deino, A.L., Hay, R.L., Swisher III, C.C., Stanistreet, I.G., Ashely, G.M., McHenry, L., J., Sikes, N.E., Van der Merwe, N.J., Tactikos, J.C., Cushing, A.E., Deocampo, D.M., Njau, J.K., Ebert, J.I., 2003. Late Pliocene *Homo* and hominid land use from Western Olduvai Gorge, Tanzania. Science 299, 1217–1221.

Blumenschine, R. J., Selvaggio, M. M. 1988. Percussion marks on bone surfaces as a new diagnostic of hominid behavior. Nature 333, 763–765.

Blumenschine, R. J., Selvaggio, M. M. 1991. On the marks of marrow bone processing by hammerstones and hyenas: their anatomical patterning and archaeological implications. In Clark, J. D. (ed.) Cultural Beginnings: Approaches to Understanding Early Hominid Life Ways in the African Savanna. R. Habelt, GMBH, Bonn, pp. 17–32.

Bobe, R., Behrensmeyer, A. K. 2004. The expansion of grassland ecosystems in Africa in relation to mammalian evolution and the origin of the genus *Homo*. Palaeogeography Palaeoclimatology Palaeoecology 207, 399–420.

Bobe, R., Behrensmeyer, A. K., Chapman, R. E. 2002. Faunal change, environmental variability and late Pliocene hominin evolution. Journal of Human Evolution 42, 475–497.

Bodenhorn, B. 2000. It's good to know who your relatives are but we were taught to share with everybody: shares and sharing among Inupiaq households. In Wenzel, G. W., Hovelsrud-Broad, G., Kishigami, N. (eds.) The Economy of Sharing: Resource Allocation and Modern Hunter-Gatherers. Senri Ethnological Studies 53, pp. 27–60.

Boesch, C., Boesch, H. 1984. Mental map in wild chimpanzees: an analysis of hammer transports for nut cracking. Primates 25, 160–170.

Bonnefille, R. 1984. Palynological research at Olduvai Gorge. National Geographic Society Research Papers 17, 227–243.

Brain, C. K. 1967. Hottentot food remains and their bearing on the interpretation of fossil bone assemblages. Scientific Papers of the Namib Desert Research Station 32, 1–11.

Brain, C. K. 1969. The probable role of leopards as predators of the Swartkrans Australopithecines. South African Archaeological Bulletin 24, 170–171.

Brain, C. K. 1981. The Hunters or the Hunted? Chicago University Press, Chicago.

Brugal, J. P. 1992. Eco-ethologie des premiers hominid'es en Afrique: Les activités de subsistance. Bulletin et Memoire de la Societé d'Anthropologie de Paris 4, 143–166.

Brugal, J. P., Fosse, P., Guadelli, J. L. 1997. Comparative study of bone assemblages made by recent and Pleistocene hyenids. Paper presented at the Eight Meeting of Working Group 1 on Bone Modification, Hot Springs, South Dakota, USA. September 1993, pp. 158–187.

Bunn, H. T. 1981. Archaeological evidence for meat-eating by Plio-Pleistocene hominids from Koobi Fora, Kenya. Nature 291, 574–577.

Bunn, H. T. 1982. Meat-eating and human evolution: studies on the diet and subsistence patterns of Plio-Pleistocene hominids in East Africa, Ph.D. Dissertation. University of California, Berkeley.

Bunn, H. T. 1983a. Evidence on the diet and subsistence patterns of Plio-Pleistocene hominids at Koobi Fora, Kenya, and at Olduvai Gorge, Tanzania. In Clutton-Brock, J., Grigson, C. (eds.) Animals and Archaeology: Hunters and their Prey. B.A.R. International Series, 163, pp. 21–30.

Bunn, H. T. 1983b. Comparative analysis of modern bone assemblages from a San hunter-gatherer camp in the Kalahari Desert, Bostwana, and froms a spotted hyena den near Nairobi, Kenya. In Clutton-Brock, J., Grigson, C. (eds.) Animals and Archaeology: Hunters and their Prey. B.A.R. International Series, 163, pp. 143–148.

Bunn, H. T. 1986. Patterns on skeletal representation and hominid subsistence activities at Olduvai Gorge, Tanzania, and Koobi Fora, Kenya. Journal of Human Evolution 15, 673–690.

Bunn, H. T. 1991. A taphonomic perspective on the archaeology of human origins. Annual Review of Anthopology 20, 433–467.

Bunn, H. T. 1994. Early Pleistocene hominid foraging strategies along the ancestral Omo River at Koobi Fora, Kenya. Journal of Human Evolution 27, 247–266.

Bunn, H. T. 1995. Reply to Tappen. Current Anthropology 36, 250–251.

Bunn, H. T. 1996. Reply to Rose and Marshall. Current Anthropology 37, 321–323.

Bunn, H. T. 2001. Hunting, power scavenging, and butchering by Hadza foragers and by Plio-Pleistocene *Homo*. In Stanford, C. B., Bunn, H. T. (eds.) Meat-Eating and Human Evolution. Oxford University Press, Oxford, pp. 199–218.

Bunn, H. T., Ezzo, J. A. 1993. Hunting and scavenging by Plio-Pleistocene hominids: nutritional constraints, archaeological patterns, and behavioural implications. Journal of Archaeological Science 20, 365–398.

Bunn, H. T., Kroll, E. M. 1986. Systematic butchery by Plio-Pleistocene hominids at Olduvai Gorge, Tanzania. Current Anthropology 27, 431–452.

Bunn, H. T., Kroll, E. M. 1988. A reply to Binford. Current Anthropology 29, 123–149.

Bunn, H. T., Kroll, E. M., Bartram, L. E. 1988. Variability in bone assemblage formation from Hadza hunting, scavenging and carcass processing. Journal of Anthropological Archaeology 7, 412–57.

Bunn, H. T., Kroll, E. M., Bartram, L. E. 1991. Bone distribution on a modern East African landscape and its archaeological implications. In Clark, J. D. (ed.) Cultural Beginnings: Approaches to Understanding Early Hominid Life Ways in the African Savanna. U.I.S.P. Monographien Band, 19, 33–54.

Capaldo, S. D. 1995. Inferring hominid and carnivore behavior from dual-patterned archaeological assemblages, Ph.D. Thesis. Rutgers University, New Brunswick.

Capaldo, S. D. 1997. Experimental determinations of carcass processing by Plio-Pleistocene hominids and carnivores at FLK 22 (Zinjanthropus), Olduvai Gorge, Tanzania. Journal of Human Evolution 33, 555–597.

Capaldo, S. D. 1998a. Methods, marks and models for inferring hominid and carnivore behavior. Journal of Human Evolution 35, 323–326.

Capaldo, S. D. 1998b. Simulating the formation of dual-patterned archaeofaunal assemblages with experimental control simples. Journal of Archaeological Science 35, 311–330.

Capaldo, S. D., Blumenschine, R. J. 1994. A quantitative diagnosis of notches made by hammerstone percussion and carnivore gnawing in bovid long bones. American Antiquity 59, 724–748.

Capaldo, S. D., Peters, C. R. 1995. Skeletal inventories from wildebeest drownings at Lakes Masek and Ndutu in the Serengeti ecosystem of Tanzania. Journal of Archaeological Science 22, 385–408.

Cavallo, J. A. 1998. A re-examination of Isaac's central-place foraging hypothesis, Ph.D. Thesis. Rutgers University, New Brunswick.

Cavallo, J. A., Blumenschine, R. J. 1989. Tree-stored leopard kills: expanding the hominid scavenging niche. Journal of Human Evolution 18, 393–399.

Cerling, T. E. 1992. Development of grasslands and savannas in East Africa during the Neogene. Palaeogeography, Palaeoclimatology, Palaeoecology 97, 241–247.

Cerling, T. E., Bowman, J. R., Oneil, J. R. 1988. An isotopic study of a fluvial lacustrine sequence: the Plio-Pleistocene Koobi Fora sequence, East Africa. Palaeogeography Palaeoclimatology Palaeoecology 63, 335–356.

Cerling, T. E., Harris, J. M., MacFadden, B. J., Leakey, M. G., Quade, J., Eisenmann, V., Ehleringer, J. R., 1997. Global vegetation change through the Miocene/Pliocene boundary. Nature 389, 153–158.

Cerling, T. E., Hay, R. L. 1986. An isotopic study of paleosol carbonates from Olduvai Gorge. Quaternary Research 25, 63–78.

Cerling, T. E., Wang, Y., Quade, J. 1993. Expansion of C^4 ecosystems as an indicator of global ecological change in the late Miocene. Nature 361, 344–345.

Child, A. M. 1995. Towards an understanding of the microbial decomposition of archaeological bone in the burial environment. Journal of Archaeological Science 22, 165–174.

Child, A. M., Gillard, R. D., Pollard, A. M. 1993. Microbiallyinduced promotion of amino acid racemization in bone: isolation of the organisms and detection of their enzymes. Journal of Archaeological Science 20, 159–168.

Clark, J. D. 1970. The Prehistory of Africa. Praeger Publishers, New YorkCleghorn, N., Marean, C. W. 2004. Distinguishing selective transport and in situ attrition: a critical review of analytical approaches. Journal of Taphonomy 2, 43–68.

Cordain, L., Miller, J. B., Eaton, S. B., Mann, N., Holt, H. A., Speth, J. 2000. Plant-animal subsistence ratios and macronutrient energy estimations in worldwide hunter-gatherer diets. American Journal of Clinical Nutrition 71, 681–692.

Creel, S., Creel, N. M. 1996. Limitation of African wild dogs by competition with larger carnivores. Conservation Biology 10, 526–538.

Cushing, A. E. 2002. The landscape zooarchaeology and paleontology of Plio-Pleistocene Olduvai, Tanzania, and their implications for early hominid ecology, Ph.D. Dissertation. Anthropology Department, Rutgers University, New Brunswick.

Cutler, A. H., Behrensmeyer, A. K., Chapman, R. E. 1999. Environmental information in a recent bone assemblage: roles of taphonomic processes and ecological change. In Martin, R., Goldstein, S., Patterson, R. T. (eds.) Fossil Taphonomy: Paleoenvironmental Reconstruction and Environmental Assessment. Palaeogeography, Palaeoclimatology, Palaeoecology 149, 359–372.

Damas, D. 1972. Central Eskimo systems of food sharing. Ethnology 11, 220–240.

Darwin, C. 1871. The Descent of Man. Random House, New York.

DeGusta, D., Beyene Y., Asfaw, B., Gilbert, H., White, T. 2003. The oldest cutmarked bones: Pliocene hominidfaunalexploitationat Bouri, Ethiopia. Paper presented at the Paleoanthropological Society Meetings, Tempe, Arizona.

Deocampo, D. M., Blumenschine, R. J., Ashely, G. M. 2002. Wetland diagenesis and traces of early hominids, Olduvai Gorge, Tanzania. Quaternary Research 57, 271–281.

Domínguez-Rodrigo, 1993. La analítica espacial como herramienta tafonómica: aplicación al entorno de una charca africana. Procesos postdeposicionales y Arqueología espacial 16–17, 82–111.

Domínguez-Rodrigo, M. 1994a. El Origen del Comportamiento Humano. Tipo, Madrid.

Domínguez-Rodrigo, M. 1994b. Dinámica trófica, estrategias de consumo y alteraciones óseas en la sabana africana: resumen de un proyecto de investigación etoarqueológico (1991–1993). Trabajos de Prehistoria 51, 15–37.

Domínguez-Rodrigo, M. 1996. A landscape study of bone conservation in the Galana and Kulalu (Kenya) ecosystem. Origini 20, 17–38.

Domínguez-Rodrigo, M. 1997a. Meat-eating by early hominids at the FLK 22 Zinjanthropus site, Olduvai Gorge, Tanzania: an experimental approach using cut mark data. Journal of Human Evolution 33, 669–690.

Domínguez-Rodrigo, M. 1997b. A reassessment of the study of cut mark patterns to infer hominid manipulation of fleshed carcasses at the FLK Zinj 22 site, Olduvai Gorge, Tanzania. Trabajos de Prehistoria 54, 29–42.

Domínguez-Rodrigo, M. 1997c. Testing meat-eating in early hominids: analysis of cut-marking processes on defleshed carcasses. Human Evolution 12, 169–182.

Domínguez-Rodrigo, M. 1999a. Flesh availability and bone modification in carcasses consumed by lions. Palaeogeography, Palaeoclimatology and Palaeoecology 149, 373–388.

Domínguez-Rodrigo, M. 1999b. Meat eating and carcass procurement by hominids at the FLK Zinj 22 site, Olduvai Gorge, Tanzania: a new experimental approach to the old hunting-versus-scavenging debate. In Ullrich, H. (ed.) Lifestyles and Survival Strategies in Pliocene and Pleistocene Hominids. Edition Archaea, Schwelm, Germany, pp. 89–111.

Domínguez-Rodrigo, M. 2001. A study of carnivore competition in riparian and open habitats of modern savannas and its implications for hominid behavioral modelling. Journal of Human Evolution 40, 77–98.

Domínguez-Rodrigo, M. 2002. Hunting and scavenging by early humans: the state of the debate. Journal of World Prehistory 16, 1–54.

Domínguez-Rodrigo, M. 2003. Bone surface modifications, power scavenging and the "display" model at early archaeological sites: a critical review. Journal of Human Evolution 45, 411–416.

Domínguez-Rodrigo, M. in press. Are all Oldowan sites palimpsests? If so, what can they tell us about hominid carnivory? In Hovers, E., Braun, D. (eds.) Interdiscipinary Approaches to Understanding the Oldowan, Springer, New York.

Domínguez-Rodrigo, M., Barba, R. 2005. A study of cut marks on small-sized carcasses and its application to the study of cut-marked bones from small mammals at the FLK Zinj site. Journal of Taphonomy 2–3, 111–124.

Domínguez-Rodrigo, M., Barba, R. 2006. New estimates of tooth marks and percussion marks from FLK Zinj, Olduvai Gorge (Tanzania): the carnivore-hominid-carnivore hypothesis falsified. Journal of Human Evolution 50, 170–194.

Domínguez-Rodrigo, M., Egeland, C. P. in press. Hominid-made bone accumulations at Olduvai

Bed Ii archaeological sites? A taphonomic study of HWK EAST 3–5, MNK MAIN, EF-HR, FC WEST, SHK and BK.

Domínguez-Rodrigo, M., Egeland, C. P., Pickering, T. R. in press. Models of passive scavenging by early hominids: problems arising from equifinality in carnivore tooth mark frequencies and the extended concept of archaeological palimpsests. In Pickering, T. R., Toth, N., Shick, K. (eds.) African Taphonomy: A Tribute to the Career of C.K. Brain. CRAFT Press, Bloomington, Indiana.

Domínguez-Rodrigo, M., Luque, L., Alcalá, L., de la Torre Sáinz, I., Mora, R., Serrallonga, Medina, V. 2002. The ST Site Complex at Peninj, West Lake Natron, Tanzania: implications for early hominids behavioural models. Journal of Archaeological Science 29, 639–665.

Domínguez-Rodrigo, M., Martí Lezana, R. 1996. Estudio etnoarqueológico de un campamento temporal Ndorobo (Maasai) en Kulalu (Kenia). Trabajos de Prehistoria 53, 131–143.

Domínguez-Rodrigo, M., Pickering, T. R. 2003. Early hominids, hunting and scavenging: a summary of the discussion. Evolutionary Anthropology 12, 275–282.

Domínguez-Rodrigo, M., Pickering, T. R., Semaw, S., Rogers, M. 2005. Cutmarked bones from Pliocene archaeological sites at Gona, Afar, Ethiopia: implications for the function of the world's oldest stone tools. Journal of Human Evolution 48, 109–121.

Domínguez-Rodrigo, M., Piqueras, A. 2003. The use of tooth pits to identify carnivore taxa in tooth-marked archaeofaunas and their relevance to reconstruct hominid carcsass processing behaviours. Journal of Archaeological Science 30, 1385–1391.

Dube, H. C. 2005. An introduction to fungi. Vikas, New Delhi.

Egeland, C. P. 2003. Carcass processing intensity and cutmark creation: an experimental approach. Plains Anthropologist 184, 39–51.

Egeland, C. P., Byerly, R. 2005. Application of return rates to large mammal butchery and transport among hunter-gatherers and its implications for Plio-Pleistocene hominid carcass forgaring and site use. Journal of Taphonomy 3(3), 135–157.

Egeland, C. P., Domínguez-Rodrigo, M and Kerbis J. In preparation. Patterns of tooth-marking and bone breakage in a hyena den from the Maasai Mara (Kenya).

Egeland, C. P., Pickering, T. R., Dominguez-Rodrigo, M., Brain, C. K. 2004. Disentangling stone age palimpsests: determining the functional independent of hominid- and carnivore-derived portions of archaeofaunas. Journal of Human Evolution 47, 343–357.

Faith, J. T., Behrensmeyer, A. K. 2006. Changing patterns of carnivore modification in a landscape bone assemblage, Amboseli Park, Kenya. Journal of Archaeological Science 12, 1718–1733.

Fernández-López, S. 2000. Temas de Tafonomía. Departamento de Paleontología, Facultad de Ciencias Geológicas.

Fernández-Jalvo, Y., Denys, C., Andrews, P., Williams, T, Dauphin, Y., Humphreys, L. 1998. Taphonomy and palaeoecology of Olduvai Bed-I (Pleistocene, Tanzania). Journal of Human Evolution 34, 137–172.

Fisher, J. W. 1995. Bone surface modification in zooarchaeology. Journal of Archaeological Method and Theory 1, 7–65.

Fosse, Ph. 1995. Les herbivores dans les gisements paléolithiques en grotte: Proies des carnivores ou gibier des hommes? Préhistoire et Anthropologie Méditerranéennes 4, 27–39.

Fosse, Ph. 1996. La grotte 1 de Lunel-Viel (Hérault, France): répaire des hyènes du Pléistocéne moyen. Paléo 8, 47–81.

Fosse, Ph., Brugal, J. L., Guadelli, P., Tournepiche, J. F. 1998. Les repaires d'Hyènes des cavernes en Europe occidentale: Présentation et comparaisons de quelques assemblages osseux. Economie Préhistorique: les comportements de subsistance au Paléolithique. XVIIIème Rencontres Internationales d'archéologie et d'Histoire d'Antibes. Éditions APDCA, Sophia Antipolis, 43–61. Gamble, C. 1986. The Paleolithic Settlement of Europe. Cambridge University Press, Cambridge.

Gamble, C. 1999. The Paleolithic Societies of Europe. Cambridge World Archaeology, Cambridge.

Gargett, R., Hayden, B. 1991. Site structure, kinship and sharing in aboriginal Australia: implications for archaeology. In Kroll, E. M., Price, T. D. (eds.) The Interpretation of Archaeological Spatial Patterning. Plenum Press, New York, pp. 11–32.

Gentry, A. W. and Gentry, A. 1978a. Fossil bovidae (Mammalia) from Olduvai Gorge, Tanzania. London, Bulletin of the British Museum 29, 29: 289–446.

Gentry, A. W. and Gentry, A. 1978b. Fossil bovidae (Mammalia) from Olduvai Gorge, Tanzania. London, Bulletin of the British Museum 29, 30: 1–83.

Gifford-Gonzalez, D. P. 1991. Bones are not enough: analogues, knowledge, and interpretive strategies in zooarcaheology. Journal of Anthropological Archaeology 10, 215–254.

Gould, S. J. 1965. Is uniformitarianism necessary? American Journal of Science 236, 223–228.

Gould, S. J. 1979. Agazziz's marginalia in Lyell's principles, or the perils of uniformity and the ambiguity of heroes. Studies in the History of Biology 3, 119–138.

Gould, R. A. 1980. Living Archaeology. Cambridge University Press, Cambridge.

Grayson, D. K. 1984. Quantitative Zooarchaeology: Topics in the Analysis of Archaeological Faunas. Academic Press, Orlando.

Greenlee, D. M. 1996. An electron microprobe evaluation of diagenetic alteration in archaeological bone. In Orna, M. V. (ed.) Archaeological Chemistry. American Chemical Society Press, Washington, D.C., pp. 334–354.

Hackett, C. J. 1981. Microscopical focal destruction (tunnels) in exhumed human bones. Medicine, Science and the Law 21, 243–265.

Haneberg, W. C. 1983. A paradigmatic analysis of Darwin's use of uniformitarianism in "The Origin of Species." The Compass of Sigma Gamma Epsilon 60, 89–94.

Harris, J. W. K., Williamson, P. G., Morris, P. J., de Heinzelin, J., Veniers, J., Helgren, D., Bellomo, R. V., Laden, G., Spang, T. W., Stewart, K., Tappen, M. J. 1990. Archaeology of the Lusso beds. In Boaz, N. T. (ed.) Evolution of the Environments and Hominidae in the African Western Rift Valley, Museum of Natural History, Martinsville, Virginia, pp. 237–272.

Hay, R. 1963. Stratigraphy of Bed I through IV, Olduvai Gorge, Tanganika. Science 139, 829–833.

Hay, R. 1971. Geologic background of Beds I and II. In Leakey, M. D. (ed.) Olduvai Gorge, Vol. 3. Cambridge University Press, London, pp. 9–18.

Hay, R. 1976. The Geology of Olduvai Gorge. Clarendon Press, Oxford.

Hay, R. L., Kyser, T. K. 2001. Chemical sedimentology and paleoenvironmental history of Lake Olduvai, a Pliocene lake in northern Tanzania. Geological Society of America Bulletin 113, 1505–1521.

Haynes, G. 1991. Mammoths, Mastodonts, and Elephants: Biology, Behavior, and the Fossil Record. Cambridge University Press, Cambridge and New York.

Hedges, R. E. M., Millard, A. R., Pike, A. W. G. 1995. Measurements and relationships of diagenetic alteration of bone from three archaeological sites. Journal of Archaeological Science 22, 201–209.

de Heinzelin, J., Clark, J. D., White, T., Hart, W., Renne, P., WoldeGabriel, G., Beyene, Y., Vrba, E. 1999. Environment and behavior of 2.5-million-year-old Bouri hominids. Science 284, 625–629.

Hill, A. 1975. Taphonomy of contemporary and late Cenozoic East African vertebrates, Ph.D. Dissertation. University of London, London.

Hill, A. 1979a. Butchery and natural dissarticulation. American Antiquity 44, 739–744.

Hill, A. 1979b. Disarticulation and scattering of mammal skeletons. Paleobiology 5, 261–274.

Hill, A. 1980. Early postmortem damage to the remains of some contemporary east african mammals. In Behrensmeyer, A. K.and Hill, A. (eds.) Fossils in the Making. University of Chicago Press, Chicago, pp. 131–152.

Hill, A., Behrensmeyer, A. K. 1984. Dissarticulation patterns of some modern East African mammals. Paleobiology 10, 366–376.

Isaac, G. L. 1976. The activities of early african hominids. In Isaac, Mc Cown (eds.) Human Origins. Benjamin, Menlo Park, California, pp. 483–514.

Isaac, G. L. 1978. The food-sharing behavior of protohuman hominids. Scientific American 238, 90–108.

Isaac, G. L. 1983. Bones in contention: competing explanations for the juxtaposition of Early Pleistocene artifacts and faunal remains. In Clutton-Brock, J. and Grigson, C. (eds.) Animals and Archaeology 1. Hunters and Their Prey. B.A.R. International Series, 163, Oxford, pp. 3–19.

Isaac, G. L. 1984. The archaeology of human origins: studies of Lower Pleistocene in East Africa, 1971–1981. Advances in World Archaeology 3, 1–87.

Isaac, G. 1997. Plio-Pleistocene archaeology. Koobi Fora Research Project. Volumen V. Cambridge University Press, Cambridge.

Isaac, G. L., Crader, D. C. 1981. To what extent were early hominids carnivorous? An archaeological perspective. In Harding, R. S. O., Teleki, G. (eds.) Omnivorous Primates. Columbia University Press, New York, pp. 37–103.

Isaac, G. L. and Marshall, F. 1981. Small is informative: the application of the study of mini-sites and least-effort criteria in the interpretation of the early Pleistocene archaeological record at Koobi Fora. In (Clark, J. D. and Isaac, G. L., eds.) Las industrias mas antiguas. Maxico, Comision VI, Xth Congress UISPP, pp. 101–119.

Jackes, M., Sherburne, R., Lubell, D., Barker, C., Wayman, M. 2001. Destruction of the microstructure in archaeological bone: a case study from Portugal. International Journal of Osteoarchaeology 11(6), 387–399.

Jacobs, B. F., Kingston, J. D., Jacobs, L. L. 1999. The origin of grass-dominated ecosystems. Annals of the Missouri Botanical Garden 86, 590–643.

Jaeger, J. J. 1976. Les rongeurs (Mammalia, Rodentia) du Pleistocene inferieur d'Olduvai Bed I (Tanzanie), 1ere partie: Les Murides. In Savage, R. J. G., Coryndon, S. C. (eds.) Fossil Vertebrates of Africa, Vol. 4. Academic Press, London, pp. 57–120.

Kappelman, J. 1984. Plio-Pleisocene environments of Bed I and lower Bed II, Olduvai Gorge, Tanzania. Palaeogeography, Palaeoclimatology, Palaeoecology 48, 171–196.

Kappelman, J. 1986. Plio-Pleistocene marine-continental correlation using habitat indicators from Olduvai Gorge, Tanzania. Quaternary Research 25, 141–149.

Kappelman, J., Plummer, T., Bishop, L., Duncan, A., Appleton, S. 1997. Bvids as indicators of Plio-Pleistocene environments in East Africa. Journal of Human Evolution 32, 229–256.

Keeley, L. H., Toth, N. 1981. Microwear polishes on early stone tools from Koobi Fora, Kenya. Nature 293, 464–465.

Kerbis, J. 1990. The roles of leopards, hyena and porcupines in ungulate carcass dispersal: implications for paleoanthropology, Ph.D. Dissertation. Department of Anthropology, University of Chicago, Chicago.

Kingston, J. D., Marino, B. D., Hill, A. 1994. Isotopic evidence for Neogene hominid paleoenvironments in the Kenya Rift Valley. Science 264, 955–959.

Klein, R. G. 1982a. Age (mortality) profiles as a means of distinguishing hunted species from scavenged ones in Stone Age archaeological sites. Paleobiology 8, 151–158.

Klein, R. G. 1982b. Patterns of ungulate mortality and ungulate mortality profiles from Langebaanweg (Early Pliocene) and Elandsfontain (Middle Pleistocene), South-western Cape Province, South Africa. Annals of the South African Museum 90, 49–94.

Klein, R. G. 1986. Reply to Bunn and Kroll (1986). Systematic butchery by Plio-Pleistocene hominids at Olduvai Gorge, Tanzania. Current Anthropology 27, 431–452.

Klein, R. G., Cruz-Uribe, K. 1984. The Analysis of Animal Bones from Archaeological Sites. University of Chicago Press, Chicago.

Kruuk, H. 1972. The Spotted Hyena. University of Chicago Press, Chicago.

Kruuk, H., Turner, M. 1967. Comparative notes on predation by lion, leopard, cheetah and wild dog in the Serengeti area, East Africa. Mammalia 31, 1–27.

Kuhn, T. S. 1962. The Structure of Scientific Revolutions. Princeton University Press, Princeton.

Lakatos, I. 1978. The Methodology of Scientific Research Programmes. Cambridge University Press, Cambridge.

Lam, Y. M. 1992. Variability in the behaviour of spotted hyaenas as taphonomic agents. Journal of Archaeological Science 19, 398–406.

Lam, Y. M., Chen, X., Pearson, O. M. 1999. Intertaxonomic variability in patterns of bone density and the differential representation of bovid, cervid and equid elements in the archaeological record. American Antiquity 64, 343–362.

Leakey, L. 1959. A new fossil skull from Olduvai. Nature 184, 491–3.

Leakey, M. 1971. Olduvai Gorge, Vol. 3. Excavations in Bed I and II, 1960–1963. Cambridge University Press, Cambridge.

Levin, N. E., Quade, J., Simpson, S. W., Semaw, S., Rogers, M. J. 2004. Isotopic evidence for Plio-Pleistocene environmental change at Gona, Ethiopia. Earth and Planetary Science Letters 219, 93–110.

Lewis, M. E. 1997. Carnivoran paleoguilds of Africa: implications for hominid food procurement strategies. Journal of Human Evolution 32, 257–288.

Ludwig, B. V. 1999. A technological reassessment of East African Plio-Pleistocene lithic artifact assemblages, Unpublished Ph.D. University of Rutgers, New Brunswick, p. 432.

Lupo, K. D. 1994. Butchering marks and carcass acquisition strategies: distinguishing hunting from scavenging in archaeological contexts. Journal of Archaeological Science 21, 827–837.

Lupo, K. D. 1998. Experimentally derived extraction rates for marrow: implications for body part exploitation strategies of Plio-Pleistocene hominid scavengers. Journal of Archaeological Science 21, 827–837.

Lupo, K. D., O'Connell, J. F. 2002. Cut and tooth mark distributions on large animal bones: ethnoarchaeological data from the Hadza and their implications for current ideas about early human carnivory. Journal of Archaeological Science 29, 85–109.

Lyman, R. L. 1984. Bone density and differential survivorship in fossil classes. Journal of Anthropological Archaeology 3, 259–299.

Lyman, R. L. 1985. Bone frequencies: differential transport, in situ destruction and the MGUI. Journal of Archaeological Science 12, 221–236.

Lyman, R. 1994. Vertebrate Taphonomy. Cambridge University Press, Cambridge.

Lyman, R. 2004. The concept of equifinality in taphonomy. Journal of Taphonomy 2, 15–26.

Lyman, R. 2006. Analyzing cut marks: lessons from artiodactyl remains in the northwestern United States. Journal of Archaeological Science 32, 1722–1732.

Lyman, R., Fox, J. 1989. A critical evaluation of bone weathering as an indication of bone assemblage formation. Journal of Archaeological Science 16, 293–317.

Maguire, J. M., Pemberton, D., Collett, M. H. 1980. The Makapansgat limeworks grey breccia: hominids, hyaenas, hystricids or hillwash? Paleontologia Africana 23, 75–98.

Manega, P. C. 1993. Geochronology, geochemistry and isotopic study of the Plio-Pleistocene hominid sites and the Ngorongoro Volcanic highland in Northern Tanzania, Ph.D. Dissertation. University of Colorado, Boulder.

Marchiafava, V., Bonucci, E., Ascenzi, A. 1974. Fungal osteoclasia: a model of dead bone resorption. Calcified Tissue Research 14, 195–210.

Marean C. W. 1989. Sabertooth cats and their relevance for early hominid diet and evolution. Journal of Human Evolution 18, 559–582.

Marean, C. W. 1998. A critique of the evidence for scavenging by Neandertal and early modern humans: new data from Kobeh Cave (Zagros, mountains, Iran) and Die Kelders Cave 1 Layer 10 (South Africa). Journal of Human Evolution 35, 111–136.

Marean, C. W., Abe, Y., Frey, C. J., Randall, R. C. 2000. Zooarchaeological and taphonomic analysis of the Die Kelders Cave 1 layers 10 and 11 Middle Stone Age larger mammal fauna. Journal of Human Evolution 38, 197–233.

Marean, C. W., Cleghorn, N. 2003. Large mammal skeletal element transport: applying foraging theory in a complex taphonomic system. Journal of Taphonomy 1, 15–42.

Marean, C. W., Domínguez-Rodrigo, M., Pickering, T. R. 2004. Skeletal element equifinality in zooarchaeology begins with method: the evolution and status of the "shaft critique". Journal of Taphonomy 2, 69–98.

Marean, C. W., Ehrhardt, C. L. 1995. Paleoanthropological and paleoecological implications of the taphonomy of a sabertooth's den. Journal of Human Evolution 29, 515–547.

Marean, C. W., Frey, C. 1997. Animal bones from caves to cities: reverse utility curves as methodological artifacts. American Antiquity 62, 698–716.

Marean, C. W., Kim, S. Y. 1998. Mousterian large mammal remains from Kobeh cave. Current Anthropology 39, 79–113.

Marean, C. W., Spencer, L. M. 1991. Impact of carnivore ravaging of bone in archaeological assemblages. Journal of Archaeological Science 18, 677–694.

Marean, C. W., Spencer, L. M., Blumenschine, R. J., Capaldo, S. 1992. Captive hyaena bone choice and destruction, the Schlepp effect and Olduvai archaeofaunas. Journal of Archaeological Science 19, 101–121.

Marlowe, F. W. 2005. Hunter-gatherers and human evolution. Evolutionary Anthropology 14(2), 54–67.

Marra, A. C., Villa, P., Beauval, L, Bonfiglio, L., Goldberg, P. 2004. Same predator, variable prey: taphonomy of two Upper Pleistocene hyena dens in Sicily and SW France. Revue de Paleobiologie 23(2), 787–801.

Marshall, F. 1986. Implications of bone modification in a Neolithic faunal assemblage for the study of early hominid butchery and subsistence practices. Journal of Human Evolution 15, 661–672.

Marshall, F. 1994. Food sharing and body part representation in Okiek faunal assemblages. Journal of Archaeological Science 21, 65–77.

Martínez, J. 1998. El modo de vida Neandertal: Una reflexión en torno a la ambigüedad en la interpretación de la subsistencia durante el Paleolítico Medio cantábrico, Tesis Doctoral. Facultad de Filosofia y Letras, Universidad Autónoma de Barcelona, Barcelona.

McClanahan, T. R., Young, T. P. 1996. East African ecosystems and their conservation. Oxford University Press, New York.

deMenocal, P. B. 1995. Plio-Pleistocene African climate. Science 270, 53–59.

deMenocal, P. B. 2004. African climate change and faunal evolution during the Pliocene-Pleistocene. Earth and Planetary Science Letters 220, 3–24.

Merrick, H. V., Merrick, J. P. S. 1976. Archaeological ocurrences of earlier Pleistocene ages from the Shungura Formation. In Coppens, Y., Howell, F. C., Isaac, G. L., Leakey, R. E. F. (eds.) Earliest man and environments in the Lake Rudolf

Basin. Chicago University Press, Chicago, pp. 574–584.
Metcalfe, D. 1989. A general cost/benefits model of the trade-off between transport and field processing. Papers presented at the 54th Annual Meeting of the Society for American Archaeology, Atlanta.
Metcalfe, D., Barlow, K. R. 1992. A model for exploring the optimal trade-off between field processing and transport. American Anthropologist 94, 340–356.
Monahan, C. M. 1996. New zooarchaeological data from Bed II, Olduvai Gorge, Tanzania: implications for hominid behavior in the Early pleistocene. Journal of Human Evolution 31, 93–128.
Monahan, C. M. 1998. The Hadza carcass transport debate revisited and its archaeological implications. Journal of Archaeological Science 25, 405–524.
Mora, R., de la Torre, I. 2005. Percussion tools in Olduvai Bed I and II (Tanzania): implications for early human activities. Journal of Anthropological Archaeology 24, 179–192.
Nilssen, P. J. 2000. An actualistic butchery study in South Africa and its implications for reconstructing hominid strategies of carcass acquisition and butchery in the Upper Pleistocene and Plio-Pleistocene, Ph.D. Dissertation. University of Cape Town, Cape Town, South Africa.
Njau, J. and Blumenschine, R. 2006. A diagnosis of crocodile feeding traces on larger mammal bone with fossil examples from the Plio-Pleistocene Olduvai Basin, Tanzania. Journal of Human Evolution 50, 142–162.
Nooter, G. 1975. Leadership and headship: changing patterns in an East Greenland hunting community. Mededelingen van het Rijksmuseum voor Volkenkunde 20.
O'Connell, J. F. 1997. On Plio-Pleistocene archaeological sites and central places. Current Anthropology 38, 86–88.
O'Connell, J. F., Hawkes, K., Blurton Jones, N. 1988. Hadza hunting, butchering and bone transport and their archaeological implications. Journal of Anthropological Research 44, 113–161.
O'Connell, J. F., Hawkes, K., Blurton Jones, N. 1990. Reanalysis of large mammal body part transport among the Hadza. Journal of Archaeological Science 17, 301–316.
O'Connell, J. F., Hawkes, K., Blurton Jones, N. 1991. Distribution of refuse-producing activities at Hadza residential base camps: implications for analysis of archaeological site structure. In Kroll, E., Price, T. D. (eds.) The Interpretation of Archaeological Spatial Patterning. Plenum Press, Nueva York, pp. 61–76.
O'Connell, J. F., Hawkes, K., Blurton Jones, N. 1992. Patterns in the distribution, site structure and assemblage composition of Hadza kill-butchering sites. Journal of Archaeological Science 19, 319–345.
O'Connell, J. F., Hawkes, K., Lupo. K. D., Blurton Jones, N. 2002. Male strategies and Plio-Pleistocene archaeology. Journal of Human Evolution 43, 831–872.
O'Connell, J., Hawkes, K., Lupo, K., Blurton-Jones, N. G. 2003. Another reply to Domínguez-Rodrigo. Journal of Human Evolution 45, 417–419.
Oliver, J. S. 1994. Estimates of hominid and carnivore involvement in the FLK Zinjanthropus fossil assemblage: some socioecological implications. Journal of Human Evolution 27, 267–294.
Pales, L., Lambert, C. 1971. Mammiferes du Quaternaire. Les membres (herbivores). Centre National de Recherche Scientifique, Paris VII.
Perez-Ripoll, M. 1992. Marcas de carnicería, fracturas intencionadas y mordeduras de carnivoros en huesos prehistoricos del mediterraneo español. Diputacion provincial de Alicante, Alicante.
Perkins, D., Daly, P. 1968. A hunter's village in Neolithic Turkey. Scientific American 219, 96–106.
Peters, C. R., Blumenschine, R. J. 1995. Landscape perspectives on possible land use patterns for Early Pleistocene hominids in the Olduvai Basin, Tanzania. Journal of Human Evolution 29, 321–362.
Petraglia, M.D. and Potts, R. 1994. Water flow and the formation of Early Pleistocene artefact sites in Olduvai Gorge, Tanzania. Journal of Anthropological Archaeology 13, 228–254.
Pickering, T. R. 2002. Reconsideration of criteria for differentiating faunal assemblages accumulated by hyaenas and hominids. International Journal of Osteoarchaeology 12, 127–141.
Pickering, T. R., Domínguez-Rodrigo, M. 2005. The acquisition and use of large mammal carcasses by Oldowan hominins in Eastern and Southern Africa: a selected review and assessment. In Toth, N., Schick, K. (eds.) The Oldowan: Studies into the Origins of Human Technology. Stone Age Institute Press, Bloomington, pp. 113–128.
Pickering, T. R., Domínguez-Rodrigo, M., Egeland, C., Brain, C. K. 2004. New data and ideas on the foraging behavior of Early Stone Age hominids at Swartkrans cave. South African Journal of Science 100, 215–220.

Pickering, T. R., Domínguez-Rodrigo, M., Egeland, C. P., Brain, C. K. 2005a. The contribution of limb bone fracture patterns to reconstructing early hominid behavior at Swartkrans Cave (South Africa): Archaeological application of a new analytical method. International Journal of Osteoarchaeology 15, 247–260.

Pickering, T. R., Domínguez-Rodrigo, M., Egeland, C., Brain, C. K., Kuman, K. 2005b. Carcass foraging by early hominids at Swartkrans Cave (South Africa): A new investigation of the zooarchaeology and taphonomy of member 3. In Pickering, T. R., Toth, N., Shick, K. (eds.) African Taphonomy: A Tribute to the Career of C.K. Brain. CRAFT Press, Bloomington, Indiana (en prensa).

Pickering, T. R., Egeland, C. P. 2006. Experimental pattern of hammerstone percussion damage on bones: implications for inferences of carcass processing by humans. Journal of Archaeological Science 33, 459–469.

Pickering, T. R., Marean, C. W., Domínguez-Rodrigo, M. 2003. Importance of limb bone shaft fragments in zooarchaeology: a response to "On in situ attrition and vertebrate body part profiles" (2002) by M. C. Stiner. Journal of Archaeological Science 30, 1469–1482.

Pickering, T. R., Wallis, J. 1997. Bone modifications resulting from captive chimpanzee mastication: implication for the interpretation of Pliocene archaeological faunas. Journal of Archaeological Science 24, 1115–1127.

Piepenbrink, H. 1984. Examples of signs of biogenic decomposition in bones long buried. Anthropologischer Anzeiger 42, 241–251.

Plummer, T. W. 2004. Flaked stones and old bones: biological and cultural evolution at the dawn of technology. Yearbook of Physical Anthropology 47, 118–164.

Plummer, T., Bishop, L. 1994. Hominid paleoecology at Olduvai Gorge, Tanzania, as indicated by bovid remains. Journal of Human Evolution 27, 47–75.

Potts, R. 1982. Lower Pleistocenes site formation and hominid activities at Olduvai Gorge, Tanzania, Ph.D. Dissertation. Harvard University, Massachussets.

Potts, R. 1983. Foraging for faunal resources by early hominids at Olduvai Gorge, Tanzania. In (Clutton-Brock, J., Grigson, C., eds.) Animals and ArchaeologyI: Hunters and their Prey. British Arcnaeological Reports International Series 163, Oxford, pp. 51–62.

Potts, R. 1984. Hominid hunters? Problems of identifying the earliest hunter/gatherers. In Foley, R. (ed.) Hominid Evolution and Community Ecology. Academic Press, London, pp. 129–166.

Potts, R. 1986. Temporal span of bone accumulations at Olduvai Gorge and implications for early hominid foraging behavior. Paleobiology 12, 25–31.

Potts, R. 1988. Early Hominid Activities at Olduvai. Aldine, New York.

Potts, R. 1991. Why the Oldowan? Plio-Pleistocene tool making and the transport of resources. Journal of Anthropological Research 47, 53–176.

Potts, R. 1993. Archeological interpretations of early hominid behavior and ecology. In Rasmussen, D. T. (ed.) The Origin and Evolution of Humans and Humanness. Jones and Bartlett, Boston. pp. 49–74.

Potts, R. 1994. Variables versus models of early Pleistocene hominid land use. Journal of Human Evolution 27, 7–24.

Potts, R., Shipman, P. 1981. Cutmarks made by stone tools from Olduvai Gorge, Tanzania. Nature 291, 577–580.

Pumarejo, P., Cabrera, V. 1992. Huellas de descarnado en el Paleolítico Superior: la cueva de El Castillo (Cantabria). Espacio, Tiempo y Forma, Serie I, Arqueología y Prehistoria 5, 139–152.

Roche, H., Delagnes, A., Brugal, J.-P., Feibel, C., Kibunjia, M., Mourre, V., Texier, P. J. 1999. Early hominid stone tool production and technical skill 2.34 Myr ago in West Turkana, Kenya. Nature 399, 57–60.

Rose, L., Marshall, F. 1996. Meat eating, hominid sociality, and home bases revisited. Current Anthropology 37, 307–338.

Rudwick, M. J. S. 1971. Uniformity and progression: reflections on the structure of geological theory in the age of Lyell. In Roller, D. (ed.) Perspectives in the History of Science and Technology. University of Oklahoma Press, Norman, pp. 209–227.

Ruiter, D. J., Berger, L. R. 2000. Leopards as taphonomic agents in dolomitic caves-implications for bone accumulations in the hominid-bearing deposits of South Africa. Journal of Archaeological Science 27, 665–684.

Schaller, G. B. 1972. The Serengeti Lion. University of Chicago Press, Chicago.

Schick, K. D. 1984. Processes of Paleolithic site formation: an experimental study. Berkeley, University of California, Ph. D. dissertation.

Schick, K. D. 1987. Modeling the formation of early stone age artifact concentrations. Journal of Human Evolution 16, 789–808.

Schick, K., Toth, N. 1993. Making Silent Stones Speak. Simon & Schuster, New York.

Scott, J. 1985. The Leopard's Tale. Elm Tree Books, London.

Scott, J. 1987. The leopard's tale. Elm Tree Books, Nairobi.

Seilacher, A., Reif, W. E., Westphal, F. 1985. Sedimentological, ecological and temporal patterns of fossil Lagerstätten. Philosophical Transactions of the Royal Society (London) 311, 5–23.

Selvaggio, M. M. 1994. Identifying the timing and sequence of hominid and carnivore involvement with Plio-Pleistocene bone assemblages from carnivore tooth marks and stone-tool butchery marks on bone surfaces, Ph.D. Dissertation. Rutgers University, New Brunswick.

Selvaggio, M. M., Wilder, J. 2001. Identifying the involvement of multiple carnivore taxa with archaeological bone assemblages. Journal of Archaeological Science 28, 465–470.

Semaw, S., Renne, P., Harris, J. W. K., Feibel, C. S., Bernor, R. L., Fesseha N., Mowbray, K. 1997. 2.5-million-year-old stone tools from Gona, Ethiopia. Nature 385, 333–336.

Shackleton, N. J. 1995. New data on the evolution of Pliocene climatic variability. In Vrba, E. (ed.) Paleoclimate and Evolution with Emphasis on Human Origins. Yale University Press, New Haven, pp. 242–248.

Sharmin, S., Kishi, F., Sano, A., Kamei, K., Nishimura, K., Miyaji, M. 2003. Direct invasion of bones by highly pathogenic fungi in an in vitro model and its ecological significance. Japanese Journal of Medical Mycology 44, 17–23.

Shea, J. H. 1982. Twelve fallacies of uniformitarianism. Geology 10, 455–460.

Shipman, P. 1975. Implications of drought for vertebrate fossil assemblages. Nature 257, 67–668.

Shipman, P. 1983. Early hominids lifestyle: hunting and gathering or foraging and scavenging. In Clutton-Brock, J., Grigson, C. (eds.) Animals and Archaeology, Vol. 1: Hunters and their Prey. BAR, 163, London, pp. 31–49.

Shipman, P. 1986. Scavenging or hunting in early hominids: theoretical framework and tests. American Anthropologist 88, 27–43.

Shipman, P., Fisher, D. C., Rose, J. 1984. Mastodon butchery: microscopic evidence of carcass processing and bone tool use. Paleobiology 10, 358–365.

Shipman, P., Harris, J. M. 1988. Habitat preference and paleoecology of Australopithecus boisei in eastern Africa. In Gribe, F. E. (ed.) Evolutionary History of the "Robust" Australopithecines. Aldin de Gruyter, New York, pp. 343–381.

Shipman, P., Phillips, J. E. 1976. On scavenging by hominids and other carnivores. Current Anthropology 17, 170–172.

Shipman, P., Rose, J. J. 1983. Early hominid hunting, butchering and carcass-processing behaviors: approaches to the fossil record. Journal of Anthropological Archaeology 2, 57–98.

Shipman, P., Rose, J. 1984. Cutmarks mimics on modern and fossil bovid bones. Current Anthropology 25, 116–117.

Sikes, N. E. 1994. Early hominid habitat preferences in East Africa: paeosol carbonate isotopic evidence. Journal of Human Evolution 27, 25–45.

Sikes, N. E. 1999. Plio-Pleistocene floral context and habitat preferences of sympatric hominid species in East Africa. In Bromage, T. G., Schrenk, F. (eds.) African Biogeography, Climate Change and Human Evolution. Oxford University Press, Oxford, pp. 301–315.

Simpson, G. G. 1970. Uniformitarianism: an inquiry into principle, theory, and method in geohistory and biohistory. In Hecht, M. K., Steere, W. C. (eds.) Essays in Evolution and Genetics in Honor of Theodosius Dobzhansky. Appleton, New York, pp.43–96.

Skinner, J. D., van Aarde, R. J. 1991. Bone collecting by brown hyaenas Hyaena brunnea in the Central Namib Desert, Namibia. Journal of Archaeological Science 18, 513–5123.

Skinner, J. D., Henshel, J. R., van Aarde, R. J. 1986. Bone-collecting habits of spotted hyenas Crocruta crocruta in the Kruger National Park. South African Journal of Tydskr. Dierk. 21, 303–308.

Smith, B. D. 1977. Archaeological inference and inductive confirmation. American Anthropologist 79, 598–617.

Sorensen, M. V., Leonard, W. R. 2001. Neandertal energetics and foraging efficiency. Journal of Human Evolution 40, 483–495.

Stanley, S. M. 1992. An ecological theory for the origin of *Homo*. Paleobiology 3, 237–257.

Stiner, M. 1990. The use of mortality patterns in archaeological studies of hominid predatory daptations. Journal of Anthropological Archaeology 9, 305–351.

Stiner, M. 1991. Food procurement and transport by human and non-human predators. Journal of Archaeological Science 18, 455–482.

Stiner, M. 1994. Honor Among Thieves: A Zooarchaeological Study of Neandertal Ecology. Princeton University Press, Princeton.

Stiner, M. 2002. On in situ attrition and vertebrate body part profiles. Journal of Archeological Science 29, 979–991.

Stiner, M. 2004. A comparison of photon-densitometry and computed tomography parameters of bone density in ungulate body part profiles. Journal of Taphonomy 2(1–4), 117–146.

Suwa, G., Nakaya, H., Asfaw, B., Saegusa, H., Amzaye, A., Kono, R. T., Beyene, Y., Katoh, S. 2003. Plio-Pleistocene terrestrial mammal assemblage from Konso, southern Ethiopia. Journal of Vertebrate Paleontology 23, 901–916.

Tappen, M. 1992. Taphonomy of a central African savanna: natural bone deposition in Parc National des Virunga, Zaire, Ph.D. Thesis. Department of Anthropology, Harvard University, Cambridge.

Tappen, M. 1995. Savanna ecology and natural bone deposition: implications for early hominidsite formation, hunting and scavenging. Current Anthropology 36, 223–260.

Todd, L. C., Rapson, D. J. 1988. Long bone fragmentation and interpretation of faunal assemblages: approaches to comparative analysis. Journal of Archaeological Science 15, 307–325.

Torre, de la, I. 2002. El Olduvayense de la Seccion Tipo de Peninj (Lago Natron, Tanzania), M.Sc. Thesis. Universidad Complutense, Madrid.

Torre, de la, I. 2004. Omo revisited. Evalating the technological skills of Pliocene hominids. Current Anthropology 45, 439–468.

Torre, de la, I. 2006. Estrategias tecnológicas en el Pleistoceno inferior de África oriental (Olduvai y Peninj, norte de Tanzania). Tesis Doctoral inédita, Universidad Complutense, Madrid.

Torre, de la, I., Mora, R. 2004. El Olduvayense de la Sección Tipo de Peninj (Lago Natron, Tanzania), Vol. 1. CEPAP, Barcelona.

Torre, de la, I., Mora, R. 2005. Unmodified lithic material at Olduvai Bed I: manuports or ecofacts? Journal of Archaeological Science 32, 273–285.

Torre, de la, I., Mora, R., Domínguez-Rodrigo, M., Alcala, L., Luque, L. (2003). The Oldowan industry of Peninj and its bearing on the reconstruction of the technological skills of lower Pleistocene hominids. Journal of Human Evolution, 44, 203–224.

Toth, N. 1982. The stone technologies of early hominids at Koobi Fora, Kenya: an experimental approach, Ph.D. Thesis. Department of Anthropology, University of California, Berkeley.

Toth, N. 1985. The Oldowan reassessed: a close look at Early Stone Age artifacts. Journal of Archaeological Science 12, 101–120.

Treves, A., Naughton-Treves, L. 1999. Risk and opportunity for humans coexisting with large carnivores. Journal of Human Evolution 36, 275–282.

Turner, C. G. 1983. Taphonomic reconstruction of human violence and cannibalism based on mass burials in the American Southwest. In LeMoine, G. M., MacEachern, A. S. (eds.) Carnivores, Human Scavengers and Predators: A Question of Bone Technology, University of Calgary Archaeological Association, Calgary, pp. 219–240.

Turner, A., Anton, M. 1997. The Big Cats and their Fossil Relatives. Columbia University Press, New York.

Verdcourt, B. 1963. The Miocene non-marine Mollusca of Rusinga Island, Lake Victoria, and other localities in Kenya. Paleontographica 121, 1–37.

Villa, P., Castel, C., Beauval, C., Bourdillat, V., Goldberg, P. 2004. Human and carnivore sites in European Middle and Upper Paleolithic: similarities and differences in bone modification and fragmentation. Revue de Paleobiologie 23(2), 705–730.

Villa, P., Mahieu, E. 1991. Breakage patterns of human long bones. Journal of Human Evolution 21, 27–48.

Walter, R. C., Manega, P. C., Hay, R. L. 1991. Laser-fusion Ar40/Ar39 dating of bed I, Olduvai-Gorge, Tanzania. Nature 354, 145–149.

Walter, R. C., Manega, P. C., Hay, R. L. 1992. Tephrochronology of Bed I, Olduvai Gorge: an application of the laser-fusion 40Ar/39Ar dating to calibrating biological and climatic change. Quaternary International 13–14, 37–46.

Wenzel, G. W. 1995. Ningiqtuq: resource sharing and generalized reciprocity in Clyde River, Nunavut. Arctic Anthropology 32(2), 43–60.

White, T. E. 1952. Observations on the butchering technique of some aboriginal peoples: no.1. American Antiquity 17, 337–338.

Wolpoff, M. H. 1999. Paleoanthropology. McGraw Hill, Madison, Wisconsin.

Wynn, J. G. 2004. Influence of Plio-Pleistocene aridification on human evolution: evidence from paleosols of the Turkana Basin, Kenya. American Journal of Physical Anthropology 123, 106–118.

Appendix A

Specimens from FLK Zinj bearing tooth marks
40
116
149
229
266
273
319
341
364
364
370
402
067/954
067/989
067/992
67/809
B135
B159
B170
B172
B180
B195
B232a
B252
B316
B385
B387
B457
B4-8
B4-8
B4-8
B4-84
B501
B529
B579
B580
B674

(*Continued*)

Specimens from FLK Zinj bearing tooth marks–cont'd

B8-12
B85
BA119
BA120
BA122
BA124
BA162
BA81
BA85
BA88
BA91
C1000
C1053
C1059
C1074
C1150
C1220
C1227a
C765
C825
C835
C868
C904
C907
C976
C983
C986
D100
D104
D112
D133
D137
D146
D164
D19
D220
D36
D44
D49
D54
D84
E115
E133
E59
E78
E94
F161
F170
F193
F216
F220
F228
F248
F360

(*Continued*)

Specimens from FLK Zinj bearing tooth marks–cont'd

G
G174
G284
G313
G318
G360a
G360c
G377
J55
K3
M110
M147
M78
T567

Appendix B

Specimens from FLK Zinj bearing percussion marks (PM) and percussion notches (PN)

06	PM, PN
80	PM
82	PM, PN
85	PM
93	PM
117	PM
180	PM
234	PM
302	PM
320	PM
365	PM
527	PM
067/4546	PM
067/992	PM
A22	PN
B0-4	PN
B12-16	PM
B12-16	PM
B140b	PM, PN
B151	PM
B161	PN
B166b	PM
B180	PM
B181	PM, PN
B264	PM
B275	PM
B283	PM
B308	PM, PN
B310	PM
B316	PM
B340	PN
B385	PN
B43	PN
B443	PM
B444	PM
B458	PM
B4-8	PN

(*Continued*)

Specimens from FLK Zinj bearing percussion marks (PM) and percussion notches (PN)—cont'd

B4-8	PM, PN
B4-8(45)	PN
B480	PM
B4-84	PM
B556	PN
B618	PM
B620	PN
b637	PM
B73	PM, PN
B75	PM
B8-12	PN
B8-12	PN
B8-12	PN
B8-12	PM
B8-12	PM
B8-12	PM
B8-12	PM
B8-12	PM, PN
B8-12	PM, PN
B8-12	PM, PN
BA122	PM, PN
BA123	PN
Ba151	PN
BA156	PM
BA162	PM
BA171	PM
Ba174	PM
BA243a	PM
BA243b	PM
Ba254	PN
BA75	PM
BA87	PN
BA88	PM, PN
C1000	PM
C1001	PM
C1003	PM, PN
C1019	PN
C1020	PM
C1022	PM
C1043	PM
C1059	PM
C1149	PM
C1152	PM
C1169	PN
C1177	PM
C1180	PN
C1181	PM
C1208	PM, PN
C1212	PM
C1215	PN
C1216	PM
C1217	PM

(*Continued*)

Specimens from FLK Zinj bearing percussion marks (PM) and percussion notches (PN)—cont'd

C1218b	PN
C1219	PM, PN
C1225	PN
C1227c	PM
C1238	PM, PN
C20-25	PM
C721	PM
C769	PN
C835	PM
C864	PM
C948	PM
C977	PM, PN
D104	PM
D105	PM
D14	PM
D164	PM
D166	PM
D171	PN
D26	PN
D28	PM
D30	PN
D38	PM
D44	PM
D58	PM, PN
D63	PM
D69	PM
D70	PN
D77	PM, PN
D96	PM, PN
E18	PM, PN
E21	PM
E57	PM
E58	PM
E7	PM
E89	PN
E92	PM
E94	PM
E96	PM
F170	PN
F172	PM
F216	PM
F228	PM
F255	PM, PN
G364	PN
J77	PN
K18	PN
K6	PM
K7	PM, PN
M132	PM
M145	PM, PN
M146	PM
M147	PM
M508	PN
M78	PM
M92	PM

Appendix C

Specimens from FLK Zinj with measured tooth pits

Catalogue no.	TM length	TM breadth
BA120	1.58	0.95
	1.30	1.07
BA91	1.31	0.80
	1.95	0.60
M81	2.09	1.08
	2.31	1.10
	2.39	1.81
	2.96	1.59
C1053	4.14	2.82
D220	2.72	1.45
	2.21	1.28
D36	1.49	1.30
	2.40	2.11
F220	7.68	5.06
	9.74	6.25
	6.15	5.27
	6.12	4.92
C1074	1.47	0.77
067/818	6.79	3.68
067/989	3.90	2.82
	4.40	4.04
	4.77	3.27
	3.44	2.55
	3.64	3.34
	3.86	2.90
	3.72	2.70
	4.27	3.40
B122	2.12	2.18
	1.75	1.11
	2.04	1.95
	2.61	2.27
BA124	5.71	3.70
D106	2.17	1.52
	2.82	2.07

(*Continued*)

Specimens from FLK Zinj with measured tooth pits—cont'd

Catalogue no.	TM length	TM breadth
F216	4.11	2.65
	1.64	1.16
	2	1.22
	2.88	2.19
	2.29	1.83
BA85	2.68	1.66
	1.35	1.10
F248	2.83	2.06
G377	2.12	1.33
	5.50	4.25
	2.40	1.64
	1.82	1.39
G	4.05	3.49
C825	3.15	2.21
	2.40	1.26
	1.94	1.27
	2.34	1.37
	2.21	1.77
D54	3.03	1.99
	4.25	4.11
	2.37	1.47
K3	2.15	1.11
	0.73	0.57
	0.88	0.70
	2.37	1.21
	2.20	1.18
	2.13	1.84
	3.65	2.55
	2.30	1.67
	1.80	1.07
116	1.82	1.02
	1.45	0.93
364	3.97	2.80
T567	1.59	0.95
B673	2.21	1.51
	1.99	1.51
	1.78	1.37
	1.10	0.81
BA88	2.25	1.02
C1059	4.73	2.10
	1.61	1.24
	1.19	0.89
	3.35	2.10

Appendix D

List of specimens in each Olduvai Bed I site bearing bone surface modifications attributed to hominids (cut and percussion marks and notches) and carnivores (tooth marks and notches).

Keys:

CM: CUT MARK
TM: TOOTH MARK
PM: PERCUSSION MARK
PN: PERCUSSION NOTCH
CN: CARNIVORE NOTCH

FLK N1-2	CM
66	
175	
175	
412	
538	
1624	
2783	
2935	
3402	
3423	
7077	
7232	
8091	
8363	
8704	
8722	
8725	
8746	
8809	
8948	
067/1163	

FLK N1-2	CM+TM
538	
8809	

FLK N1-2	PM
140	
289	
416	
1528	
3037	
3420	
3431	
3444	
3469	
3471	
3840	
8571	

FLK N1-2	TM	FLK N1-2	CN
000		5	
5		124	
7		140	
9		159	
10		167	
15		174	
21		234	
27		308	
28		319	
36		454	
39		540	
50		595	
58		997	
65		1025	
66		1056	
67		1301	
70		1338	
78		1500	
83		1528	
84		1551	
88		1569	
94		1577	
103		6258	
104		7176	
114		7178	
115		7291	
117		7660	
130		7677	
131		7701	
133		8068	

(*Continued*)

FLK N1-2	TM	FLK N1-2	CN
134		8341	
135		8358	
153		8359	
154		8366	
163		8443	
170		8486	
174		8488	
195		8510	
234		8514	
245		8556	
246		8568	
250		8632	
260		8697	
261		8704	
266		8777	
273		8778	
289		8846	
292		8850	
296		8901	
308		9009	
311		9011	
316		9016	
319		9028	
323		9037	
326		9056	
329		9103	
331		9114	
341		9127	
386		9148	
392		9151	
398		10036	
401		10055	
402		10062	
406		10065	
413		10074	
415		10081	
417		067/521	
428		067/795	
438		067/803	
439		8962A	
452		91b	
454		x2774	
461		x2947	
462		x2971	
462		x3011	
488		x3017	
507		x3037	
538		x3423	
539		x3431	
553			
557			
559			
580			
603			
611			

(*Continued*)

FLK N1-2	TM	FLK N1-2	CN
612			
613			
659			
927			
999			
1003			
1013			
1016			
1031			
1033			
1046			
1056			
1059			
1097			
1098			
1102			
1103			
1117			
1241			
1246			
1247			
1248			
1253			
1253			
1256			
1260			
1267			
1272			
1274			
1280			
1290			
1295			
1298			
1301			
1331			
1338			
1339			
1343			
1351			
1353			
1485			
1487			
1500			
1501			
1507			
1512			
1516			
1528			
1529			
1530			
1531			
1550			
1551			
1552			
1554			

(*Continued*)

FLK N1-2	TM	FLK N1-2	CN
1564			
1568			
1569			
1569			
1577			
1583			
1587			
1591			
1601			
1606			
1607			
1615			
1619			
1641			
6293			
7012			
7018			
7024			
7038			
7039			
7043			
7044			
7049			
7050			
7054			
7055			
7058			
7066			
7067			
7068			
7070			
7071			
7074			
7076			
7078			
7160			
7165			
7168			
7169			
7176			
7178			
7193			
7198			
7203			
7205			
7213			
7221			
7230			
7245			
7246			
7257			
7262			
7267			
7268			
7274			

(*Continued*)

FLK N1-2	TM	FLK N1-2	CN
7278			
7286			
7297			
7336			
7354			
7358			
7372			
7401			
7413			
7413			
7419			
7423			
7459			
7462			
7478			
7579			
7621			
7624			
7625			
7652			
7658			
7659			
7660			
7664			
7667			
7668			
7670			
7674			
7675			
7680			
7686			
7695			
7701			
7709			
7720			
7722			
7729			
8002			
8004			
8009			
8017			
8022			
8024			
8064			
8066			
8068			
8069			
8071			
8074			
8075			
8077			
8091			
8099			
8316			
8321			

(*Continued*)

FLK N1-2	TM	FLK N1-2	CN
8333			
8335			
8337			
8343			
8344			
8356			
8358			
8374			
8378			
8380			
8382			
8393			
8400			
8407			
8412			
8414			
8422			
8428			
8433			
8436			
8439			
8444			
8464			
8479			
8482			
8486			
8488			
8504			
8510			
8516			
8531			
8538			
8546			
8547			
8573			
8632			
8661			
8671			
8672			
8689			
8697			
8701			
8717			
8737			
8741			
8749			
8760			
8768			
8777			
8783			
8792			
8799			
8809			
8812			
8828			

(*Continued*)

FLK N1-2	TM	FLK N1-2	CN
8842			
8846			
8851			
8856			
8875			
8878			
8879			
8882			
8891			
8894			
8896			
8899			
8915			
8935			
8938			
8942			
8958			
8969			
8974			
8981			
8994			
9003			
9009			
9014			
9028			
9039			
9056			
9058			
9061			
9089			
9090			
9106			
9112			
9114			
9120			
9131			
9149			
9151			
10001			
10004			
10013			
10023			
10024			
10031			
10032			
10036			
10038			
10041			
10044			
10058			
10062			
10063			
10064			
10065			
10067			

(*Continued*)

FLK N1-2	TM	FLK N1-2	CN
10077			
10081			
10088			
10100			
10110			
10112			
10116			
10121			
(66)183			
(66)184			
(66)190			
(66)192			
(66)195			
(66)204			
067/1000			
067/1011			
067/1047			
067/1049			
067/1050			
067/1074			
067/1120			
067/1143			
067/1150			
067/1151			
067/1153			
067/1157			
067/1254			
067/1532			
067/1544			
067/1574			
067/1625			
067/1655			
067/3275			
067/3369			
067/3370			
067/3386			
067/3391			
067/3404			
067/3880			
067/4284a			
067/4284b			
067/4284c			
067/4284d			
067/4290b			
067/4482			
067/4743			
067/4793			
067/4794			
067/4796			
067/4797			
067/4798			
067/4830			
067/500			
067/506			
067/508			

(*Continued*)

FLK N1-2	TM	FLK N1-2	CN
067/516			
067/518			
067/521			
067/596			
067/618			
067/638			
067/638			
067/647			
067/656			
067/661			
067/670			
067/678			
067/763			
067/803			
067/804			
067/8296			
067/8299			
067/8312			
067/8322			
067/8336			
067/8475			
067/8505			
067/8726			
067/8841			
067/9137			
067/995			
067/996			
1052B			
1079-1094-1350			
23?			
7704A			
8395A			
8557A			
91b			
Lb			
x2673			
x2709			
x2766			
x2767			
x2773			
x2774			
x2776			
x2777			
x2785			
x2788			
x2794			
x2798			
x2800			
x2802			
x2803			
x2807			
x2809			
x2814			
x2829			
x2840			

(*Continued*)

FLK N1-2	TM	FLK N1-2	CN
x2842			
x2871			
x2892			
x2928			
x2931			
x2932			
x2976			
x2981			
x2982			
x2986			
x2993			
x2994			
x2995			
x3011			
x3012			
x3017			
x3019			
x3021			
x3025			
x3027			
x3039			
x3043			
x3052			
x3065			
x3088			
x3182			
x3191			
x3251			
x3318			
x3344			
x3344			
x3349			
x3361			
x3408			
x3415			
x3416			
x3420			
x3424			
x3429			
x3430			
x3431			
x3444			
x3447			
x3452			
x3461			
x3468			
x3469			
x3471			
x3484			
x3488			
x3489			
x3840			
x3844			

FLK N3	TM	FLK N3	CN
30		1364	
712		1370	
1027		1413	
1121		1415	
1129		1701	
1134		3019	
1137		5052	
1146		5073	
1148		6046	
1157		6075	
1161		6081	
1164		6091	
1169		6119	
1377		6200	
1378		6239	
1380		7104	
1389		7105	
1394		7108	
1396		7115	
1414		7499	
1579		7499	
1648		7501	
1685		7504	
1689		7788	
1692		8111	
1712		9183	
1721		9219	
1723		9228	
1729		9239	
1730		9255	
1735		9257	
1740		10170	
3019		067/1141	
3298		7766&7749	
5009		x2127	
5023		x2194	
5031		x2201	
5036		x2209	
5041		x2311	
5047		x2319	
5074		x2322	
5142		x2329	
5149		x2362	
5171			
6000			
6006			
6014			
6019			
6030			
6034			
6039			
6043			
6046			
6046			
6047			

(*Continued*)

FLK N3	TM	FLK N3	CN
6050			
6054			
6057			
6062			
6065			
6066			
6069			
6071			
6075			
6076			
6081			
6091			
6092			
6093			
6094			
6115			
6116			
6118			
6121			
6123			
6200			
6243			
6254			
6271			
6280			
6284			
6285			
6288			
6291			
6292			
7092			
7097			
7104			
7105			
7108			
7305			
7307			
7307			
7309			
7317			
7476			
7481			
7482			
7484			
7491			
7499			
7502			
7505			
7507			
7508			
7509			
7510			
7730			

(*Continued*)

FLK N3	TM	FLK N3	CN
7735			
7742			
7760			
7764			
7768			
7769			
7776			
7788			
7790			
7792			
7797			
7850			
8102			
8173			
9103			
9162			
9169			
9185			
9189			
9199			
9205			
9219			
9236			
9237			
9247			
9251			
9254			
9256			
9257			
9401			
10128			
10157			
10165			
10167			
067/1051			
067/1101			
067/1133			
067/1135			
067/1136			
067/1137			
067/1139			
067/1142			
067/1168			
067/3297			
067/3340			
067/497			
067/499			
067/667			
067/668			
067/700			
067/738			
067/768			
067/999			
7767&7759			
x2040			

(*Continued*)

FLK N3	TM	FLK N3	CN
x2091			
x2195			
x2197			
x2200			
x2201			
x2202			
x2207			
x2209			
x2224			
x2299			
x2300			
x2310			
x2311			
x2320			
x2321			
x2322			
x2324			
x2326			

FLK N4	TM	FLK N4	CN
1182		1429	
1184		5082	
1189		6314	
1423		7124	
1425		x3889	
1428		x3891	
1429			
1737			
1738			
1764			
1780			
5082			
5087			
5092			
5101			
5108			
5114			
5116			
5131			
5188			
6126			
6139			
6145			
6204			
6209			
6295			
6297			
6301			
6325			
6328			
6330			
6332			

(*Continued*)

FLK N4	TM	FLK N4	CN
7326			
7326			
7520			
7534			
7536			
8127			
8128			
8130			
8133			
8142			
9264			
9267			
9268			
10181			
(66)488			
(66)588			
067/1069			
067/2130			
067/2133b			
067/2135			
067/245			
067/2614			
067/3804			
067/3825			
067/3836			
067/3866			
067/3872			
067/4525			
067/498			
067/504			
067/52			
067/53			
067/637			
067/637			
067/733			
067/735			
067/752			
067/776			
067/781			
x3866			

FLK N5	CM	TM
1207		
1231		
1443		
1450.2		
1454		
1458		
1807		
1810		
1813		
4155		
5135		

(*Continued*)

FLK N5	TM
6151	
7131	
7133	
7135	
7148	
7333	
7338	
7546	
7570	
7571	
7574	
7818	
7821	
7827	
7835	
7858	
8137	
8144	
8154	
8184	
8193	
8203	
8212	
9281	
9297	
9320	
9325	
9348	
9353	
9362	
9363	
9387	
9399	
10227.1	
10246	
10250	
10252	
10258	
10259	
10261	
10264	
10271	
10278	
10287	
067/479.3	
067/505	
067/507	
067/571.1	
067/626.3	
067/648	
067/657	
067/676	
067/682	
067/686	
067/732	
067/782	

(*Continued*)

FLK N5	TM
067/783	
067/798	
067/1109	
067/1110	
067/1172	
067/3307	
067/3336	
067/4280.1	
067/4280.2	
067/4280.4	
067/4280.5	
067/4280.6	
067/4280.7	
067/4280.8	
067/4280.9	
067/4280.10	
067/4280.11	
067/4280.13	
067/4280.14	
067/4280.15	
067/4280.17	
067/4280.19	
067/4280.20	
067/4280.21	
067/4280.23	
067/4280.282	
067/4298.1	
067/4298.2	
067/4298.3	
067/4298.4	
067/4298.6	
067/4298.7	
067/4298.8	
067/4298.9	
067/4298.10	
067/4298.11	
067/4298.12	
067/4298.13	
067/4298.14	
067/4298.15	
067/4298.17	
067/4298.18	
067/4298.20	
067/4298.22	
067/4298.23	
067/4298.24	
067/4298.27	
067/4298.46	
067/4298.76	
067/4298.119	
067/4299.3	
067/4299.4	
067/4299.6	
067/4299.7	

(*Continued*)

FLK N5	TM
067/4299.8	
067/4299.10	
067/4299.11	
067/4299.12	
067/4300.1	
067/4300.3	
067/4786	

FLK N5	CM
10066	

FLK N6	TM
828	
8249	
8256	
9461	
10416	
10417	
10448	
067/496	
067/610	

FLK NN1	TM	FLK NN1	CN
3		9	
5		546	
20		818	
30			
40			
56			
102			
275			
540			
547			
549			
552			
558			
804			
807			
810			
818			
821			
822			
834			
838			
1017			
1049			
067/1212			
067/3742			
067/5439			

FLK NN2	TM
CATNUM	
1	
11	
175	
341	
342	
343	
344	
355	
365	
368	
369	
370	
459	
480.12	
499	
499.1	
500	
505	
521	
562	
567	
577	
578	
585	
611	
678	
690	
692	
700	
730	
732	
734	
850	
853	
882	
891	
892	
897	
903	
907	
912	
956	
962.1	
964	
967	
970	
970.1	
996	
997	
1004	
1005	
1015	
1025	
067/6969.7	

FLK N3	TM	FLK N3	CN
30		1364	
712		1370	
1027		1413	
1121		1415	
1129		1701	
1134		3019	
1137		5052	
1146		5073	
1148		6046	
1157		6075	
1161		6081	
1164		6091	
1169		6119	
1377		6200	
1378		6239	
1380		7104	
1389		7105	
1394		7108	
1396		7115	
1414		7499	
1579		7499	
1648		7501	
1685		7504	
1689		7788	
1692		8111	
1712		9183	
1721		9219	
1723		9228	
1729		9239	
1730		9255	
1735		9257	
1740		10170	
3019		067/1141	
3298		7766&7749	
5009		x2127	
5023		x2194	
5031		x2201	
5036		x2209	
5041		x2311	
5047		x2319	
5074		x2322	
5142		x2329	
5149		x2362	
5171			
6000			
6006			
6014			
6019			
6030			
6034			
6039			
6043			
6046			
6046			

(*Continued*)

FLK N3	TM	FLK N3	CN
6047			
6050			
6054			
6057			
6062			
6065			
6066			
6069			
6071			
6075			
6076			
6081			
6091			
6092			
6093			
6094			
6115			
6116			
6118			
6121			
6123			
6200			
6243			
6254			
6271			
6280			
6284			
6285			
6288			
6291			
6292			
7092			
7097			
7104			
7105			
7108			
7305			
7307			
7307			
7309			
7317			
7476			
7481			
7482			
7484			
7491			
7499			
7502			
7505			
7507			
7508			
7509			
7510			
7730			
7735			

(*Continued*)

FLK N3	TM	FLK N3	CN
7742			
7760			
7764			
7768			
7769			
7776			
7788			
7790			
7792			
7797			
7850			
8102			
8173			
9103			
9162			
9169			
9185			
9189			
9199			
9205			
9219			
9236			
9237			
9247			
9251			
9254			
9256			
9257			
9401			
10128			
10157			
10165			
10167			
067/1051			
067/1101			
067/1133			
067/1135			
067/1136			
067/1137			
067/1139			
067/1142			
067/1168			
067/3297			
067/3340			
067/497			
067/499			
067/667			
067/668			
067/700			
067/738			
067/768			
067/999			
7767&7759			
x2040			

(*Continued*)

FLK N3	TM	FLK N3	CN
x2091			
x2195			
x2197			
x2200			
x2201			
x2202			
x2207			
x2209			
x2224			
x2299			
x2300			
x2310			
x2311			
x2320			
x2321			
x2322			
x2324			
x2326			

DK2

CATNUM	MARK
10	TM
15	TM
17	TM
19	TM
19.2	TM, CM
31.2	TM
31.4	TM
38.3	TM
43	TM
43.1	TM
43.2	TM
50	TM
69	TM
71	TM
74	TM
78	TM
79	TM
82	TM
86	TM
90	TM
91	TM
91	TM
109	TM
125	TM
129	TM
141	TM
141	TM
143	TM
159	TM
163.2	TM
163.3	TM

(*Continued*)

DK2–*cont'd*

CATNUM	MARK
163.4	TM
163.5	TM
163.6	TM
163.7	TM
163.8	TM
163.9	TM
163.11	TM
163.12	TM
163.16	TM
163.20	TM
163.21	PM
163.22	TM
163.24	TM
163.25	TM
163.39	TM
163.52	TM
164	TM
166	TM
168	TM
172	TM
175	TM
175	TM
175.1	TM
189	TM
195	TM
198	TM
217	TM
284	TM
289	TM
290	TM
295	TM
295.1	TM
299	TM
299.1	TM
305	TM
310	TM
324	TM
338	TM
352	TM
409	TM
415	TM
419	TM
451	TM, CM
453	TM
457	TM
471	TM
486	TM
575	TM, CM
576	TM, PM
577	TM
583	TM
612	TM
639	TM
704	TM

(*Continued*)

DK2–*cont'd*

CATNUM	MARK
780	TM
788	TM
3044.1	TM
3051	TM
3052	TM
3053A	TM
3072	TM
3073	TM
3079	TM
3090	TM
3474	TM
4121	TM
4137	TM
4137.1	TM
4137.2	TM
4137.3	TM
4137.5	TM
4137.6	TM
4137.8	TM
4137.9	TM
4137.10	TM
4137.11	CM
4137.13	CM
4137.14	TM
4137.17	TM
4137.21	TM
4141	TM
4157	TM
4181.1	TM
4215	TM
4216	TM
4225	TM
4240	TM
4270	TM
4304	TM
4451	TM
4536	TM
5072.1	TM
5504	TM
5562	TM
7934	TM
067/3093	TM
067/3427	TM
067/3465	TM
067/3471	TM
067/3493	TM
067/4106	PM
067/4107	TM
067/4109	TM
067/4127.1	TM
067/4127.2	TM
067/4127.4	TM
067/4135	TM

(*Continued*)

DK2–*cont'd*

CATNUM	MARK
067/4140	TM
067/4159.2	PM
067/4159.4	PM
067/4159.8	TM
067/4159.10	TM
067/4159.17	TM
067/4191.1	TM
067/4191.2	TM, CM
067/4191.3	TM
067/4191.79	TM
067/4193	TM, CM
067/4197.2	TM
067/4240.1	TM
067/4245.1	TM
067/4245.2	TM
067/4245.3	TM
067/4245.4	TM
067/4245.5	TM
067/4245.7	TM
067/4245.8	TM
067/4245.9	TM
067/4245.10	TM
067/4245.11	TM
067/4245.14	TM
067/4245.15	TM
067/4245.17	TM
067/4245.26	TM
067/4245.29	TM
067/4245.31	TM
067/4260.1	TM
067/4260.2	TM
067/4260.4	TM
067/4260.12	TM
067/4260.19	TM
067/4260.24	TM

DK3

CATNUM	MARK
131	TM
132	TM
134	TM
136	TM
144	TM
153	TM
160	TM
180	TM
183	TM
188	TM
203	TM
201	PM
212	TM

(*Continued*)

DK3–*cont'd*

CATNUM	MARK
217	TM
218	TM
222	TM
223	TM
226	TM
227	TM
228	TM
231	TM
238	TM
243	TM
248	TM
257	TM
517	TM
541	TM
601	TM
649	TM
690	TM
777	TM
864	TM
866	TM
879	TM
962A	TM
1673	TM
1708	TM
3107	TM
3110	TM
3111	TM
3118	TM
3154	TM
3201	TM
3230	TM
3239	TM
3284	TM
3288.2	TM
3294	TM
3309	TM
3340	TM
3343	TM
3375	TM
3394	TM
3396	TM
4131.1	TM
4229	TM
4277	TM
4424	TM
4565	TM
5342	TM
067/3070	TM
067/3071	TM
067/3074	TM
067/3075	TM
067/3080	TM
067/4087	TM

(*Continued*)

DK3–*cont'd*

CATNUM	MARK
067/4156	TM
067/4176.2	TM
067/4176.18	TM
067/4176.20	TM
067/4176.21	TM
067/4176.98	TM
067/4179	CM
067/4184	TM
067/4198.1	TM
067/4198.2	TM
067/4198.4	TM
067/4198.11	TM
067/4198.22	TM
067/4198.24	TM
067/4210.1	TM
067/4210.2	TM
067/4210.3	TM
067/4210.7	TM
067/4210.16	TM
067/4210.17	TM
067/4211.1	TM

Index

Vertebrate Paleobiology and Paleoanthropology

Published and forthcoming titles:

Neanderthals Revisited: New approaches and perspectives
Edited by K. Harvati and T. Harrison
ISBN: 978-1-4020-5120-3, 2006
A Volume in the Max-Planck Institute Sub-series in Human Evolution

The Evolution and Diversity of Humans in South Asia: Interdisciplinary studies in archaeology, biological anthropology, linguistics and genetics
Edited by M. Petraglia and B. Allchin
ISBN: 978-1-4020-5561-4, 2007

Dental Perspectives on Human Evolution: State of the art research in dental paleoanthropology
Edited by S.E. Bailey and J-J. Hublin
ISBN: 978-1-4020-5844-8, 2007
A Volume in the Max-Planck Institute Sub-series in Human Evolution

Hominin Environment in the East African Pliocene: An Assessment of the Faunal Evidence
Edited by R. Bobe, Z. Alemseged and A.K. Behrensmeyer
ISBN: 978-1-4020-3097-0, 2007

Deconstructing Olduvai: A Taphonomic Study of the Bed I Sites
By M. Domínguez-Rodrigo, R. Barba Egido and C.P. Egeland
ISBN: 978-1-4020-6150-9, 2007

MIX
Papier aus verantwortungsvollen Quellen
Paper from responsible sources
FSC® C105338

If you have any concerns about our products,
you can contact us on
ProductSafety@springernature.com

In case Publisher is established outside the EU,
the EU authorized representative is:
Springer Nature Customer Service Center GmbH
Europaplatz 3, 69115 Heidelberg, Germany

Printed by Libri Plureos GmbH
in Hamburg, Germany